T0389719

Data Driven Mathematical Modeling in Agriculture: Tools and Technologies

RIVER PUBLISHERS SERIES IN MATHEMATICAL, STATISTICAL AND COMPUTATIONAL MODELLING FOR ENGINEERING

Applied mathematical techniques along with statistical and computational data analysis has become vital skills across the physical sciences. The purpose of this book series is to present novel applications of numerical and computational modelling and data analysis across the applied sciences. We encourage applied mathematicians, statisticians, data scientists and computing engineers working in a comprehensive range of research fields to showcase different techniques and skills, such as differential equations, finite element method, algorithms, discrete mathematics, numerical simulation, machine learning, probability and statistics, fuzzy theory, etc.

Books published in the series include professional research monographs, edited volumes, conference proceedings, handbooks and textbooks, which provide new insights for researchers, specialists in industry, and graduate students.

Topics included in this series are as follows:-

- Discrete mathematics and computation
- Fault diagnosis and fault tolerance
- Finite element method (FEM) modeling/simulation
- Fuzzy and possibility theory
- Fuzzy logic and neuro-fuzzy systems for relevant engineering applications
- Game Theory
- Mathematical concepts and applications
- Modelling in engineering applications
- Numerical simulations
- Optimization and algorithms
- Queueing systems
- Resilience
- Stochastic modelling and statistical inference
- Stochastic Processes
- Structural Mechanics
- Theoretical and applied mechanics

For a list of other books in this series, visit www.riverpublishers.com

Data Driven Mathematical Modeling in Agriculture: Tools and Technologies

Editors

Sabyasachi Pramanik
Haldia Institute of Technology,
India

Sandip Roy
Department of Computer Science & Engineering,
School of Engineering, JIS University,
Kolkata, India

Rajesh Bose
Department of Computer Science & Engineering,
School of Engineering, JIS University,
Kolkata, India

NEW YORK AND LONDON

Published 2024 by River Publishers
River Publishers
Alsbjergvej 10, 9260 Gistrup, Denmark
www.riverpublishers.com

Distributed exclusively by Routledge
605 Third Avenue, New York, NY 10017, USA
4 Park Square, Milton Park, Abingdon, Oxon OX14 4RN

Data Driven Mathematical Modeling in Agriculture: Tools and Technologies / Sabyasachi Pramanik, Sandip Roy and Rajesh Bose.

Routledge is an imprint of the Taylor & Francis Group, an informa business

ISBN 978-87-7004-100-3 (hardback)
ISBN 978-87-7004-182-9 (paperback)
ISBN 978-10-4012-097-2 (online)
ISBN 978-1-003-51588-3 (ebook master)

Contents

Preface

Overview of the Subject Matter

This book examines how likely and how often technology components are used to embrace precision agriculture methods. Zero-inflated Poisson and negative binomial count data regression models were used to detect farmers' preferences for more accurate technologies. The count data analysis of a random sample of farm operators found that farm dimension, farmer demographics, soil texture, urban impacts, farmer liability position, and farm state position were significantly associated with precision farming technology approval severity and likelihood.

A Description of Where My Topic Fits in the World Today

Farm management information systems (FMISs) have become more complicated as they have adopted new technologies, including the Internet. Few FMISs have really tapped into the Internet's capabilities, and the newly developed notion of precision agriculture gets little or no support in the FMIS being marketed. Precision agriculture FMIS must fulfill additional requirements, which increases the technical complexity of its adoption. The authors investigated cutting-edge web-based technologies to build an FMIS that meets these additional criteria. The purpose was to identify precision agriculture's FMIS needs.

Description of the Target Audience

Primary audience

The primary audience includes mathematicians, statisticians, and computer scientists working in a variety of areas, including agricultural developments, biometrics, econometrics, data analysis, graphics, simulation, algorithms, knowledge-based systems, and Bayesian computing.

Secondary audience

The secondary audience includes lecturers and tutors: They will find it handy to impart quality technical knowhow in the relevant fields to the undergraduate/graduate and post-graduate students.

Benefits to Audience

The authors in Chapter 1 titled "Use of CNNs and Their Frameworks for the Detection of Fungal Herb Disease" describe that 85% of plants are infected by fungus, bacteria, and viruses, which damage the ecosystem. One of the most common signs of fungal infection is plant color. In literature, several rule-dependent algorithms and image processing methods are used to diagnosis fungal plant diseases. Old approach offers poor sickness identification accuracy. Convolutional neural networks may be used for plant pathology image identification and classification. This chapter examines LeNet, VGGNet, ResNet, and ZFNet CNN architectures for plant disease sensing, along with their structures and advantages. ResNet and ZFNet are efficient in accuracy and error rate.

In the research "Technologies based on the IoT and Artificial and Natural Intelligence for Sustainable Agriculture," the authors examine blockchain technology's promises and new challenges for the Russian digital economy. He says that the blockchain innovator group actively debates fewer concerns and offers more solutions. The Russian study found that most of these problems were not addressed by Russian researchers. Private or consortium blockchains may address some of these instead of an open network. End-user help (difficult to use and understand) and developer support (inadequacy of tools) are now minor technical impediments. The most recent advancements in developer assistance include the deployment of blockchain using model-based approaches, sliding intellectual control and transfer, and technical problem tools. The chapter shows avatars communicating using several methods for smart agriculture and sustainable farming.

The authors of Chapter 3 "IoT for Smart Farming Technology: Practices, Methods, and Future" specify that farming research has changed as a result of the IoT. IoT should be thoroughly investigated before being employed in agriculture since it is still in its early stages. The writers here talk about IoT applications and challenges in applying technology to better agriculture. Implementations of IoT devices and wireless transmission technologies in agriculture are thoroughly studied to focus on specific demands. We investigate sensor-based IoT frameworks that provide intelligent agriculture practices. Use cases analyze IoT-based solutions by deployment variables from various organizations, individuals, and categories. Although there are problems with these solutions, they also highlight opportunities for development and an IoT-based action plan.

In "Integrating Artificial Intelligence into Pest Management," the authors specify that the farmers face unpredictable bug outbreaks owing to climate change, increased global trafficking of contaminated commodities, and pest management challenges. These challenges may be solved using integrated pest management (IPM), a sustainable pest control method that uses natural resources efficiently. IPM considers climate, economic thresholds, plant sensitivity, and reproduction. The most crucial part of IPM is skilled workers. The expert designs systems, monitors ecological factors, and makes decisions. Sustainable pest management may employ AI to monitor biological and ecological elements and choose the optimal timing and strategy. This chapter will explain IPM's use of artificial intelligence and its tools, algorithms, and methods.

In "Practices of Deep Learning in Farming: What Deep Learning Can Do in Intelligent Agriculture," the authors tell that machine learning, which is a subset of deep learning, is a cutting-edge technology for processing images and analyzing vast volumes of data. Many sectors improve performance using DL approaches. Since current agricultural processes cannot keep up with population growth, DL approaches may benefit the agriculture business. This chapter covers the latest deep learning breakthroughs in agriculture and associated research topics. Additionally, the implemented models will be compared to other models in use.

In the research work "Building a Solar-powered Greenhouse having SMS and Web Information Framework," the authors specify that the greenhouse crops cost a lot for heating and cooling. This has made many reconsider energy-saving methods, including solar energy as a realistic, sustainable, and renewable greenhouse farming alternative. This article proposes a solar-powered framework that uses GSM and IoT to detect, manage, and maintain greenhouse weather. The proposed system automatically calculates and tracks temperature, humidity, soil moisture, and light intensity. The design framework offers users online and SMS access to regular parameters regardless of location. The study also contains a method to self-adjust the greenhouse's environment for plant growth. A method like this may boost greenhouse crop productivity and quality. Tests show the system meets design objectives.

In the chapter "Agriculture using Digital Technologies," the authors introduce agriculture's latest technology, including cloud computing, ML, and AI. The agricultural cycle – crop choice, soil production, seed choice, seed planting, irrigation, crop development, fertilization, and harvesting – is then briefly reviewed. Digital technology for the agricultural cycle is also discussed in this chapter. The rest of the chapter will explain how farming will change and how digital technology will be integrated into crop cycles.

The research titled "Agriculture Digitization: Perspectives on the Networked World" begins with agricultural digitization. It discusses the need for new Agtech companies that use cutting-edge digital technologies. The chapter discusses agriculture's digital benefits and post-COVID effects. Telematics, precision agriculture, blockchain, AI, and other agricultural digitalization subjects are covered. Connection, interoperability, portability, and public–private partnership were the last significant problems discussed.

The authors of the work "Cucumber in PH Disease Monitoring using IoT-based Mobile App" show that agriculture is a given in most countries since it creates employment and generates cash. Modern agriculture includes the unique and popular PH farming. Using a PH that provides water and nutrients in the proper amounts and under controlled circumstances may boost yield. Since stagnant air and lack of air circulation promote bug proliferation and material loss, PH crops must be regularly maintained. Thus, this chapter proposes a farm/PH IoT-dependent disease-tracking prototype. To identify cucumber disease early, a prototype is constructed and tested. This study initially focuses on identifying dangerous cucumber infections in PHs using a NodeMCU and Raspberry-Pi hardware paradigm. Farmers will be notified of their options and any major changes in identified metrics via a custom smartphone app.

In the chapter "New Technologies for Sustainable Agriculture," the authors found that agriculture has made great strides in technology in recent decades. Technology is transforming agriculture into digital agriculture. These innovations transformed conventional agriculture into intelligent farming. Smart farmers collect data and analyze it to fertilize and care for their crops. The smart agriculture system costs less and provides more accurate crop forecasting and protection. This technology makes agriculture more lucrative, efficient, waste-free, and sustainable. This chapter discusses cutting-edge agricultural technologies and proposes a smart agriculture paradigm.

In the research titled "Agriculture Automation," the authors showed that to increase agricultural yields using agricultural automation, new skills and technology are required. Intelligent agriculture tracks, monitors, and analyzes agricultural activities using technology. IoT hardware and software, cameras, drones, edge devices, cloud servers, sensors, actuators, and communication technologies make up the system. This chapter covers agricultural automation hardware and software. Four segments make up the chapter. A smart agriculture and precision farming overview is offered. The second half reviews current studies' literature. The third section examines IoT, sensors, cloud, and edge computing for smart agriculture. The following section presents smart agricultural IoT case studies. Thc chapter will discuss the IoT framework solution for agricultural automation.

The authors of the work "Food 4.0: A Survey" finds that the 4th Industrial Revolution may replace people and physical work with robots and AI. Humanity uses the 4IR concept in all aspects. Academics are studying the change as corporate executives prepare for its inevitable and rapid adoption in energy, automotive, communications, facilities, security, medical, and other sectors. The agricultural and food sector – also known as "Food 4.0" – is the biggest employer of human resources and is expected to benefit considerably from 4IR as it enters a new development phase.

In the chapter "Crop Monitoring in Real Time in Agriculture," the authors found that many countries, including India, employ traditional agriculture. Our farmers' lack of knowledge worsens the agricultural business. Farmers endure huge losses due to weather predictions and estimates, which might be inaccurate, resulting in debt and even deaths. Soil moisture, quality, air quality, and irrigation all affect crop yield and cannot be neglected. Global population growth is frightening, and agricultural production cannot keep up. By 2050, the global population is expected to exceed nine billion, requiring a 70% increase in agricultural supplies. To do this, plant growth must be monitored from sowing to culture.

In the research work "Smart Farming Utilizing Wireless Sensor Network and Internet of Things," the authors found that Indian agriculture is suffering, and crop yield is falling every day. To increase productivity, a solution must be found and implemented. To develop agriculture, smart technologies are used. The agriculture industry is increasingly using Internet of Things, big data, cloud-based capabilities, and GPS. Demand is rising for crop analysis accuracy, real-time field data processing, and automated farming for future progress. With these smart procedures, a smart agriculture company is expected. This chapter discusses the benefits and downsides of the Internet of Things (IoT) and various data collecting sensors.

In the chapter titled "Intelligent Agriculture using Autonomous UAVs," the researchers cover intelligent farming, robotic frameworks, ML techniques, and agricultural robot development and effect. The goal is to collect communicated autonomous unmanned aerial vehicles and UGVs to reduce target time. It investigated numerous stop-number techniques using particle swarm optimization. The ideal time to harvest the apples distributed to the stalls was computed using deterministic, binary mixed (0-1) integer modeling and *K*-means. This modeling determines which UAV may harvest the apple and how to detect whether it did using 0-1 binary modeling. The 2-opt, closest neighbor, and nearest insertion methods constructed the unmanned UGV's route.

In the research "Agriculture using Smart Sensors," the authors found that the people's main occupation is agriculture. Urbanization poses new issues and causes a mass exodus of village residents to the metropolitan due to increased ecological pollution, climate change, soil and water degradation, population growth, farm profit decline, etc. New technologies that use smart sensors to construct smart farms using IoT approach offer a solution. Smart agriculture improves crop productivity, animal tracking, soil moisture tracking, remote water tank level monitoring, temperature and humidity sensing, farmland security, environmental condition tracking, and tool monitoring. This supports farmers in remote property monitoring and protection. The new smart farming system incorporates IoT-based sensors. A camera with a microcontroller, Wi-Fi, a variety of interfacing sensing nodes for service, internet service, smart remote devices, or PCs with internet track sensor node function in an IoT-dependent smart agricultural system. These sensors include temperature, soil moisture, PIR sensors, which detect people, animals, and objects in fields, and GPS-dependent remote control robots for sprinkling, weeding, security, and moisture sensing. This chapter follows: an introduction to the agriculture sector, including its current issues and a newer approach to address them; the need for IoT in farming; a detailed examination of the relationship between IoT technology and WSNs; IoT-dependent systems, applications, and advantages.

The chapter titled "Technologies that Work Together for Precision Agriculture" finds that precision agriculture (PA) helps farmers optimize inputs to increase yield and uniform harvests while lowering costs and environmental impact. Due to extensive farm lands and mechanized agricultural production techniques, industrialized countries associate with precision agriculture. Data collection, analysis, and graphing on yield, soil standard parameters, and ecological factors at different field locations determine the farm's water, nutrient, and fertilizer needs. Most developing countries lack precision agriculture technologies. The fields are smaller, and technology, knowledge, and funds are few. However, developing country farmers continue to seek ways to increase agricultural production and output.

In the chapter "Utilizing Smart Farming Methods to Reduce Water Scarcity," the researchers found that due to exponential population expansion and increased demand for food and industrial goods, many countries are suffering a water crisis due to restricted water supplies. About 70% of water is used for irrigation. Thus, untreated wastewater is commonly used for irrigation, increasing health concerns. Studies suggest that employing information technology to enhance irrigation may help farmers consume less water. Smart agricultural approaches employ information technology to manage water supply, reduce waste, and measure water quality. Intelligent agriculture methods may help solve the world's biggest problems and achieve sustainable development.

In the chapter "Real-Time Irrigation Optimization for Horticulture Crops using WSN, APSim, and Communication Models," the researchers show that prescriptive agriculture is a new field of research in agriculture that uses WSNs, IoTs, and advanced technologies. Prescriptive agriculture enforces precision agriculture (PA), which monitors, computes, and reacts to inter- and intra-farm field variability. In this chapter, WSN, APSim, and communication models inspired a novel irrigation optimization strategy in agriculture. A GSM module sends temperature and relative humidity data from sensors to a server. Inter- and intra-farm field conditions were considered while assessing crop datasets. Finally, way2SMS and WebHost server destroy irrigation data.

The chapter titled "Greenhouse Gas Discharges from Farming Modeled Mathematically for Various End Users" finds that abstract models approximate real systems. There is a practicable constraint to model outputs that is partly defined by the intended end use, even if theoretically there is no limit to statistical framework refinement and facts, which allows more detailed representation of the physical system. The collaborative model studies in this chapter range from basic empirical models to large simulation models. The authors provide examples of agricultural greenhouse gas (FHG) model applications and their end customer amounts. The authors will demonstrate that stakeholder communities must employ many models and methods to develop knowledge. We conclude that a model's value should be based on both its accuracy and utility. Communicating information to many parties has led to several remarkable advancements in recent years. The ecosystem land-usage model (ELUM) is one of many recent enticing innovations to make these methodologies more accessible to laypeople, despite the fact that process-based models often employ conventional procedures.

List of Contributors

Aadil, Mir, *Department of Computer Science and IT, School of CS and IT, Jain (Deemed to be) University, India*

Afaq, Ahmar, *Symbiosis Law School, Nagpur Campus, Symbiosis International (Deemed University), India*

Amesho, Kassian T.T., *National Sun Yat-sen College of Engineering, Taiwan*

Awasthi, Aishwary, *Department of Mechanical Engineering, Sanskriti University, India*

Awasthi, Ashutosh, *College of Agriculture Science, Teerthanker Mahaveer University, India*

Bargavi, S. K. Manju, *Department of Computer Science and IT, School of CS and IT, Jain (Deemed to be) University, Bangalore, India*

Beemkumar, N., *Department of Mechanical Engineering, Faculty of Engineering and Technology, JAIN (Deemed to be University), India*

Benedict, Abhishekh, *Symbiosis Law School, Nagpur Campus, Symbiosis International (Deemed University), India*

Bhardwaj, Shambhu, *College of Computing Science and Information Technology, Teerthanker Mahaveer University, India*

Bhatia, Himanshi, *Symbiosis Law School, Nagpur Campus, Symbiosis International (Deemed University), India*

Bhuvana, J., *Department of Computer Science and IT, School of CS and IT, Jain (Deemed to be) University, India*

Bui, Thanh, *Data Analytics and Artificial Intelligence Laboratory, School of Engineering and Technology, Thu Dau Mot University, Vietnam*

Cengiz, Korhan, *University of Fujairah, UAE*

Chandra, Rushil, *Symbiosis Law School, Nagpur Campus, Symbiosis International (Deemed University), India*

Dari, Sukhvinder Singh, *Symbiosis Law School, Nagpur Campus, Symbiosis International (Deemed University), India*

Dhabliya, Dharmesh, *Department of Information Technology, Vishwakarma Institute of Information Technology, India*

Dhabliya, Dharmesh, *Department of Information Technology, Vishwakarma Institute of Information Technology, India*

Dhage, Prashant, *Symbiosis Law School, Nagpur Campus, Symbiosis International (Deemed University), India*

Elngar, Ahmed, *Beni-Suef University, Egypt*

Fatima Rizvi, Nuzhat, *Symbiosis Law School, Nagpur Campus, Symbiosis International (Deemed University), India*

Ganesh, D., *Department of Computer Science and IT, School of CS and IT, Jain (Deemed to be) University, India*

Ghayathri, J., *Department of Computer Science and IT, School of CS and IT, Jain (Deemed to be) University, Bangalore, India*

Godbole, Aditee, *Symbiosis Law School, Nagpur Campus, Symbiosis International (Deemed University), India*

Gupta, Ankur, *Department of Computer Science and Engineering, Vaish College of Engineering, India*

Gupta, Mohan Vishal, *College of Computing Science and Information Technology, Teerthanker Mahaveer University, India*

Gupta, Namit, *College of Computing Science and Information Technology, Teerthanker Mahaveer University, India*

Gupta, Rupal, *College of Computing Science and Information Technology, Teerthanker Mahaveer University, India*

Gupta, Sachin, *Department of Management, Sanskriti University, India*

Hadar, Ofer, *Ben-Gurion University of the Negev, Israel*

Jaison, Feon, *Department of Computer Science and IT, School of CS and IT, Jain (Deemed to be) University, India*

Jose, Jimmy, *Symbiosis Law School, Nagpur Campus, Symbiosis International (Deemed University), India*

Joshi, Manish, *College of Computing Science and Information Technology, Teerthanker Mahaveer University, India*

Kalnawat, Aarti, *Symbiosis Law School, Nagpur Campus, Symbiosis International (Deemed University), India*

Kalra, Nuvita, *Symbiosis Law School, Nagpur Campus, Symbiosis International (Deemed University), India*

Karthikeyan, M.P., *Department of Computer Science and IT, School of CS and IT, Jain (Deemed to be) University, India*

Khatun, Ayesha, *Symbiosis Law School, Nagpur Campus, Symbiosis International (Deemed University), India*

Khubalkar, Deepti, *Symbiosis Law School, Nagpur Campus, Symbiosis International (Deemed University), India*

Kishor, Kaushal, *Department of Agriculture, Sanskriti University, India*

Krishnamoorthy, Ramkumar, *Department of Computer Science and IT, School of CS and IT, Jain (Deemed to be) University, India*

Kumar, Pawan, *Department of Computer Science and IT, School of CS and IT, Jain (Deemed to be) University, India*

Kumar, Rajeev, *Department of Agriculture, Sanskriti University, India*

Kumar, Sunil, *College of Agriculture Science, Teerthanker Mahaveer University, India*

Le, Dac-Nhuong, *Faculty of Information Technology, Haiphong University, Vietnam*

Maurya, Kuldeep, *Department of Agriculture, Sanskriti University, India*

Michael, Gabriela, *Symbiosis Law School, Nagpur Campus, Symbiosis International (Deemed University), India*

Mishra, Manoj Kumar, *Department of Agriculture, Sanskriti University, India*

Mishra, Umesh Kumar, *Department of Agriculture, Sanskriti University, India*

Murugan, R., *Department of Computer Science and IT, School of CS and IT, Jain (Deemed to be) University, India*

Nair, Rakesh, *Symbiosis Law School, Nagpur Campus, Symbiosis International (Deemed University), India*

Nait-Abdesselam, Farid, *School of Science & Engineering, University of Missouri Kansas City, USA*

Natesan, Gobi, *Department of Computer Science and IT, School of CS and IT, Jain (Deemed to be) University, India*

Noor, Ayman, *College of Computer Science and Engineering, Taibah University, Saudi Arabia*

Obaid, Ahmed J., *University of Kufa, Iraq*

Ojha, Ananta, *Department of Computer Science and IT, School of CS and IT, Jain (Deemed to be) University, India*

Pandey, Rajendra P., *College of Computing Science and Information Technology, Teerthanker Mahaveer University, India*

Pareek, Vaidehi, *Symbiosis Law School, Nagpur Campus, Symbiosis International (Deemed University), India*

Polke, Nikhil, *Symbiosis Law School, Nagpur Campus, Symbiosis International (Deemed University), India*

Pramanik, Sabyasachi, *Department of Computer Science & Engineering, Haldia Institute of Technology, India*

Rabie, Khaled M., *Department of Engineering, Manchester Metropolitan University (MMU), UK*

Raghavendra, R., *Department of Computer Science and IT, School of CS and IT, Jain (Deemed to be) University, India*

Raj, Saurabh, *Symbiosis Law School, Nagpur Campus, Symbiosis International (Deemed University), India*

Ramya, D. Janet, *Department of Computer Science and IT, School of CS and IT, Jain (Deemed to be) University, India*

Ranka, Siddharth, *Symbiosis Law School, Nagpur Campus, Symbiosis International (Deemed University), India*

Rastogi, Ajay, *College of Computing Science and Information Technology, Teerthanker Mahaveer University, India*

Rizvi, Nuzhat Fatima, *Symbiosis Law School, Nagpur Campus, Symbiosis International (Deemed University), India*

Samanta, Debabrata, *Rochester Institute of Technology, Europe*

Sanjaya, Karun, *Symbiosis Law School, Nagpur Campus, Symbiosis International (Deemed University), India*

Saxena, Abhilash Kumar, *College of Computing Science and Information Technology, Teerthanker Mahaveer University, India*

Saxena, Ashendra Kumar, *College of Computing Science and Information Technology, Teerthanker Mahaveer University, India*

Saxena, Shakuli, *College of Agriculture Science, Teerthanker Mahaveer University, India*

Saxena, Vineet, *College of Computing Science and Information Technology, Teerthanker Mahaveer University, India*

Shah, Pradeep Kumar, *College of Computing Science and Information Technology, Teerthanker Mahaveer University, India*

Sharma, Anu, *College of Computing Science and Information Technology, Teerthanker Mahaveer University, India*

Sharma, Parth, *Symbiosis Law School, Nagpur Campus, Symbiosis International (Deemed University), India*

Sharma, Shilpa, *Symbiosis Law School, Nagpur Campus, Symbiosis International (Deemed University), India*

Sherazi, Hafiz Husnain Raza, *School of Computing and Engineering, University of West London, UK*

Sikchi, Pratieek, *Symbiosis Law School, Nagpur Campus, Symbiosis International (Deemed University), India*

Singh, Akanchha, *College of Agriculture Science, Teerthanker Mahaveer University, India*

Singh, Devendra Pal, *College of Agriculture Science, Teerthanker Mahaveer University, India*

Singh, Dinesh, *Department of Agriculture, Sanskriti University, India*

Singh, Praveen Kumar, *College of Agriculture Science, Teerthanker Mahaveer University, India*

Singhal, Priyank, *College of Computing Science and Information Technology, Teerthanker Mahaveer University, India*

Srikanth, V., *Department of Computer Science and IT, School of CS and IT, Jain (Deemed to be) University, India*

Srivastava, Gautam, *Computer Science, Brandon University*

Thiruvenkdam, Department of Computer Science and IT, *School of CS and IT, Jain (Deemed to be) University, India*

Tiwari, Pankaj Kumar, *Department of Agriculture, Sanskriti University, India*

Tripathi, Sachin, *Symbiosis Law School, Nagpur Campus, Symbiosis International (Deemed University), India*

Tripathi, Umesh Kumar, *Department of Agriculture, Sanskriti University, India*

Upasana, College of Agriculture Science, *Teerthanker Mahaveer University, India*

Yadav, Vinaya Kumar, *Department of Agriculture, Sanskriti University, India*

Zaidi, Taskeen, *Department of Computer Science and IT, School of CS and IT, Jain (Deemed to be) University, India*

Zubair Khan, Mohammad, *Department of Computer Science and Information, Taibah University Medina Saudi Arabia*

List of Figures

List of Tables

List of Abbreviations

4IR	Fourth Industrial Revolution
AI	Artificial intelligence
AIoT	Agricultural IoT
AlexNet	A convolutional neural network
ANN	Artificial neural network
API	Application-programming interface
AR	Augmented reality
AVR	Automatic voltage regulator
AWS	Amazon Web Services
BIM	Building information modeling
BMP	Bitmap
CDMA	Code-division multiple access
CFS	Container freight station
CIMMYT	International Maize and Wheat Improvement Center
CNN	Convolutional neural network
CPS	Cyber–physical system
CVSM	Crop variety selection method
CWSI	Crop water stress index
DD	Degree-days
DENSENET	Dense convolutional network
DGPS	Differential global positioning system
DL	Deep learning
DSS	Decision support system
EC	Emulsifiable concentrate
ECOSSE	Estimation of Carbon in Organic Soils – Sequestration and Emission
ELUM	Ecosystem land-usage model
ERP	Enterprise Resource Planning
FAA	Food and Agriculture Agency
FDD	Food and Dairy Division
FFD	Full-function device
FGW	Farm gate way

FHG	Farming greenhouse gas
FMIS	Farm management information system
FPGA	Field programmable gate array
FSTSP	Flying sidekick traveling salesman problem
GHG	Greenhouse gas
GIS	Geographical information systems
GLCM	Gray-level co-occurrence matrix
GNSS	Global navigation satellite system
GoogLeNet	22-layer deep convolutional neural network
GRID	Global resource information database
GSM	Global system for mobile communications
GUI	Graphic user interface
HAB	Harmful algae bloom
HSI	Hue, saturation, and intensity
HTTP	HyperText transfer protocol
HVAC	Heating, ventilation, and air conditioning
HYV	High yielding variety
ICT	Information and Communication Technologies
IIDF	Internet Initiatives Development Fund
IIoT	Industrial Internet of Things
Industry 4.0	Fourth Industrial Revolution
IoT	Internet of Things
IoTA	IoT Association
IPM	Integrated pest management
ISM	Information security management
KNN	K-nearest neighbor
LAN	Local area network
LDR	Light dependent resistor
LeNet	LeNet-5 CNN
LIDAR	Light detection and ranging
LOPCOW	LOgarithmic percentage change-driven objective weighting
LoRa	Long range
LP	Linear programming
LPWA	Low power wide area
LSTM	Long short-term memory
LTE	Long-term evolution
M2M	Machine-to-machine
MAI	Moisture Adequacy Index
MILP	Mixed integer linear programming
ML	Machine learning

MSP	Minimum support price
NDVI	Normalized difference vegetation index
NEE	Net ecosystem exchange
NIR	Near infrared
NLP	Natural language processing
NN	Neural network
NPP	Net primary productivity
NTC	Negative temperature coefficient
NTI	National Technology Initiative
OGC	Open geospatial consortium
PA	Precision agriculture
PAN	Personal area coordinator
PC	Protected cultivation
PCB	Polychlorinated biphenyl
PH	Potential of Hydrogen
PSO	Particle swarm optimization
PwC	PriceWaterhouseCoopers
QoS	Quality of Service
q-ROFN	q-rung orthopair fuzzy number
RAV	Robotic agricultural vehicle
RDF	Refuse-derived fuel
ReLu	Rectified linear unit
RESNET	Residual network
RF	Radio frequency
RFD	Reduced-function device
RFID	Radio frequency identifier
RGB	Red, green, and blue
RNN	Recurrent neural network
RPA	Robotic process automation
SaaS	Software as a service
SFAIS	Smart agricultural Automated Irrigation System
SGDM	Stochastic gradient descent with momentum
SiPs	School of Integrative Plant Science
SLPM	Semantic learning process mining
SMG	Smart hydroponic greenhouse
SMTP	Stachybotrys microspora triprenyl phenol
SNMP	Simple network management protocol
SOA	School of Agriculture
SoCs	Soil organic carbon
SOM	Soil organic matter

SQUEEZENET	CNN 18 layer deep
SSA	Smart sustainable agriculture
SSCM	Site-specific crop management
SSD	Seed Systems Development
SSE	Sum of squared error
SVG	Scalable vector graphics
SysML	Systems Modeling Language
TCP	Transfer control protocol
TGR	Third Green Revolution
TOPSIS	Technique for Order Preference by Similarity to Ideal Solution
TSP	Traveling salesman problem
UAS	Autonomous aircraft systems
UAV	Unmanned aerial vehicle
UGV	Unmanned ground vehicle
UML	Unified Modeling Language
VGGNet	Visual Geometry Group
VIKOR	VlseKriterijumska Optimizacija I Kompromisno Resenje
VPFA	Vodafone Precision Fertilizer
VR	Virtual reality
VRA	Variable rates applications
VRP	Vehicle routing problem
VRT	Variable-rate technology
WEF	World Economic Forum
WSN	Wireless sensor network
WWW	World Wide Web
XML	Extensible Markup Language
ZFnet	Zeiler and Fergus network

1

Use of CNNs and Their Frameworks for the Detection of Fungal Herb Disease

R. Raghavendra[1], Ashendra Kumar Saxena[2], Aishwary Awasthi[3], Sukhvinder Singh Dari[4], Jimmy Jose[4], Debabrata Samanta[5], and Farid Nait-Abdesselam[6]

[1]Department of Computer Science and IT, School of CS and IT, Jain (Deemed to be) University, India
[2]College of Computing Science and Information Technology, Teerthanker Mahaveer University, India
[3]Department of Mechanical Engineering, Sanskriti University, India
[4]Symbiosis Law School, Nagpur Campus, Symbiosis International (Deemed University), India
[5]Rochester Institute of Technology, Europe
[6]School of Science and Engineering, University of Missouri, USA

Email: r.raghavendra@jainuniversity.ac.in; ashendrasaxena@gmail.com; aishwary@sanskriti.edu.in; director@slsnagpur.edu.in; jimmyjose@slsnagpur.edu.in; debabrata.samanta369@gmail.com; naf@umkc.edu

Abstract

Eighty-five percent of plants suffer from illnesses brought on by fungi, bacteria, and viruses, which wreak havoc on the environment. The plant color is one of the most frequent indicators that a plant is infected with a fungal illness. To diagnose the fungal plant diseases, numerous conventional rule-dependent algorithms and common image processing approaches are employed in literature. The old method, however, has low illness identification precision. One possible deep learning neural network utilized for picture identification and classification in plant pathology is the convolutional neural network (CNN). Some possible CNN designs for herb disease sensing, including LeNet, VGGNet, ResNet, and ZFnet, are explored in this chapter along with their structures and benefits. In accordance with accuracy and error rate, ResNet and ZFnet are shown to have high efficiency.

1.1 Introduction

The natural ecology is severely harmed by illnesses brought on by fungi that impact around 85% of plants. Fungi, bacteria, and viruses are the microbes that usually cause fungal diseases in plants. The presence of the illnesses might be recognized early on by carefully observing how the plant appears. According to [1], the most typical clues given by plants suffering from fungal diseases include fading of herb color, decline of leaf dimension, yellow color of branches, rusting of stems, drying of seeds, development of white patterns, curved stem borders, crumpling of leaves, and others.

The fungus in plants with fungal damage consumes all of the nutrients and energy from the plants, severely harming both the plants and their byproducts. By harming the plants' cells and tissues, the damage stresses the plants. Animals, dirt, equipment, human workers, weeds, seeds, and other things are some of the origins of fungus-related illnesses. The fungus is produced from these sources and enters the plant either by the stomata, which are the plant's natural opening, or past a crack that is made over the herb by pests or various automated devices used during plantation. Cottony rot, browning, death of foliage, rusts, and other fungal plant diseases are the most prevalent ones, according to [2]. The identification of fungal plant diseases in literature uses a variety of conventional rule-dependent approaches and common image processing approaches. The traditional approach, on the other hand, has a number of drawbacks, including poor disease estimation accuracy, unsuitable data preprocessing, repetitious difficult tasks, inability to control unrestricted data capturing circumstances, bottlenecks in segmentation, shortage of scalability, inability to operate data transition, and so forth. It prompted the development of automatic artificial-intelligence-dependent systems in locating infectious fungi that cause plant illnesses. In a back-propagation neural network, DL, supervised learning, reinforcement learning, unsupervised learning, CNN [3], RNN [4], and so on are some of the commonly used AI methodologies in recognizing fungal herb infections. To train the algorithms for disease detection, the automation systems can employ numerical data or photos obtained throughout the vast range of fungal infected plants. By effectively addressing issues with real-time operation, processing resources, inadequate data collection, distribution of training data, and complexity in calculation, data transfer, and other issues, automated artificial intelligence techniques perform exact operations.

When it comes to recognizing fungal plant diseases, image-based automated systems outperform all other types of automated approaches in terms of output quality. One possible deep learning neural network utilized in plant pathology for picture identification and classification is CNN. Convolutional, pooling, and fully connected layers make up CNN, which may properly account for dependencies associated with the spatial and temporal characteristics of pictures. LeNet, AlexNet, GoogLeNet, and ZFnet are a few of the sophisticated CNN architectures utilized for image analysis.

1.2 LeNet

LeNet is an acronym for LeNet5, which Yann LeCun developed in 1998 at Bell Labs. It consists of two sets of convolutional layers and a pooling layer. It is one of the first CNN models created for deep learning purposes, and it was the first to be used with a back-propagation-based learning method. LeNet was initially used to categorize the issue of handwritten character or digit recognition. Besides input and output layers, LeNet5 has six layers. The parameters of a single layer are utilized to train the parameters of subsequent levels. Six 5 × 5 convolutional kernels make up the first layer (C1), and a 28 × 28 feature map is used for the feature mapping. Each layer's features are 14 × 14 in size, while the second layer, S2, does sub-sampling. The third layer, or C3, is a convolutional layer with kernels that are 24 × 6 × 6 in size. S4 is the fourth layer, comparable to S2 of size 4 × 4, and it produces 18 samples. The fifth layer, C5, is a convolutional layer that has 120 kernels of 5 × 5 dimensions. The sixth layer, F6, is completely coupled to the C5 layer and outputs graphs with 84 characteristics. Figure 1.1 shows a typical LeNet5 design.

The frequent occurrence of class imbalance ratio, suffering from over-fitting complication, improper future estimation, huge collection of error gradients, network instability because of overtraining, more training period, explosion of gradient problems, geometric variation because of distortions, traction in mission-critical issues, inappropriate structural mapping, sluggish progressive learning, and other issues are few difficulties faced by the LeNet5 architecture.

The LeNet5 [5] architecture has a number of potential benefits, including: it is simple to understand because it has fewer layers and shallower depth; it can produce trained architectures that are widely accepted because all of its modules are optimized together; it responds well and robustly to

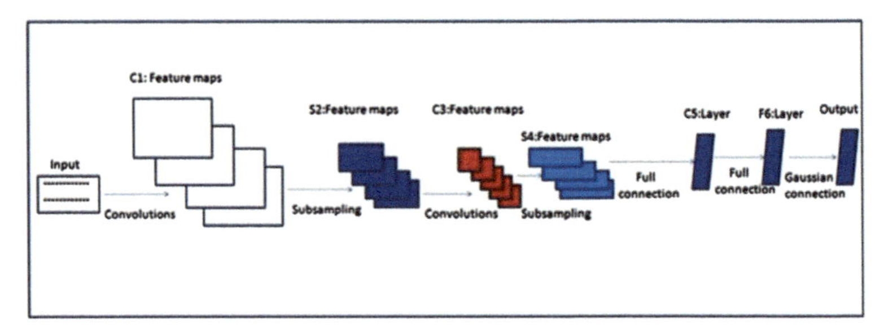

Figure 1.1 LeNet5's typical architecture.

spatially distributed input patterns because it masters through filters; and it attains higher accuracy than various shallow frameworks because filters [6] are combined to e-applications of the LeNet5 architecture in plant pathology that include the detection of degraded components, categorization of plant images, identification of disease symptoms, classification of diseases, and analysis of plant diseases.

1.3 AlexNet

Greater than LeNet, AlexNet is a tree of layers where the first levels are made up of nodes and the later layers are made up of computation units. A classic AlexNet [7] design consists of 50 million parameters and 650K neurons, and it takes five to six days to train the framework. The broader framework of AlexNet contrasted to LeNet compels it to be suitable to do computer-vision-associated [8] tasks. It has eight layers, the first five of which are visible are fully linked layers, while the last three layers are convolutional layers. Figure 1.2 displays an example of AlexNet's design.

The AlexNet architecture faces a number of difficulties, including performance degeneration in extensive image processing, repeated bottlenecks in challenging image classification issues, over-fitting issues, difficulty in training larger networks, accuracy issues when concerned with low-resolution images, lack of concurrent framework transparency, erupting gradient issues in training, and errors when fine-tuning the network.

Some potential benefits of the AlexNet architecture include: using rectified linear units [9] rather than that of the tanh function, which constitute it to be faster and lowers the error rate by 25% contrasted to LeNet; allowing training the network utilizing various graphical processing units [10], which shortens the time-span required to train the huge-size network; and exhibiting the capability to pool the adjacent deposit of neurons of little to no overlapping, which avoids the emergence of new connections.

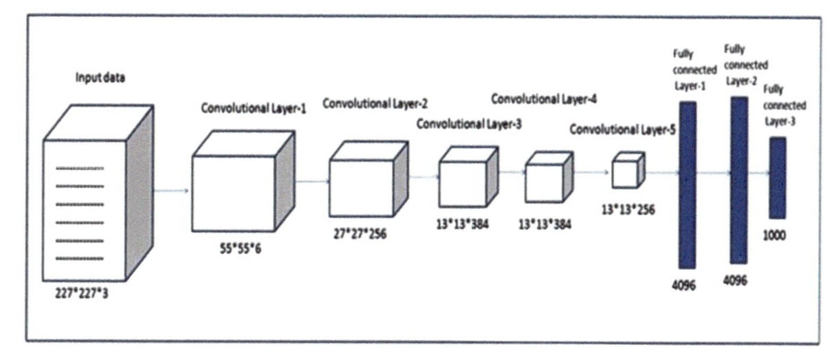

Figure 1.2 Typical AlexNet architecture.

Herb image vision, disease estimation, fine-grained leaf image segmentation, earlier aster yellows detection, video feature extraction of bacterial wilt, bacterial bite recognition, birds-eye spot recognition, image processing of chlorosis, abnormality recognition in seedling images, greensickness approval, and other implementations of AlexNet framework in herb pathology are a few of them.

1.4 VGGNet

VGGNet is an improved variant of AlexNet that uses 3 × 3 kernel-sized filters in lieu of the large-sized kernel. It is a uniform kind of network with 16 convolutional layers and has a highly alluring appearance. Given that AlexNet has more filters and 3 × 3 convolutional layers than other networks, there is a great deal of similarities between them. VGGNet [11], which is made up of 128 neurons and 138 parameters, takes two to three weeks to train on a normal 4-GPU machine. Rather than having to boost the depth of learning, a single big kernel is combined with several smaller kernels [12]. The network's breadth starts at 64 and grows by a factor of 2 when each sampling layer or pooling layer is added. Figure 1.3 displays a typical VGGNet design.

Some potential benefits of the VGGNet architecture include the ease with which eight configurations can be implemented, the high accuracy attained in extracting the features of the images, the reduced training time required due to the low quantity of parameters, the elimination of the vanishing gradient issue by bypassing the configurations in the network, the ability to identify incorrect object classifications at multiple scales, and the low cost of learning complex structures.

Plant cell wall categorization, stomata visualization, soft rot recognition, determining the plant's acid tolerance level, leaf rust recognition,

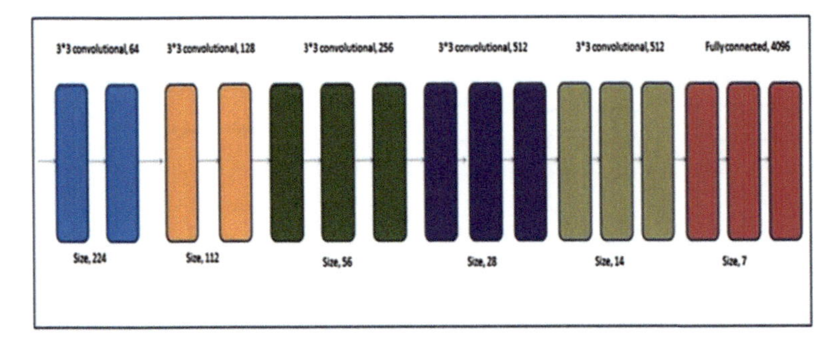

Figure 1.3 Typical VGGNet framework.

stem rust identification, computerized recognition of white mold development, detection of bacterial oozes, fruit spot scrutiny, canker segmentation, and other implementations of VGGNet framework in herb pathology are just a few.

1.5 GoogLeNet

For the deep nesting and object identification in Google's planned image recognition operations, there are 22 layers in the GoogLeNet [13] network. As the quantity of parameters analyzed decreases from 50 to 5 million, the design of GoogLeNet differs from that of AlexNet and ZFnet. It includes an auxiliary classifier, global average pooling, an inception module, and 1×1 convolution to facilitate the creation of deeper model architecture. The 7×7 pictures are entered into the global average pooling layer, which averages them down to 1×1. The inclusion of a bottleneck layer results in a significant decrease in the amount of computation necessary to run the network. Figure 1.4 depicts the usual design of GoogLeNet.

The GoogLeNet architecture may have some benefits, including the reduction of training parameters due to the use of smaller-level convolutions, the depth of the framework's use of 1×1 convolution [14], the high trainable accuracy attained due to the application of global average pooling at last, the ability to drive I/P images of various sizes with superior scalability due to the use of convolutional filters of various sizes, and the ability to perform precise classification.

Mosaic leaf design, wrinkled leaves identification, processing of herb stunts, leaf blight categorization, canker finding, stem rot estimation, fruit rot identification, dampening of seed plant diagnosis, pest bite estimation, plum assault recognition, soaked lesion improvement monitoring, concentric ring spotting, tuber disease inspection, atrophy fraud recognition, and dwarfing

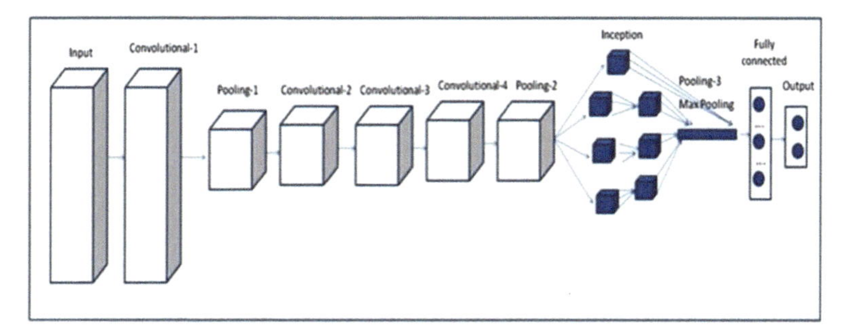

Figure 1.4 Typical GoogLeNet architecture.

categorization are fewer implementations of GoogLeNet framework in herb pathology.

1.6 ResNet

The pyramid-shaped cells in the cerebral cortex serve as the foundation for Kaiming's residual kind of neural network, known as ResNet, which works by bypassing numerous repeating levels. There are many types of ResNet designs, but ResNet50 and ResNet101 [15] are the most often used types. It has two 3 × 3 convolutional layers and the ability to learn around 11 million parameters at once. The network can accommodate input photos of various sizes since the pooling layers are situated at the network's beginning and end. Residual block that plays the function of learning and is recurrent across the network is the building block of the ResNet. The weights allocated to midway layers are set to 0 and progressively increased to 1 for achieving optimum learning. It is possible to create two different types of mapping: non-trainable mapping, which performs padding of zero, and trainable mapping, where a convolutional layer is in charge of mapping the I/Ps to the O/Ps. Figure 1.5 displays a representative ResNet design.

The vanishing gradient problem, auxiliary loss from extra supervision, structured regression issue, uneven network dimension, under-fitting, exponential rise in the quantity of free parameters, failure to determine the multi-class image classification [16] problem, inadequate learning rate, incorrect data segmentation, incorrect handling of gradient descent in deep networks, and unequal design are some of the difficulties encountered in ResNet.

Some potential benefits of the ResNet architecture include its ability to train large NNs with thousands of neurons because it is less complex than conventional networks; the vanishing gradient problem is ably managed because it allows for multiple levels of nesting; the likelihood of misclassification

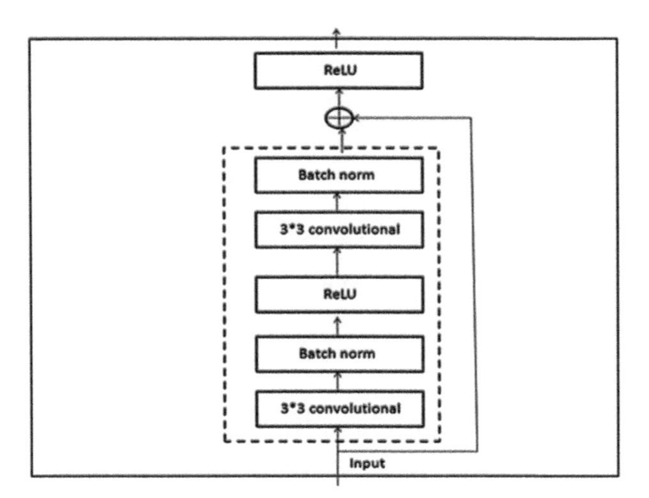

Figure 1.5 Typical ResNet architecture.

errors is reduced by enlarging the depth of the network; and the ability to execute quick network training because the ReLu layer remains over the top layer.

Powdery mildew estimation, internal obligate pattern recognition, leaf coating categorization, fungus spore tracking system, imaging of herb shot holes, damages developing in stems, segmentation of affected tissues, swelling recognition, abnormal expansion in plant organ tracking, damage of turgidity recognition, automated tracking of smoky leaves, downy mildews tracking, and white conidia for white mold are fewer implementations of ResNet framework in herb pathology.

1.7 ZFnet

Zeiler and Fergus created ZFnet, sometimes known as Zeiler Furgus Net, an upgraded form of AlexNet, in 2013. To avoid losing the image's pixel information, ZFnet shrinks the size of the filter. It is primarily made up of three fully connected feed-forward layers, dropout layers, max-pooling layers, and five convolutional layers in shared form. The first layer uses a 7 × 7 filter with a reduced stride value, while the final layer uses a softmax filter. The deconvolutional layer is utilized to see and comprehend convolutional networks on a fundamental level. The third, fourth, and fifth convolutional layers now include more activation maps, which increases the network's capacity for defect detection.

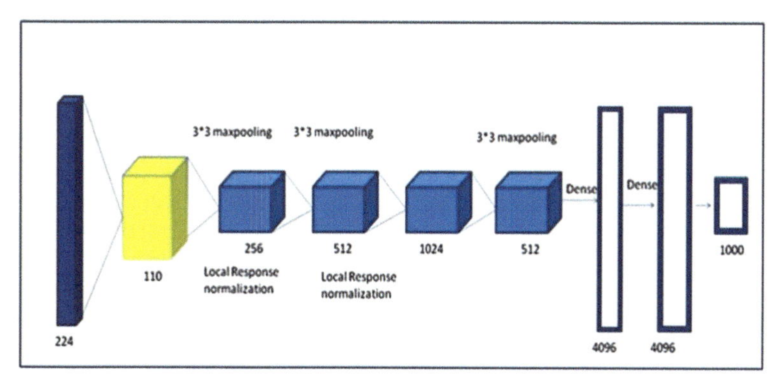

Figure 1.6 ZFnet's usual architecture [12].

ZFnet faces a number of difficulties, including max layer overlap, an increase in image recognition error rates, an incompetence to handle various sources, small receptive fields, better debugging issues in hidden layers, an increase in processing time for larger documents, imperfect accuracy when processing distorted images of high quality, ineffective segmentation of ultrasound networks, an incompetency to handle noise in the I/P data, and an inability to extract baseline features in node detection. The accuracy [17] attained by ZFnet is greater than AlexNet's because it employs 7 × 7 filters, and the architecture's significant reduction in error rate has two possible benefits. Enhanced parameterization and use of deep learning mechanism enhances visualization approach, increased classification accuracy makes it suitable for application in cutting-edge technology to assist humanity. ZFnet uses 256 depth-wise feature maps of the input photos to mitigate the issue of over-fitting and similar problems.

Leaf drooping identification, herb pathogen processing, herb health assessment, stem disease administration, digital herb identification, spotted leaf spot recognition, leaf curl segmentation, oak wilt premature identification, microbes ooze categorization, mutilated herb tissue spotting, gray mold recognition, brown marks parsing, and dropping leafs estimation are fewer implementations of ZFnet [18] framework in herb pathology.

1.8 Efficiency Assessment

The comparison of the effectiveness attained by different CNN categories, i.e., Table 1.1 compares the performance of AlexNet, VGGNet, and ZFnet in terms of metrics like the quantity of parameters, error rate, depth, and accuracy. From Table 1.1, it can be seen that ZFnet and ResNet [19] operate at a high efficiency level due to their high precision and low mistake rate.

Table 1.1 Comparison of the effectiveness of CNN kinds.

CNN categories	Quantity of parameters	Error rate	Depth	Accuracy	Favorable feature
LeNet	70,000	25%	8	80.3	Elementary-layer composition
AlexNet	50×10^6	18%	10	78.6	Deeper
VGGNet	140×10^6	8.5%	20	92.4	Certain size kernel
GoogLeNet	5×10^6	7.5%	25	61.5	Wide parallel kernels
ResNet	25×10^6	4.2%	160	94.6	Shortcut connections
ZFnet	70×10^6	15.3%	10	97.4	Visualization layers

1.9 Conclusion

This chapter gives a quick overview of plant fungal infections and the harm they do to the production of the plant. The identification of fungal plant diseases is compared using conventional and automated methods. It describes several of the probable convolutional neural network frameworks for identifying herb illnesses, including AlexNet, VGGNet, ResNet, and ZFnet. The effectiveness of the aforementioned CNN designs is examined in accordance with performance metrics including precision, error rate, depth, and the quantity of parameters used in training. When compared to other CNN architectures that were taken into consideration for study in order to identify plant diseases, ZFnet and ResNet had higher efficiency.

References

[1] Arif, Y., Bajguz, A. & Hayat, S. (2023). *Moringa oleifera* Extract as a Natural Plant Biostimulant. *J Plant Growth Regul* 42, 1291–1306 https://doi.org/10.1007/s00344-022-10630-4

[2] Batista, E., Lopes, A., Miranda, P. *et al.* (2023). Can species distribution models be used for risk assessment analyses of fungal plant pathogens? A case study with three Botryosphaeriaceae species. *Eur J Plant Pathol* 165, 41–56. https://doi.org/10.1007/s10658-022-02587-7

[3] Praveenkumar, S., Veeraiah, V., Pramanik, S., Basha, S.M., Lira Neto, A.V., De Albuquerque, V. H. C. and Gupta, A. (2023). Prediction of Patients' Incurable Diseases Utilizing Deep Learning Approaches, ICICC 2023, Springer

[4] Chellam, V. V., Veeraiah, V., Khanna, A., Sheikh, T. H., Pramanik, S. and Dhabliya, D. (2023). A Machine Vision-based Approach for Tuberculosis Identification in Chest X-Rays Images of Patients, ICICC 2023, Springer

[5] Mondal, D., Ratnaparkhi, A., Deshpande, A., Deshpande, V., Kshirsagar, A. P. and Pramanik, S. (2023). Applications, Modern Trends and Challenges of Multiscale Modelling in Smart Cities, in Data-Driven Mathematical Modeling in Smart Cities, IGI Global, 2023, DOI: 10.4018/978-1-6684-6408-3.ch001

[6] Pramanik, S. (2023). An Adaptive Image Steganography Approach depending on Integer Wavelet Transform and Genetic Algorithm, Multimedia Tools and Applications, 2023, https://doi.org/10.1007/s11042-023-14505-y.

[7] Chandan, R. R., Soni, S., Raj, A., Veeraiah, V., Dhabliya, D., Pramanik, S., Gupta, A., (2023). Genetic Algorithm and Machine Learning, in "Advanced Bioinspiration Methods for Healthcare Standards, Policies, and Reform", Hadj Ahmed Bouarara, IGI Global, DOI: 10.4018/978-1-6684-5656-9

[8] Mandal, A., Dutta, S., Pramanik, S. (2021). Machine Intelligence of Pi from Geometrical Figures with Variable Parameters using SCILab", in Methodologies and Applications of Computational Statistics for Machine Learning, D. Samanta, R. R. Althar, S. Pramanik and S. Dutta, Eds, IGI Global, 2021, pp. 38–63, DOI: 10.4018/978-1-7998-7701-1.ch003

[9] Pramanik, S. and Bandyopadhyay, S. (2023). Identifying Disease and Diagnosis in Females using Machine Learning, in Encyclopedia of Data Science and Machine Learning, Eds. John Wang, IGI Global, 2023, DOI: 10.4018/978-1-7998-9220-5.ch187.

[10] Bansal, R., Obaid, A. J., Gupta, A., Singh, R., Pramanik, S. (2021). Impact of Big Data on Digital Transformation in 5G Era", 2nd International Conference on Physics and Applied Sciences (ICPAS 2021), doi:10.1088/1742-6596/1963/1/012170, 2021.

[11] Jayasingh, R., Kumar J., R.J.S, D. B. Telagathoti, Sagayam, K. M. and Pramanik, S. (2022). Speckle noise removal by SORAMA segmentation in Digital Image Processing to facilitate precise robotic surgery, International Journal of Reliable and Quality E-Healthcare, vol. 11, issue 1, DOI: 10.4018/IJRQEH.295083, 2022

[12] Veeraiah, V., Talukdar, V., Manikandan K., Talukdar, S.B., Solavande, V.D., Pramanik, S., Gupta, A. (2023). Machine Learning Frameworks in Carpooling, in Handbook of Research on AI and Machine Learning Applications in Customer Support and Analytics, Eds, Md S. Hossain, R. C. Ho and G. Trajkovski, IGI Global, 2023.

[13] Pramanik, S., Obaid, AJ, N M, and Bandyopadhyay, S. K. (2023). Applications of Big Data in Clinical Applications", Al-Kadhum 2nd

International Conference on Modern Applications of Information and Communication Technology, *AIP Conference Proceedings* 2591, 030086 (2023), https://doi.org/10.1063/5.0119414

[14] Dushyant, K., Muskan, G., Gupta, A. and Pramanik, S. (2022). Utilizing Machine Learning and Deep Learning in Cyber security: An Innovative Approach", in Cyber security and Digital Forensics, M. M. Ghonge, S. Pramanik, R. Mangrulkar, D. N. Le, Eds, Wiley, 2022, https://doi.org/10.1002/9781119795667.ch12

[15] Meslie, Y., Enbeyle, W., Pandey, B. K., Pramanik, S., Pandey, D., Dadeech, P., Belay, A., Saini, A. (2021). Machine Intelligence-based Trend Analysis of COVID-19 for Total Daily Confirmed Cases in Asia and Africa", in Methodologies and Applications of Computational Statistics for Machine Learning, D. Samanta, R. R. Althar, S. Pramanik and S. Dutta, Eds, IGI Global, 2021, pp. 164–185, DOI: 10.4018/978-1-7998-7701-1.ch009

[16] Bandyopadhyay, S., Goyal, V., Dutta, S., Pramanik, S. and Sherazi, H. H. R. (2021). Unseen to Seen by Digital Steganography", in Multidisciplinary Approach to Modern Digital Steganography, S. Pramanik, M. M. Ghonge, R. Ravi and K. Cengiz, Eds, IGI Global, pp. 1–28, 2021, DOI: 10.4018/978-1-7998-7160-6.ch001

[17] Pramanik, S., Galety, M. G., Samanta, D., Joseph, N. P (2022). Data Mining Approaches for Decision Support Systems, 3rd International Conference on Emerging Technologies in Data Mining and Information Security, 2022.

[18] Bhattacharya, A., Ghosal, A., Obaid, A. J., Krit, S., Shukla, V. K., Mandal, K. and Pramanik, S. (2021). Unsupervised Summarization Approach with Computational Statistics of Microblog Data", in Methodologies and Applications of Computational Statistics for Machine Learning, D. Samanta, R. R. Althar, S. Pramanik and S. Dutta, Eds, IGI Global, pp. 23–37, DOI: 10.4018/978-1-7998-7701-1.ch002

[19] Anand, R., Singh, J., Pandey, D., Pandey, B. K., Nassa, V. K. and Pramanik, S. (2022). Modern Technique for Interactive Communication in LEACH-Based Ad Hoc Wireless Sensor Network", in Software Defined Networking for Ad Hoc Networks, M. M. Ghonge, S. Pramanik and A. D. Potgantwar, Eds, Springer, 2022, https://doi.org/10.1007/978-3-030-91149-2_3

2

Technologies based on the IoT and Artificial and Natural Intelligence for Sustainable Agriculture

S. K. Manju Bargavi[1], Mohan Vishal Gupta[2], Rajeev Kumar[3], Ahmar Afaq[4], Pratieek Sikchi[4], and Dac-Nhuong Le[5]

[1]Department of Computer Science and IT, School of CS and IT, Jain (Deemed to be) University, Bangalore, India
[2]College of Computing Science and Information Technology, Teerthanker Mahaveer University, India
[3]Department of Agriculture, Sanskriti University, India
[4]Symbiosis Law School, Nagpur Campus, Symbiosis International (Deemed University), India
[5]Faculty of Information Technology, Haiphong University, Vietnam

Email: b.manju@jainuniversity.ac.in; mvgsrm@indiatimes.com; rajeev.ag@sanskriti.edu.in; ahmar@slsnagpur.edu.in; nhuongld@hus.edu.vn

Abstract

The authors discuss the key prospects and new problems presented by blockchain technology for the Russian digital economy in this chapter. The authors mention that blockchain innovator group vigorously debate fewer issues and provide a wide range of possible results. The study in Russia revealed that the bulk of these concerns were not addressed by the Russian research fraternity. Instead of a completely open network, some of these may be solved utilizing private or consortium blockchain. In terms of end-user assistance (difficult to utilize and comprehend) as well as developer support (inadequacy of tools), technical obstacles are now rather minor. The efforts toward model-based implementations of blockchain utilizations, as well as the sliding mode in intellectual control and transmission, as well as the tools that have been developed to assist with technical issues, are some of the most recent developments in developer assistance. The chapter demonstrates how avatars may interact with one another using a range of communication techniques for smart agriculture and sustainable farming.

2.1 Introduction

Unquestionably, the degree of computerization and the broad use of newer technology contributed to the globe entering a new digital age. High-end technologies are rapidly advancing all aspects of our lives. Throughout this time utilization of 3D printing [1], the IoTs, VR and AR, cloud technologies, and the "Industry 4.0" is being driven by the development of quantum technologies, robots, and other technologies.

Since there were no farms using sensor technology before the turn of the century, it is anticipated that the number would rise to 650 million by 2025 and up to 2 billion intelligent agro-sensors by 2050. Additionally, it is anticipated that during the years 2018–2022, the agro-industrial complex's overall average yearly growing rate for the IoT demands (agricultural IoT or AIoT) will be 16%–17%. The minimal economic impact of implementing IoT technology into the agro-industrial complex by optimizing labor expenditures and lowering harvest damages, fuels, and lubricants by 2027 may be 500 billion rubles, in accordance with the PwC prediction.

The long-term effects of IoT technology adoption in agriculture will largely be characterized by considerable material and resource savings, and, as a consequence, cost optimization for agricultural operations will take place. Additionally, new technology will boost yield, which will raise income. All of this will, in the future, directly affect the competitiveness and marginality of businesses. According to projections for the global food market, consumption will naturally shift in favor of agricultural goods by 2050 due to an increase in the Earth's population of around 2.3 billion people (or about 33%) and an improvement in its standard of living. Moreover, the Russian Federation has the greatest potential for expanding the amount of agricultural land required for food yield. Thus, one of the most important responsibilities is to find and integrate new techniques and technologies in the agro-industrial complex while keeping in mind current developments and the key trends in global growth. This problem is being actively explored and resolved at the state level. Currently being developed as portion of the National Technology Initiative is the idea of a FoodNet market strategy. The FoodNet market is a market for the manufacture and trading of food and finished food commodities (personalized and general, depending on conventional resources and their alternatives), and associated IT recommendations (for instance, offering logistics and selection services for individualized nutrition). Smart farming, enhanced selection, new raw material sources, reasonably priced organics, and individualized nutrition will be the major market areas. According to NTI, the ultimate goal is to develop these services and solutions that, depending on the market niche, will occupy somewhere between 5% and 15% of the worldwide market. The utilization of IoT technology served as the foundation for the growth of the aforementioned FoodNet market sectors.

As part of carrying out the instructions of the Deputy Prime Minister of the Russian Federation and the talks with Russian President Vladimir Putin dated May 19, 2018 “On initiatives to apply the state research and development policy in the motivations of farming progress” and “For implementing the approach for enhancing the grade of food products until 2035,” dated June 5, 2018, an outline plan was built for the growth of the IoT in the Russian agricultural–industrial complex. Having the assistance of the Open Government, the IoT Association (IoTA), the Department of IT, and the Russian Ministry of Farming, specialists from the Internet Initiatives Development Fund (IIDF) developed a paper. According to the proposed strategy, domestic element-based developments for tools in the IoTs should account for 15% of advancements by 2020 (up from 7% now), and 30% of agricultural firms should be employing IoT solutions. Additionally, at the same time, a minimum of 25 main projects for the use of IoT technology in the agriculture sector are expected to be implemented.

2.2 Methods and Materials

Because of the technical potentials and scalability for multiple use cases, the blockchain technology attracted the attention of several people and businesses. These use cases make it abundantly evident that system and experienced designers should acquire the necessary skills and continue honing them as technical advances. After the progress brought about by this quick establishment, there are now three forms of blockchain technology [2]: “public,” “private,” and “federated.” These blockchains basically have identical features. They mostly depend on the use cases, authorization levels, and security when it comes to discrepancies. The blockchain has several advantages for businesses, including time savings over work processes, reduced expenses, less risk, and increased trust. We designers will be able to predict how this technology will change the way our clients' companies operate if we are aware of its values and advantages, in addition to having the knowledge and assurance we require to lead our clients and provide the best answers for their problems. But for that to occur, the company has to have some kind of network to provide a strong basis for a suitable blockchain use case. The adage "with great technology comes great responsibility" matters concerning technology. The blockchain eliminates reliance on centralized authority and outside parties by returning control and possession of information to its correct holder.

2.3 Background

One of the important technologies that, at this time, have the power to fundamentally change both the global and national wealth is the IoTs. In light of multiple factors and overall trends, the agricultural–industrial complex is the most crucial sector where these technologies may and should be used. Fundamental societal changes are primarily determined by the fast progress of ICTs [3] and their global usage. The growing priority of human capital as the cornerstone of the knowledge economy is another feature of the new social realities. The advancement of research and education is crucial for the creation of social capital. An individual today requires being able to properly handle in quick altering habitats, have the correct decisions quickly, and gain experience for being a fortunate and an extremely trained person. This is only feasible if the person continuously refreshes and modifies his mastery, expands his expertise, and acquires experience. This implies that his education extends beyond the boundary of a school and that he will get the proper certificate. A person just has to continue learning throughout his or her life, and the information era offers many possibilities for this. Lifelong learning is a concept that is particularly pertinent. Without the use of ICTs, which may offer the greatest flexibility and various structures, it is impossible to obtain continuing education, which is a basic mode in the progress of education, including the facts, individuality, difference, and diverseness. Let us look more closely at how the advancement of ICT affected knowledge and resulted in changes to different training methods. It is important to note that representations of educational materials, such as visual aids and technological teaching aids, are changing more than the actual training methods themselves.

The modifications made to the way educational data is shown reflect the advancement of technology tools. Thus, one can follow the evolution of visual materials from paper to electronic form and finally to the utilization of different technological presentation instruments. The growth of the Internet greatly increased the options in sharing instructional data, initially via email, then chat, and, social networks, giving educators the chance to interact with many students at once. It is possible to distinguish between the subsequent categories of ICTs: electronic, computer, multimedia, remote, the Internet, and video conferencing technologies. The combination of personal computers into the educational technique is linked to the development of educational information technology. Electronic innovations from the 1970s and 1980s were predicated on the usage of computers and information retrieval systems, but by the 1980s, computer technologies had already supplanted them. During this time, the personal computer evolved as a tool for education.

An examination of global experience with industry transformation into the digital economy and the shift to CPSs for the IoTs to the Fourth Industrial Revolution, or Industry 4.0 [4], is being implemented as a portion of the change to the digital economy and the initiation of intelligent manufacturing, digital manufacturing, internet of manufacturing, and open manufacturing technologies, according to an analytics of global event in modifying industry into the digital economy and the alteration to IIoT [5].

For a very long time, research publications have laid out the fundamentals of the shift to digital economy methods for using technology (IoT, Smart City, and big data) effectively. It was stated, for instance, that combining building information modeling (BIM) with GIS [6] technology is the key to creating systems that function well throughout the design, construction, and usage of a facility. Leading international specialists and practitioners have come to this conclusion.

Cyber–physical systems are fundamentally a connection between software and electronic systems and the physical production processes or other processes (such as the control of electric power transmission and distribution), which necessitate the feasible execution of continual command in real time. In Russian literature, this subject has not received much attention. Nevertheless, its significance is clear.

Multidimensionality, structural complexity, and functional complexity are characteristics of cyber–physical systems. This field of study is still in its infancy and is at the nexus of many other academic fields. All of this establishes the need for creating appropriate procedures for their creation. The model-based strategy shows the greatest promise. Review of the CFS modeling, integration, and design processes is provided, along with indications of their utilization. The variety of illustrations of these systems, which include physical, cybernetic, and communication components, need a consistent technique to the explanation that will facilitate the reuse of components, facilitate portability, and enable interoperability. The languages may be used to some degree to meet these needs. The drawback of UML [7], SysML, and XML is again a concentration on the display of syntactic information or constrained semantics. Technicalities based on the semantic web provide considerably more opportunities. Ontology-based descriptions, for instance, allow for the presentation of both syntactic and semantic data. Ontological rezoning is another option, and it may be helpful in the analysis, verification, and validation of the ontological model.

The prototypes and approaches put forth in the chapter are in line with the state-of-the-art in scientific research, as evidenced by the rise in recent years in the quantity of research publications in the area of designing and developing intelligent cyber–physical frameworks like "Intelligent Energy Grids," "Smart Roads," and "Smart Cities" [8]. These frameworks are based on automations in decentralized large data processing methodologies for machine learning and intellectual analysis, M2M interaction of cyber–physical tools in the industrial Internet of Things network, etc.

The Smart Road is an illustration of a smart environment for a smart city. As a result, research and development in this field take up a lot of space. A smart transport facility tracking system – the winner of the Geospatial World Excellence Award is Imagem and Antea Group's Smart City Road Monitor – is a key component for such settings. The creation of a road traffic modeling framework in predicting vehicle events with coordinate recommendations to digital map layers is an instance of experimentation on the survey of streaming information and predicting the evolution of circumstances in this field. It is suggested to employ sensor network nodes (a kind of "foggy" computation) to handle sensory data in cyber–physical systems (CPS). Since wireless sensor networks are already extensively employed in many aspects of human life, scientists and researchers are very interested in using them to generate new, cutting-edge technologies. Leading manufacturers of hardware and software, like NXP, provide creative solutions for combining various network frameworks and protocol stacks, such as the sensor node's coupled ZigBee and Bluetooth modules. As a variant of the cloud computation concept, the fog computation platform uses sensor nodes instead of servers as the computation nodes in decentralized data processing. These nodes have constrained computation and energy capabilities. A semantic methodology depending on the usage of models is proposed in system architecture, scrutinizing needs, simulation, and evaluation of CPS [9]. Ontologies and ontological models have just lately started to be employed in modeling CPSs. Domain ontologies are utilized in this research in computation and decision-making. Work is involved in creating the knowledge frameworks that will facilitate the proper "correct-system." This chapter offers fresh ontological understanding; decision-taking for CPS may be supported by a foundational and logical conclusion. This extends the approach to the design of CPS based on model management by enabling the construction of deterministic, verifiable, and viable frameworks of CPS backed by dependable semantics.

A method is put forward to create a digital representation of every piece of information that can be obtained from and about an item, which might be an H/W platform or S/W platform. Semantic knowledge granting methods like RDF, RDF Schema, and OWL are the foundation of digital display. Here, the authors moreover offer the idea of the Semantic I4.0 element that makes use of semantic technologies to address the issues of comprehension and communication in Industry 4.0 situations.

The overview supplied above provides evidence for the use of ontological models in modeling cyber–physical system contexts, cyber–physical system structures, and cyber–physical system project verification systems, choice-making throughout the design phase, and informational display of an item. Nevertheless, it is also true that, as of right now, there are no tasks that utilize dynamic ontological frameworks in which modifications will be included into their semantics.

The ontological method may eventually totally replace model-oriented design. Ontology-dependent S/W evolution, in general, assigns to novel approaches that use ontologies to enhance models, methodologies, and software development procedures. Benefits include enhanced degrees of inter-linkage and combination between SW system elements, optimum verification of program code, and reuse of artifacts.

Automations in knowledge extraction and intelligent investigation of large data include anticipating the acts of CPS and techniques as well as management technique creation. Big data supports a variety of characteristics. Higher speed data production and processing in real time enables one to decide how particular influences on the control process will affect it in the best way possible. Diversity in a broad span of data produced from multiple sources in different forms, having varying size and organization, categorized into numerous categories pertaining to various facets of the management technique, permit for the preparation of classifications, groups, and correlations. A thorough and heterogeneous data processing approach is necessary due to the complication of processing and data management, as well as the diversity of data gathered from multiple sources.

The evolution of the technique of deep analysis of data (data mining) may be seen in the method of deep analysis of processes (process mining), but process mining produces an O/P which explains the dynamics of the framework. Wil van der Aalst is the forerunner of the process mining [10] technique. The group using this methodology is actively developing it. The fundamental tenets of the process mining methodology are explained. In tracking frameworks, the process mining technique starts to be diligently applied. As a consequence of investigating and analyzing the framework utilizing the process mining technique and contrasting the results with the formal description of the tested framework, a novel method for the automated production of trust attributes is presented.

It is suggested to use the process mining approach to both a conventional and online audit system. Additionally, a continuous information monitoring system is suggested, which, by keeping an eye on risk variables in businesses and organizations, may anticipate and avoid dangers in the big data [11] world. This chapter's goal is to create an initial risk factor confirmation framework utilizing real-world sales audit cases. The growth of the semantic component's process mining approach represents a potential direction. There was an effort in that direction. In order to assist real-time process identification, monitoring, and improvement, this research investigates the learning technique – how data from diverse technique domains may be retrieved, semantically processed, and translated into workable mining forms. Additionally, the suggested approach enables individual pattern/behavior prediction via additional semantic analysis of the produced frameworks. The study suggests formalizing the supposed "semantic learning process mining (SLPM)," which is technologically realized as a "fuzzy semantic miner."

According to researchers, an examination of the review work on the application of the process mining technique in tracking frameworks revealed that the expansion of the process mining technique's semantic element depending on the usage of ontologies is a potential but underexplored area.

As a final point, the authors would like to point out that a review of the literature survey and current design research works in this area of study revealed a number of issues, including the shortage of suitable mathematical frameworks in the analytics of big sensory data and the detailed examination of techniques in CPS systems, the shortcomings of frameworks for detecting hidden patterns in time series and event logs, accounting for the impact of random fluctuations and outside influences on the nature of the CPS, and the complexity of technology to obtain my decision-making in the technique of tracking and process control in CPS. It supports the project's significance, which aims to provide fresh frameworks and approaches in conceptualizing and creating CPS.

2.3.1 The principal scientific rivals

With the help of the government, research institutions and universities are conducting the most extensive studies in the area of knowledge connected to the developing of CPS and technicalities for the IoT in the USA, China, Europe, and Korea. We take a look at some of the efforts in this field.

1. In order to improve distributed computing performance, we simplify the feasible usage of cloud computations, and boost the security and solidity of decentralized computing frameworks; Intel is creating concepts, methodologies, and models for cloud and GRID computation. The creation of tools and solutions in cloud data centers is the primary emphasis.
2. Partly of the execution of the "Intelligent Human Fraternity" idea, Mitsubishi is employed in the area of mixing cloud computation, big data, and intelligent automations to assist the research of the energy sector, medicine, and facilities.
3. With the use of a network architecture made up of several sensors and dispersed data processing frameworks built on a fog and cloud computation paradigm, Cisco is creating Internet of Things technologies. In order to manage multimodal transportation and smart cities, Mitsubishi and Cisco are collaborating on development into the IoTs [12], machine communications, fog, cloud computing [13], and ubiquitous wireless internet. The major objective is to improve the productivity, functional capabilities, and efficiency of technological processes in industry, transportation, and the city environment. The foundation is the creation of the Internet of Everything by fusing Toshiba Group technology for network point management with Cisco fog computing network architecture. This will enable one to design distributed computing solutions and monitor and manage globally dispersed devices.

The following individuals do scientific study on the design and implementation of an industrial IoT network in CPS, intelligent energy grids, and the usage of fog computation:

1. A group of scientists from the University of Ottawa in Canada and Deakin University in Burwood, Australia, under the direction of Professor Ivan Stojmenovic: They talked about how to design hybrid control frameworks in decentralized tracking frameworks, create linked cars and intelligent traffic signals, and establish smart roads for UAVs [14] (smart automobile).

2. The idea of erecting a computation environment dependent on mobile transmission tools that will concurrently resolve decentralized data processing jobs in the circumstances is put forth by a scientific team of researchers under the direction of Prof. Alfredo Alexander-Katz (MIT, USA) based on permanently installed PCs with an internet connection. Depending on the movement of cloud and fog computing frameworks, a batch of academics is creating a migration framework for decentralized data processing. Relying on the tools available at the moment when altering the capacity to access tools, altering bandwidth, etc., this enables one to employ cloud–fog tools in a changing manner, for instance, to do computations again in the cloud or in a foggy environment.
3. The group of scientists led by Luca Daniel: He is presently a professor at MIT, USA teaching on CPS in design and manufacturing. He oversaw the Industry 4.0 area of expertise at the UC San Diego in California, USA, where he also served as a part-time professor.
4. Prof. Justin Soloman (MIT, USA) research group: His primary duties consist of the planned growth of automation programs for the smart grid and the verification of development projects.

Research rivals in this area of knowledge might moreover consist of:

1. Prof. Pieter Abbeel teaches at the University of California at Berkeley in USA. His development focuses on the utilization of ICTs to industrial systems and automation. At University of California, he is the director of the FAST lab, whose objective is to combine human and machine intelligence. He has participated in several US R&D projects like Arrowhead, C2NET: Cloud Collaborative Manufacturing Networks, Movus: S ICT Cloud-based platform and mobility services: available, universal and safe for all users.
2. The following are Mark Austin's current research areas: a) system framework and combination of CPS depending on ontological frameworks; b) modeling, survey, and implementation of CPS; c) semantic techniques to modeling the habit of decentralized CPS. Mark Austin is an associate professor at the University of Maryland in the United States.
3. Prof. Amay Karkare teaches at the India Institute of Technology Kanpur. His research focuses on the design of cyber–physical control systems, formal specifications of CPS, data mining and the synthesis of frameworks depending on intelligence, multi-agent systems, and

programmable controllers. He also models decentralized embedded and production frameworks and develops computational technologies for production, automates production, and creates cyber–physical control systems.

4. Prof. Edward A. Lee of the University of California, Berkeley, in the United States, specializes in the creation, modeling, and implementation of real-time cyber–physical frameworks. Both the Berkeley Ptolemy Project and the Berkeley Computer Research Center are under his direction.

5. German Professor Michael Georg teaches at the Technical University of Munich, Germany. He is the leader of a research group that focuses on data science and thorough process analysis. Data science, detailed process survey, business management, and simulation are some of his key areas of interest.

2.4 Center of the Chapter

Provisions, disagreements, concerns, issues, answers, and suggestions:

Works by Russian and international authors on the analytical and scholarly provisions of employing IoTs in the agro-industrial system, as well as disclosure on the subject of study in the monthly press and online, served as the foundation for this work. The traditional general scientific research procedures used in this methodology include analysis, synthesis, induction, deduction, generalization, categorization, and equivalent and methodical investigation. The idea behind the IoT is that objects can recognize themselves by utilizing tags (like barcodes, for example), meaning that every component has data about its origin, intended use, and other details. The phrase "Internet of Things" first originated in 2009, when there were more gadgets online than people on the planet.

When the Internet of Things (IoT) industry is discussed, products that facilitate machine-to-machine analogy, like telemetry or tracking the position of manufacturing provisions, are often mentioned as examples of these technological phenomena. These solutions are closed systems with a strong industrial focus that are often used with specialized hardware and software. A broad range of technologies and solutions from several providers that are a part of the IIoT market's ecosystem support all phases of the IoT's growth and the interlinks between frameworks.

Corresponding to methodology findings from numerous consultation firms, the adoption of the IoTs in the agricultural–industrial complex would help to increase production ability, improve population standard of life, and address environmental issues, ensuring the long-term viability of the sector.

The farming industry has historically been susceptible to a wide range of issues, from the use of incorrect planting and harvesting techniques and faulty forecast data to inappropriate irrigation and inferior soil standards. Naturally, the entire thing has a detrimental impact on the performance as a whole. By receiving high-precision data in real time from the fields where they are situated, the Internet of Things in agriculture may greatly lower such hazards. Specialists are able to make important choices, such as when to irrigate, harvest, etc., based on the information they get. This system reduces essential threats and enables farmers to decide wisely, in the production and the planning phase.

Precision agriculture [15], often called "satellite farming" is a method for regulating agricultural yield that is depending on the utilization of a variety of satellites and computer technologies.

As predicted, the typical agricultural farm in 2018 has just 200,000 data points available to it. As a result, "connected farms" will expand every year as rapidly as they can. In light of this, a greater focus on data collection and processing will result in more precise monitoring and management of the required parameters (soil characteristics, plant health, level of disease and pest infestation, state of agricultural equipment, etc.). In this instance, big data technology's ("Big Data") rising popularity is under discussion.

In the agro-industrial complex, the employment of UAVs, or agricultural drones [16], that are employed and build precise 3D field maps is tending to be more and more significant, specifically the 3D modeling [17] technique. Continuous oversight and control are very important. The worth of the worldwide agro-industry is already worth more than $ 32 billion and it is anticipated to grow significantly over the next five years.

A significant element that has a detrimental impact on agricultural output is improper water management. According to research, oversaturation, pollution, and other associated issues result in the waste of over 60% of the water allotted to agriculture crop damage because inadequate or excessive watering occurs often. In order to control and improve the process, these issues are fairly successfully resolved utilizing OGC data services from sensors that gather data from tank-filling activities. Along with information on the soil's moisture content and acidity, the received information is also analyzed. All of these factors work together to make the limited water resources used more effectively. Various sources claim that the usage of the aforementioned technologies results in an annual water savings of 30–50 billion gallons. The agro-industrial complex's use of IoT technology has created new

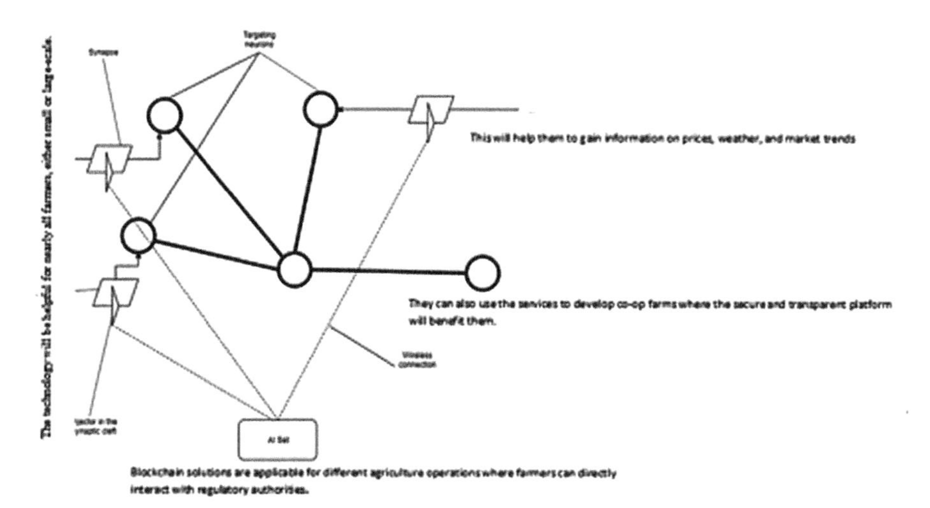

Figure 2.1 Additional used other than collecting land and Automated tractors.

possibilities in the growth of intensive agriculture, which is described by little crop rotation and a lot of resource usage. By building comparable settings based on the data acquired, the open-source platform enables the quick gathering and distribution of data from a single medium that is then utilized as a "climate recipe" for its further growth. As a result, farmers are able to artificially produce environments that are favorable for the development of any certain sort of crop. Self-propelled tractors and other equipment are already present on many contemporary farms. Intelligent tractors are yet quite new and might tend to be stronger in the near future in addition to gathering land (Figure 2.1).

2.4.1 Methodology

Agriculture has long benefited from the use of artificial intelligence (AI) and the Internet of Things (IoT), in addition to other cutting-edge computer technologies.

However, the use of these smart technologies is increasingly receiving more attention. For many thousands of years, agriculture has been a significant source of food for humans. It has also led to the development of suitable agricultural techniques for various crop varieties. New, cutting-edge IoT technologies may make it possible to monitor the agricultural environment and guarantee high-quality output. However, there is still a dearth of research and development on smart sustainable agriculture (SSA), along with complicated challenges brought on by the fragmentation of agricultural processes, such as the management and control of IoT/AI devices, interoperability, data sharing and management, and massive volumes of data analysis and storage. This research looks at the IoT/AI technologies that have already been used for SSA in the first place and then finds the IoT/AI technological architecture that can support the creation of SSA platforms. This study adds to the body of information already in existence, evaluates the research and development that has been done in SSA, and offers an IoT/AI architecture as a way to create a platform for smart, sustainable agriculture.

In addition to doing the literature review, the present researcher added to the study by conducting informal interviews with seasoned farmers. The goal of the project is to create an IoT/AI SSA architecture and investigate the possibility of using IoT and AI as the foundation for an SSA platform. In order to accomplish the aforementioned goal, a review was used to find, evaluate, and examine important books, journals, reports, and white papers. This study adds to the body of knowledge by building an IoT/AI framework for the adoption of smart technologies, in order to create smart sustainable agriculture practices, since there are not many previous studies in this field.

2.5 Outcomes

This part is broken down into three sections: (A) domains of smart sustainable agricultural model; (B) proposed IOT/AI SSA platform as a solution; and (C) proposed IoT/AI technical architecture for SSA platform. This section is based on the study goal described in Section 2.4.1.

A. The Smart Sustainable Agricultural Model's Domains

When implementing the smart agriculture model, a number of domains need to be taken into account, according to the findings of the literature review. The complexity and interdependence of data flow across several smart sustainable agriculture areas are shown in Figure 2.2.

B. The Suggested IoT/AI SSA Platform

This platform would show to be a useful tool for SSA domains to share and exchange data. Although many academics have created IoT designs, their work has often focused on certain IoT/AI domains, such as weather monitoring systems or sensors. The present study suggests an all-encompassing IoT/AI platform that can handle every aspect of an SSA environment and carry out the following functions: (a) manage and control the flow of data between SSA domains; (b) make it easier for the various SSA architecture components to integrate; (c) address interoperability problems brought on by the use of disparate tools and software; (d) offer user-friendly interfaces for

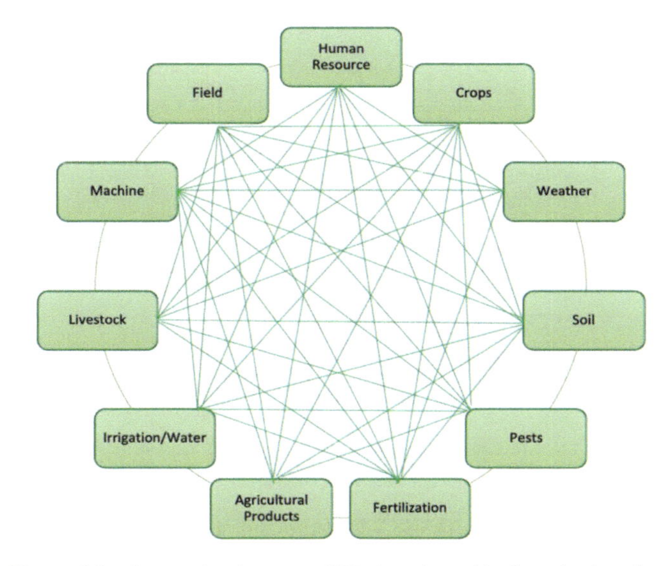

Figure 2.2 Interaction between SSA domains with chaos in data flow.

interaction; (e) enable the generation of reports based on real-time data and keep them up to date; (f) store generated data in a sustainable storage location (like the Cloud), allowing it to bc permanently recorded for future reuse; (g) isolate different layers to improve the development process in the future; and (h) the platform should have multiple nodes so that, in the event of a failure, other nodes can keep the system operational. The use of the SSA-IoT/AI platform at the heart of the SSA domain to support data sharing and business process in an intelligent, sustainable agricultural environment is shown in Figure 2.3.

C. The SSA Platform's IoT/AI Technical Architecture

The whole AI/IoT technology architecture for SSA is shown in Figure 2.2. It is made up of two primary parts: The first component is the SSA layers and AI/IoT technologies; the second is the data lifecycle inside the SSA architecture and the location of the data process. To provide more light on the framework, the description that follows provides further information on each component:

1. The first element Layers of Smart Sustainable Agriculture (SSA) and AI/IoT technology for SSA

   The following layers make up the AI/IOT framework, which is the first element of smart, sustainable agriculture (SSA): (a) physical hardware

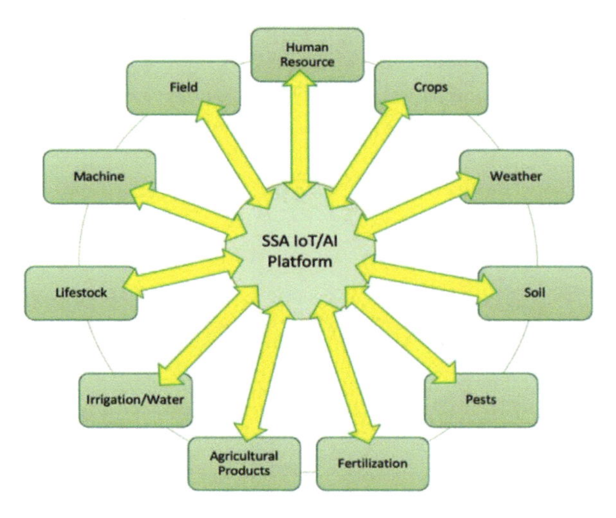

Figure 2.3 The SSA-IoT/AI platform is proposed.

and storage layer; (b) layer for AI and data management and governance; (c) network layer; (d) layer for security; (e) layer for application; (f) layer for IoT and sensing; and (g) layer for SSA domain.

2.6 Future Directions for Research

Moreover, there are more applications for the Internet of Things technology other than just growing crops. Farmers are able to acquire precise and trustworthy data and build more fruitful choices thanks to the real-time quality control systems for water, food, and livestock mentioned above. The Internet of Things, for instance, enables prompt diagnosis of animal ailments and prompt help. Farmers may employ technical sensors that evaluate farming herb growth issue regions with pests to prevent crop losses brought on by different pests.

2.7 Conclusion

The main engineering of the "Industry 4.0" period was recognized as a consequence of the study, and its attributes and utilization roles were provided. Consequences are drawn on the effect that these technologies' introduction will have positively on investment, employment, productivity, and revenue growth. The resultant full explanation of the many uses of the IoT in farming organizations' operations is given in the conclusion. The approach permits us to draw the conclusion that the digitization of the agriculture industry will effect in the establishment of better goods. Moreover, Industry 4.0 will result in the evolution of more adjustable frameworks, whose supporters will interact through the Internet, hugely compelling worker productiveness and minimizing manufacturing expenditures. Digitalization is a completely natural process that affects every sector of the economy, including marketing, retail, and services. Neural networks and modern information systems will be able to assess more variables and greatly improve the efficiency of any commercial activity. Naturally, this also applies to farming. Some farming builder in the cut-throat scenario has two key responsibilities: to reduce the expenditure of production and raise the resultant net income, while facilitating a continuously higher level of product quality. All phases of the manufacturing process must be completely controllable and transparent in order to address them. For each unit of manufacturing, for instance, you must clearly and progressively monitor the value chain. At the agricultural firm, a single information environment is being developed where high-tech machinery, implementation software, and management IT frameworks continuously interchange data. The research demonstrated how using IoT technologies might fundamentally alter agricultural management. Traditional farms may be transformed into newer development farms, or "smart farms," by introducing different types of sensors, big data technologies, and the usage of UAVs and self-propelled tractors and machinery. Blockchain technologies may be used in a variety of agricultural businesses where farmers may communicate rightly with management bodies. They will be able to learn more about pricing, the environment, and market trends as a result. The safe and open platform will help them when they utilize the services to build cooperative farms. Almost every farmer, whether they operate on a small or big scale, will benefit from the technology. The technology will then be utilized for supply-chain-dependent contracts that monitor food.

This chapter has shown how crucial it is for the agricultural industry to use cutting-edge computer technology, especially AI and IoT. It is believed that agriculture is essential to human existence. The performance, quality, and volume of produce may all be increased by using modern IoT and AI technologies to support conventional agricultural techniques. The current IoT/AI technologies covered in the major agricultural research publications have been studied for this study. Moreover, it classified the primary areas of intelligent and sustainable agriculture, including fields, crops, weather, soil, pests, fertilizer, agricultural goods, irrigation/water, machinery, and cattle. This chapter's primary contribution is to the AI/IoT technological architecture for SSA, which emphasizes the need to build a unified AI/IoT platform for the region in order to effectively address problems brought on by the fragmented structure of the agricultural industry. The process of deploying AI/IoT technologies for SSA will be investigated in future study by using the suggested AI/IoT technological architecture – a unified platform prototype – on actual test cases. In order to make additional improvements and enhancements, this will identify the pertinent strengths and shortcomings.

References

[1] Kechagias, J.D., Vidakis, N., Ninikas, K. *et al.* Hybrid 3D printing of multifunctional polylactic acid/carbon black nanocomposites made with material extrusion and post-processed with CO_2 laser cutting. *Int J Adv Manuf Technol* 124, 1843–1861 (2023). https://doi.org/10.1007/s00170-022-10604-6

[2] V. Jain, M. Rastogi, J. V. N. Ramesh, A. Chauhan, P. Agarwal, S. Pramanik, A. Gupta, FinTech and Artificial Intelligence in Relationship Banking and Computer Technology, in AI, IoT, and Blockchain Breakthroughs in E-Governance, Eds, K. Saini, A. Mummoorthy, R. Chandrika and N.S. Gowri Ganesh, IGI Global, 2023.

[3] S. Pramanik and S. Bandyopadhyay, Identifying Disease and Diagnosis in Females using Machine Learning, in Encyclopedia of Data Science and Machine Learning, Eds. John Wang, IGI Global, 2023, DOI: 10.4018/978-1-7998-9220-5.ch187.

[4] A. Bhattacharya, A. Ghosal, A. J. Obaid, S. Krit, V. K. Shukla, K. Mandal and S. Pramanik, "Unsupervised Summarization Approach with Computational Statistics of Microblog Data", in Methodologies and Applications of Computational Statistics for Machine Learning, D. Samanta, R. R. Althar, S. Pramanik and S. Dutta, Eds, IGI Global, 2021, pp. 23–37, DOI: 10.4018/978-1-7998-7701-1.ch002

[5] P. T. Khanh, T. H. Ngọc, and S. Pramanik, Future of Smart Agriculture Techniques and Applications, in Advanced Technologies and AI-Equipped IoT Applications in High Tech Agriculture, Eds, A. Khang, IGI Global, 2023.

[6] R. Bansal, A. J. Obaid, A. Gupta, R. Singh, S. Pramanik, Impact of Big Data on Digital Transformation in 5G Era", 2nd International Conference on Physics and Applied Sciences (ICPAS 2021), doi: 10.1088/1742-6596/1963/1/012170, 2021.

[7] R. Jayasingh, J. Kumar R.J.S, D. B. Telagathoti, K. M. Sagayam and S. Pramanik, "Speckle noise removal by SORAMA segmentation in Digital Image Processing to facilitate precise robotic surgery",

International Journal of Reliable and Quality E-Healthcare, vol. 11, issue 1, DOI: 10.4018/IJRQEH.295083, 2022

[8] P. K. Mall, S. Pramanik, S. Srivastava, M. Faiz, S. Sriramulu and M. N. Kumar, FuzztNet-Based Modelling Smart Traffic System in Smart Cities Using Deep Learning Models, in Data-Driven Mathematical Modeling in Smart Cities, IGI Global, 2023, DOI: 10.4018/978-1-6684-6408-3.ch005

[9] T. H. Ngọc, P. T. Khanh and S. Pramanik, Smart Agriculture using a Soil Monitoring System, in Advanced Technologies and AI-Equipped IoT Applications in High Tech Agriculture, Eds, A. Khang, IGI Global, 2023.

[10] D. Samanta, S. Dutta, M. G. Galety and S. Pramanik, A Novel Approach for Web Mining Taxonomy for High-Performance Computing, The 4th International Conference of Computer Science and Renewable Energies (ICCSRE'2021), 2021, DOI: 10.1051/e3sconf/202129701073.

[11] S. Pramanik and S. Bandyopadhyay, Analysis of Big Data, in Encyclopedia of Data Science and Machine Learning, Eds. John Wang, IGI Global, 2023, DOI: 10.4018/978-1-7998-9220-5.ch006

[12] Reepu, S. Kumar, M. G. Chaudhary, K. G. Gupta, S. Pramanik and A. Gupta, Information Security and Privacy in IoT, in Handbook of Research in Advancements in AI and IoT Convergence Technologies, Eds, J. Zhao, V. V. Kumar, R. Natarajan and T. R. Mahesh, IGI Global, 2023.

[13] S. Pramanik, "An Effective Secured Privacy-Protecting Data Aggregation Method in IoT", in Achieving Full Realization and Mitigating the Challenges of the Internet of Things, Eds, M. O. Odhiambo and W. Mwashita, IGI Global, 2022, DOI: 10.4018/978-1-7998-9312-7.ch008

[14] V. Veeraiah, V. Talukdar, Manikandan K., S. B. Talukdar, V. D. Solavande, S. Pramanik, A. Gupta, Machine Learning Frameworks in Carpooling, in Handbook of Research on AI and Machine Learning Applications in Customer Support and Analytics, Eds, Md S. Hossain, R. C. Ho and G. Trajkovski, IGI Global, 2023

[15] S. Choudhary, V. Narayan, M. Faiz and S. Pramanik, "Fuzzy Approach-Based Stable Energy-Efficient AODV Routing Protocol in Mobile Ad hoc Networks", in Software Defined Networking for Ad Hoc Networks, M. M. Ghonge, S. Pramanik and A. D. Potgantwar, Eds, Springer, 2022, https://doi.org/10.1007/978-3-030-91149-2_6

[16] G. Taviti Naidu, KVB Ganesh, V. Vidya Chellam, S Praveenkumar, D. Dhabliya, S. Pramanik, A. Gupta, Technological Innovation Driven by Big Data, in "Advanced Bioinspiration Methods for Healthcare

Standards, Policies, and Reform", Hadj Ahmed Bouarara IGI Global, 2023, DOI: 10.4018/978-1-6684-5656-9

[17] D. Mondal, A. Ratnaparkhi, A. Deshpande, V. Deshpande, A. P. Kshirsagar and S. Pramanik, Applications, Modern Trends and Challenges of Multiscale Modelling in Smart Cities, in Data-Driven Mathematical Modeling in Smart Cities, IGI Global, 2023, DOI: 10.4018/978-1-6684-6408-3.ch001

3

IoT for Smart Farming Technology: Practices, Methods, and the Future

Pradeep Kumar Shah[1], Manoj Kumar Mishra[2], J. Ghayathri[3], Jimmy Jose[4], Nuvita Kalra[4], and Thanh Bui[5]

[1]College of Computing Science and Information Technology, Teerthanker Mahaveer University, India
[2]Department of Agriculture, Sanskriti University, India
[3]Department of Computer Science and IT, School of CS and IT, Jain (Deemed to be) University, Bangalore, India
[4]Symbiosis Law School, Nagpur Campus, Symbiosis International (Deemed University), India
[5]Data Analytics and Artificial Intelligence Laboratory, School of Engineering and Technology, Thu Dau Mot University, Vietnam

Email: pradeep.rdndj@gmail.com; manoj.ag@sanskriti.edu.in; ghayathri.j@jainuniversity.ac.in; jimmyjose@slsnagpur.edu.in; nuvitakalra@slsnagpur.edu.in; tuhungphe@gmail.com

Abstract

The IoT has shifted farming research. Since IoT is developing, it should be extensively examined prior to being used in agriculture. Here, the authors discuss IoT applications and the obstacles of using it for superior agriculture. IoT devices and wireless transmission technologies in agriculture implementations are rigorously researched to concentrate on particular needs. Sensor-based IoT frameworks that deliver smart agricultural methods are explored. Use cases examine IoT-based solutions from diverse organizations, people, and categories by deployment factors. These solutions have issues, but they also reveal areas for improvement and an IoT-based action plan.

3.1 Introduction

Updated farming and society need more food to nourish the world. Newer technology and remedies are being used in agriculture to obtain and analyze information more efficiently while increasing net productivity. Modern agriculture and farming need new and better methods due to climate change and water shortages. For this objective, automation and smart decision-making are getting increasingly vital. The IoT, ubiquitous computing, WSNs, radio frequency identifier, cloud computing, remote sensing, etc., are growing more popular.

3.1.1 Motivation

IoT is largely used in agriculture to improve conventional agricultural practices. The rapid development of nanotechnology in the recent time has authorized small, affordable sensors. Self-contained operation, modular hardware systems, and scalable and cost-efficient automations have made the IoT a possible instrument for self-industry, decision-making, and automation in farming. Precision farming automated irrigating arranging, plant growing optimization, farms tracking, greenhouse tracking, and farming yield technique management are essential uses. IoT is still developing; hence, it has restrictions including interoperability, heterogeneity, memory-constrained hardware systems, and security. These constraints make agricultural IoT application design difficult. In farming, most IoT apps target diverse applications like IoTs for ecological environment tracking with soil nutrient data prediction, crop health, and yield quality over time. IoTs forecast irrigation schedule by tracking soil moisture and weather. By adding sensor nodes to an IoT application, its performance may be scaled to monitor more metrics. Such applications have challenges with device interoperability, technological heterogeneity, security, measurement interval, and routing protocols.

In general, IoT-dependent farming solutions should be cheap for end customers to purchase. However, population growth is exponentially increasing food-grain consumption. A new analysis cautions that food grain output is growing slower than population. This has pushed researchers to seek better technology to enhance productivity. According to the Food and Farming Agent, worldwide food grain needs will arrive at 3 billion tones by 2040. Newer and innovative technology is examined for numerous farming uses to attain the goal. IoT should be used to make agriculture smart and precise to accelerate this pace.

3.1.2 Contribution

This chapter highlights major agricultural applications and discusses how IoT might increase performance and production. Presenting IoT characteristics, analyzing usable hardware platforms, WSN standards, and IoT cloud services for agriculture, this document also lists sensor-based IoT solutions. The author also analyzes IoT installations in many fields. This chapter's contributions are summarized below.

- Study of the IoT concept, farming systems, and uses. Analysis of IoT installations' communication technologies, hardware frameworks, and cloud services.
- Investigation of present agricultural applications' challenges using worldwide IoT case studies.
- Highlighting key criteria for improving current situations and a future road plan.

The chapter consists of sections as follows. IoT and agricultural applications are covered in Section 3.2. IoT technologies are covered in Section 3.3. Section 3.4 reviews IoT-dependent farming sensor frameworks. Few case studies are covered in Section 3.5. Section 3.6 analyzes problems and outlines a future road map. Section 3.7 ends the chapter.

3.2 IoT Framework and Agricultural Applications

3.2.1 How to define IoT

IoT [1] implementation of machine-to-machine (M2M) [2] connections is the key to current digital industry expansion. IoT is characterized by several terms, including "a dynamic global network infrastructure with self-configuring capabilities based on standard and interoperable communication protocols where physical and virtual 'Things' have identities, physical attributes, and virtual personalities, use intelligent interfaces, and are seamlessly integrated into the information network." "A concept: anytime, anywhere, any media, resulting in 1:1 radio-human ratio."

IoT connects persons, techniques, devices, and automations having sensors and actuators. The combination of IoT having human communications, cooperation, and technology analytics allows real-time decision-making. Smart places communicate with identities and virtual personalities employing smart interfaces to interact and transmit in society, ecology, and user settings.

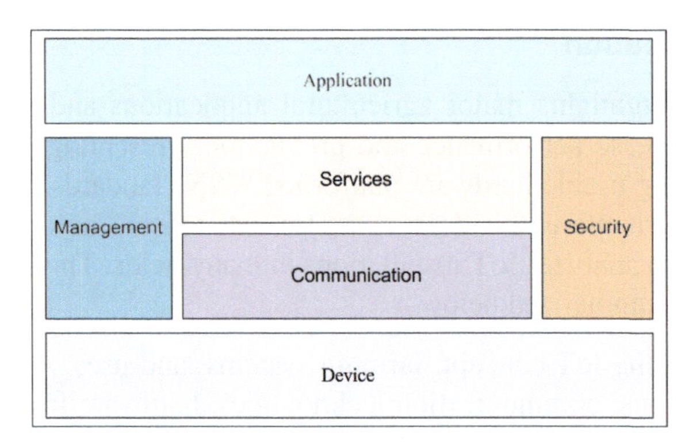

Figure 3.1 IoT building pieces in action.

3.2.2 Functional blocks of IoT

IoT systems use functional blocks for sensing, tracking, actuation, communication, and management. These functional blocks are shown in Figure 3.1.

Device: IoT systems use sensing, actuation, control, and monitoring devices. IoT devices may share data with various connected devices and apps, gather data from various devices, and analyze it locally or transfer it to central servers or cloud-dependent implementations.

Depending on temporal and space limits (memory, processing capacities, transmission latencies, speeds, and deadlines), certain operations are done locally and others in IoT infrastructure. The device structure is shown in Figure 3.2. IoT devices may have wired and wireless interfaces for communication. They comprise sensor I/O interfaces [3], internet connectivity interfaces, memory [4] and storage interfaces [5], and audio/video interfaces [6]. IoT gadgets consist of wearable sensors [7], smart watches [8], LED lights [9], cars, and industry machinery. Most IoT devices generate data that, when processed by data analytics systems, can guide local or remote actions. For example, sensor data from a soil moisture tracking device in a garden may assist to determine the best watering schedule.

Communication: The communication unit connects devices and distant servers. Data connection, network, transport, and application layers are where IoT communication methods function. The procedures are covered in Section 3.2.3.

Services: IoT systems provide several services including device modeling, control, publishing, analytics, and discovery.

Management: Management unit offers several applications to control an IoT framework to explore its governance.

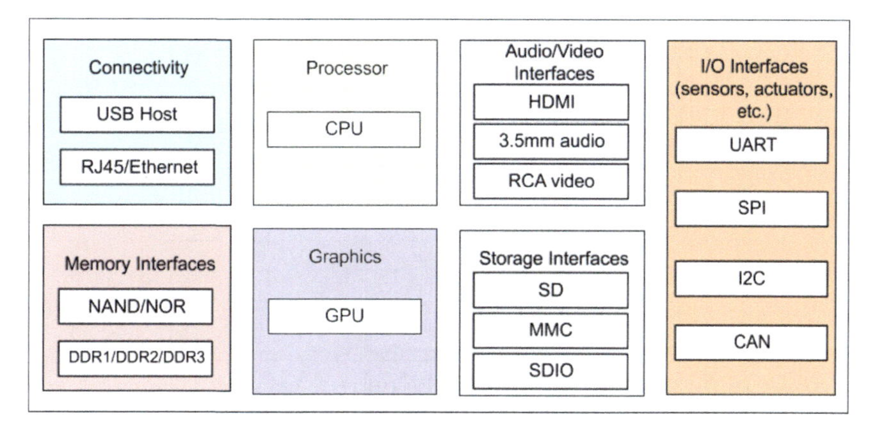

Figure 3.2 Block diagram of IoT.

Security: The security unit safeguards the IoT framework by facilitating authentication, privacy, message reliability, content reliability, and data security.

Application: The application layer is beneficial to users since it offers modules to manage and monitor the IoT framework. Implementations enable users to see and explore system status and estimate future possibilities.

3.2.3 IoT farming system

This section details an IoT system for full-scale farming applications (Figure 3.3). The six-layer system covers hardware, internet and allied transmission technologies, IoT middleware, IoT-based cloud facilities, big data analysis, and farmer experience.

Physical layer: This layer contains sensors, actuators, microcontroller modules, gateways, routers, switches, and various network equipment. Here, environment parameters are sensed, tasks are executed, and ground-level jobs are processed. This layer's microcontroller oversees networking and various operations.

Sensors and actuators: This layer's principal job is to transfer refined root-level data to high abstraction levels. The network layer includes the Internet and other transmission technologies. In agriculture, Wi-Fi [10], GSM [11], CDMA [12], and LTE (4G) [13] technologies are common. ZigBee [14] is a good long-range enabler when GSM/CDMA/LTE is unavailable. HTTP, WWW, and SMTP are suitable for agricultural Internet pavement.

Middleware layer: IoT middleware manages devices, contexts, interoperation, platform portability, and security. Middleware includes HYDRA,

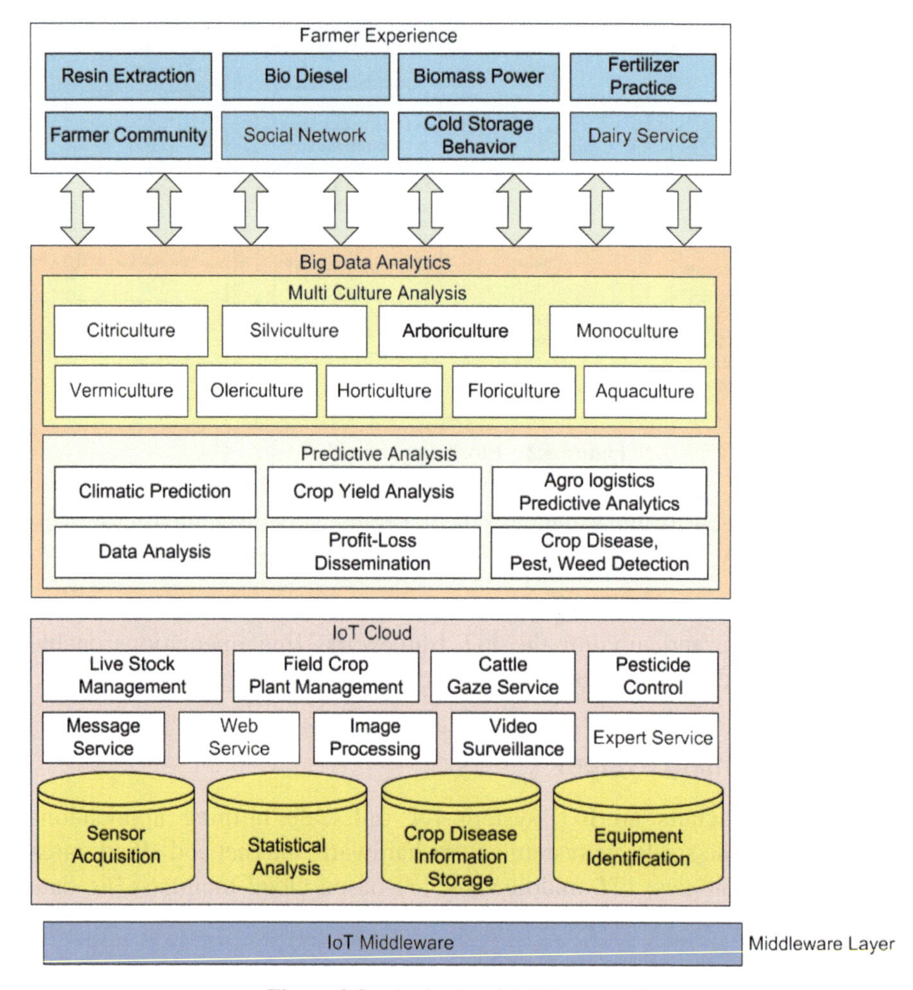

Figure 3.3 Agricultural IoT framework.

UBIWARE, UBIROAD, and the finest context-aware functionality is SMEPP, whereas SOCRADES, GSN, SIRENA, etc., include security and user privacy in their design.

Service layer: IoT cloud aided service layer is vital for agricultural cloud [15] storage and SaaS [16]. Sensor data collecting, equipment tracking, crop disease information repository, and statistical analytical techniques enable sensing, actuating, and disease detection. Additionally, livestock management, pesticide control, and automated cow gaze monitoring facilities provide value from agricultural data. Web, messaging, and expert services provide farmers with information. Data image and video analysis aids real-time on-demand service monitoring. Farmers want to enquire regarding cattle grazing in the farms, soil condition through virtual images, insect infiltration, etc. Web-based, user-friendly control panel solves all farmer needs with several services.

Analytics layer: Big data processing requires predictive and multi-cultural analytics. Prediction [17] measures next-season crop productivity probabilities. Farmers may anticipate soil moisture, temperature, heat, light intensity, rain fall, and other weather conditions.

Taking precautions to safeguard agricultural field, this detection capacity predicts crop disease incidence based on prior data. Farmers may observe insect and weed behavior in the farms. Agro-logistics service predicts optimized maintenance costs for vehicles like tractors and how they are utilized to boost profit margins by selling items in the market. This affects crop and vegetable retention since much decay is owing to improper use. Prediction may also predict next season's profit or loss. Thus, big data analytics may scientifically reduce agricultural risk factors. To create, process, and effectively manage a few agricultural methods, multi-culture analytics is included. To measure water-featured botanic growth, aquaculture may use big data analytics. This might be used to forecast fish breeding and growth in aquaculture. Big data analytics may help horticulture, floriculture, and viticulture make seasonal growth, pest management, and profit margin analysis decisions for fruits, flowers, and citrus fruits like lemons. Earthworms are raised in vermiculture. Vermicompost is organic fertilizer made from vermiculture. Silviculture can effectively regulate forest formation, growth, constituent, health, and quality to meet population demands. Big data may improve forest growth by analyzing environmental data. Arboriculture is a horticulture method for herbs, vineyards, and various perennial woody trees. Big data analysis might be used to statistically model how these plants develop and adapt to their surroundings. Olericulture can anticipate vegetable plant development for human consumption. Big data analytics can simulate water, temperature, and fertilization to boost herbaceous plant output.

User experience layer: The top layer is geared for farmers' experiences. This layer allows farmers to use social media to educate and share information about agricultural areas including politics and economics. Cold storage stores crops for multi-season usage. Behavioral and pattern analysis of cold storage might boost farmer profits. The farmer may determine the fruitful selection for crop development. Some industrial implementations are also gathered with this layer. Resin extraction from plants may be connected to the

IoT system to track resin production, which might lead to scientific collaboration and significant income. Milk production, filtration, marketing, animal disease management, and other dairy services will be integrated with the IoT framework to maximize profits. Trees and crops now provide power instead of hydro, thermal, and nuclear sources. IoT frameworks can link, manage, and monitor power use utilizing smart grids for biomass power production. Bio diesel is a renewable, biodegradable fuel. Bio diesel may be obtained from vegetable oils and animal fats. Bio diesel production, consumption, and distribution can be tracked and regulated via IoT.

3.2.4 Uses of IoT

Key usefulness aspects of IoT are listed below.

Dynamic and self-adapting: IoT devices and frameworks must be able to adjust to altering contexts and behave depending on their operational circumstances, user context, or sensing surroundings. Consider a surveillance system [18] with many cameras. The security cameras may switch modes. The day or night determines infrared night modes. When motion is spotted, cameras may convert to better resolution and inform adjacent cameras. In this scenario, the surveillance system adapts to context and dynamic situations.

Self-configuring: IoT devices can self-configure, enabling several devices to operate together for weather tracking. These devices may configure themselves, setting up networking, and download software updates without human intervention.

Interoperable communication protocols: IoT appliances may connect with different devices and the infrastructure via a variety of protocols. IoT devices have distinct identities and identifiers like IP addresses or URLs. Intelligent interfaces in IoT systems may adapt to the situation and communicate with people and the environment. In conjunction with control, configuration, and administration infrastructure, IoT device interfaces enable users to query, track, and remotely control devices.

IoT devices are frequently incorporated into the information network to interact and interchange data with different devices and frameworks. IoT devices may be dynamically detected by other devices and/or networks and explain themselves to various devices or user applications. A weather tracking node may define its functionalities to various attached nodes to communicate and share data. The collective intelligence of devices working with the infrastructure makes IoT solutions "smarter" when integrated into the infrastructure network. So, data from several weather tracking IoT nodes may be combined and explored to anticipate weather.

Context-awareness: Sensor nodes learn about their surroundings through physical and surrounding characteristics. The sensor nodes' subsequent opinions are context-aware.

Intelligent decision-making: This function boosts energy efficiency and network longevity in wide areas. Multiple sensor nodes come together and make the ultimate conclusion using this functionality.

3.2.5 IoT agriculture apps

This section lists IoT-implemented agriculture, farming, and related applications.

System for managing irrigation: Modern agriculture demands a better irrigation management system to maximize water use. Smart irrigation systems integrate real-time weather tracking data, allow farmers to direct their systems from anywhere in the globe utilizing their homes, use Wi-Fi and Ethernet, sync with moisture sensors in their yards, and reduce their yearly costs when conserving water. IoT is becoming more common in irrigation management systems worldwide.

Pest and disease control: Managed pesticide and fertilizer use improves crop quality and lowers agricultural costs. However, we must monitor agricultural pests to reduce pesticide use. To forecast this, we require IoT infrastructure to gather disease and insect pest data from sensor nodes, analyze and mine it, etc. A three-layered IoT architecture can recognize sickness and identify the bug that caused it. Farmers can indicate the needed medication to rescue the company.

Monitor cattle movement: A field full of cats may be tracked using IoT. Any livestock may be tracked in real time. IoT-dependent cloud systems like Connecterra [19] are popular for smart dairy monitoring. It can detect animal heat and health.

Water quality tracking: Sensor nodes with WSNs assist monitor water quality. A new article uses IoT to check water quality in real time. The system monitors water temperature, pH, turbidity, conductivity, and dissolvable O_2. The Internet displays sensor data utilizing cloud services.

Monitoring greenhouse conditions: Agriculture and greenhouses are linked. Greenhouse gases raise climatic temperature, affecting agriculture. Greenhouse gas emissions depend on pH, temperature, CO_2, etc. HarvestGeek [20] uses IoT cloud facilities to remotely track and regulate greenhouse conditions.

Tracking soil quality is important for agriculture. Data of soil improves corps output. Gupta et al. [21] use LoWPAN [22] and IoT to remotely aggregate soil conditions using sensor nodes' real-time network monitoring using SNMP.

Precision agriculture by UAV: Advanced technology like UAVs [23] and drones [24] may improve agricultural productivity. PrecisionHawk uses UAV, GIS, and sensor-based IoT cloud platforms to station AI for in-air flight route computations to identify meteorological conditions. In-flight diagnostics and monitoring processes continually monitor its own status while in flight and counts on the operational weather limitations, land mapping, and real time analytics supports.

Management of agricultural production supply chains: Operating efficiency on agricultural goods is important for farmers to profit. IoT can monitor agri-product supply chain management. Mondal et al. [25] examine how IoT affects product supply chain business processes and how it drives agricultural product adoption. It also offers a reference framework for product chain node firms to consider ramifications.

Traffic categorization research, which eliminates pay-load inspection, has advanced in the previous five years. Different approaches vary in features (some requiring packet payload inspection), supervised or unsupervised classification methods, and traffic types, making it difficult to compare them. Comparisons across research are complicated by the fact that classification performance depends on classifier training and accuracy test data. Unfortunately, there is no global test traffic data to compare classifiers.

3.3 Supported IoT Technology in Agriculture

We explore IoT hardware frameworks and WSN [26] technologies utilized in farming implementations in this section. Various IoT cloud service providers, which are popular for these applications, are also explored.

3.3.1 Hardware platform

Several agricultural IoT-supported hardware platforms exist. Communication protocols power IoT systems and allow network connection and application coupling. Communication protocols enable devices for network data sharing. The protocols establish data interaction formats, data encoding, device addressing, and packet routing. Sequence, flow, and lost packet retransmission are further protocol functions. IoT cloud solutions provide pay-as-you-go real-time data gathering, visualization, analytics, decision-making, device management, and more distant cloud servers. Cloud services are growing in popularity in agriculture and farming.

3.4 IoT-based Agricultural Sensor Systems

A standalone sensor gadget and back-end web service automates home plant cultivation in Bitponics. Entering the plant and hydroponic system type to grow starts the system. The program uses this information to create a bespoke growth plan that includes: (1) the plants' daily light needs and when to modify them, (2) safe pH range for the plant, and (3) water pump operation. Bitponics monitors the setup utilizing sensors like water, air, temperature, humidity, light, pH, etc. It then records this data to the user's online account for evaluation and changes. The gadget has two internet-controlled power plugs for managing settings. The free service allows 50 logs per day, 500 picture uploads and shares, email alerts, and six months of storage.

The concept allows neglected plants to phone and SMS for help. Users connect through Twitter status updates. The plant sends a message to its friend on social media when it needs water. The AT Mega368 microcontroller allows DIY construction.

The system uses a water valve that connects to the sprinkler line to automatically manage moisture levels without overwatering. The attached soil probe measures soil parameters after planting. The Edyn cloud service streams the data via the Internet. The cloud analyzes and compares plant databases to discover their optimal growth circumstances. The gardener may use a smartphone app to get cloud findings and monitor its health in real time. Water is added to the valve to adjust moisture.

3.4.1 Parrot: Flower power

Parrot is a Bluetooth-4.0-based sensor system that tracks and analyzes four vital plant parameters: sunlight, ambient temperature, fertilizer, and soil moisture. A plastic stem called Parrot detects plant temperature, humidity, and soil salinity precisely. It then delivers this info to the user's phone or tablet.

A base station attached to the gardener's router receives measured data. The framework has automated watering valves. The system sends data to a PlantLink cloud server through Zigbee from the router. The cloud helps gardeners check plant conditions, arrange automated watering, and get watering alarms. Gardener receives all late alerts through email and SMS.

Soil moisture is measured every 10–15 minutes. It adjusts to local weather patterns and optimizes your plant watering schedule based on a database of plant watering demands you may choose from HarvestGeek.

HarvestBot monitors the garden's vitals and sends data to cloud servers using sensors like air temperature, soil moisture, ambient light, CO_2 ppm level, etc. Gardeners can get real-time data via HarvestBot. Each unit is specialized for its particular duty and may be employed in big commercial indoor operations. HarvestBot must control actuators after server analysis. Email and push notifications on smartphones and tablets educate gardeners of different garden details.

Pandey et al. [27] automate sprinkler controllers. The Wi-Fi-controlled smart gadget may be managed remotely from anywhere in the globe. The user may also use automated settings modifications depending on local weather stations or the Internet.

Spruce's temperature and soil moisture sensors deliver real-time data to the cloud server. Trees get enough water as needed. The smartphone user may check the status anywhere. Open Garden is a DIY gadget, which works in residences, greenhouses, gardens, and hydroponics. The heart of Open Garden is a gateway that uploads data to a web server through Wi-Fi, GPRS, or 3G. Nodes communicate through 433-MHz wireless radios. Fixing sensors to nodes allows the user to collect data on ambient temperature, humidity, light levels, soil wetness or hydroponic growth medium temperature, pH, and conductivity. The system displays real-time sensor data online. The device is connected to water pump and dropper (indoor), electric valve and sprinkler (outdoor), O_2 pump, and grow light (hydroponics). Web programming skills may let users schedule the system automatically.

Koubachi is an APP-controlled automatic sprinkler system that optimizes garden plant watering. It has sensors for air, soil, ambient light, and soil moisture. Koubachi taps and sprinkles water from the water valves to the hose lines as needed by the soil. The gardener's home gateway links to the Koubachi cloud servers through 6LoWPAN, a low-power personal communications protocol. Users may mesh as many soil sensors or watering controls as they like. Each device may be a transceiver in this architecture. Real-time cloud data controls valves automatically. The garden's state may be checked on a smartphone or tablet from anywhere in the globe.

Indoor smart hydroponic system Niwa may be installed in a container box at home. No garden or gardener is needed here. Wi-Fi connects Niwa to cloud servers through the Internet. The user may tell plants what they desire.

3.5 Case Studies

We discuss a few situations where IoT has greatly improved agricultural and farming quality. IoT is popular in beekeeping, greenhouse facility platforms, wheat disease estimation, tomato pest management, UAV-dependent precision farming, etc.

3.5.1 Case study 1: Monitoring honey bee hives

Recently, IoT has been used for beekeeping. This project is related to open energy tracking for measuring sustainable energy. This project equips electronic components using a national crown board. The monitoring system has the items listed in Table 5. Project supplies include microcontroller, wireless communication module, power supply, display unit, and sensors. Beekeepers may see real-time bee hive temperature and battery power. The Arduino-based RF receiver is linked to a Raspberry Pi gateway, which logs, graphs, and analyzes data every 60 seconds. Figure 3.4 demonstrates the National Crown Board's monitoring rationale. The graphic shows the board and additional sensor and electrical goods.

Figure 3.4 illustrates the feasible frame construction. Cons: Only temperature sensor placed.

3.5.2 Case study 2: IoT-cloud green home service framework

Recently installed is an IoT-based wheat disease, pest, and weed monitoring and control system. IoT in real-time-based approach uses ZigBee and Wi-Fi to provide remote collection and controller systems (Figure 3.5). Sensors feed environmental data to the collector module that processes it and sends it through the gateway. The gateway sprinkles water and turns lights on and off as per user request. The gateway and data processing centers are wirelessly connected. The data center and web server at the monitoring (Figure 3.6) center store and distribute wheat disease, pest, and weed data. Data is kept permanently in a huge database that provides categorization and matching information as needed. Pros: Wireless gateway adoption is effective. Cons: Fewer collector items.

Precision farming with drones and IoT is happening worldwide. The drone is exploring the use of aerial imagery to deliver farmers demand-based agricultural information. Agribotix uses its fixed-wing Hornet and Enduro drones, depending on RangeVideo's RV Jet airframe, with Canon S200 and

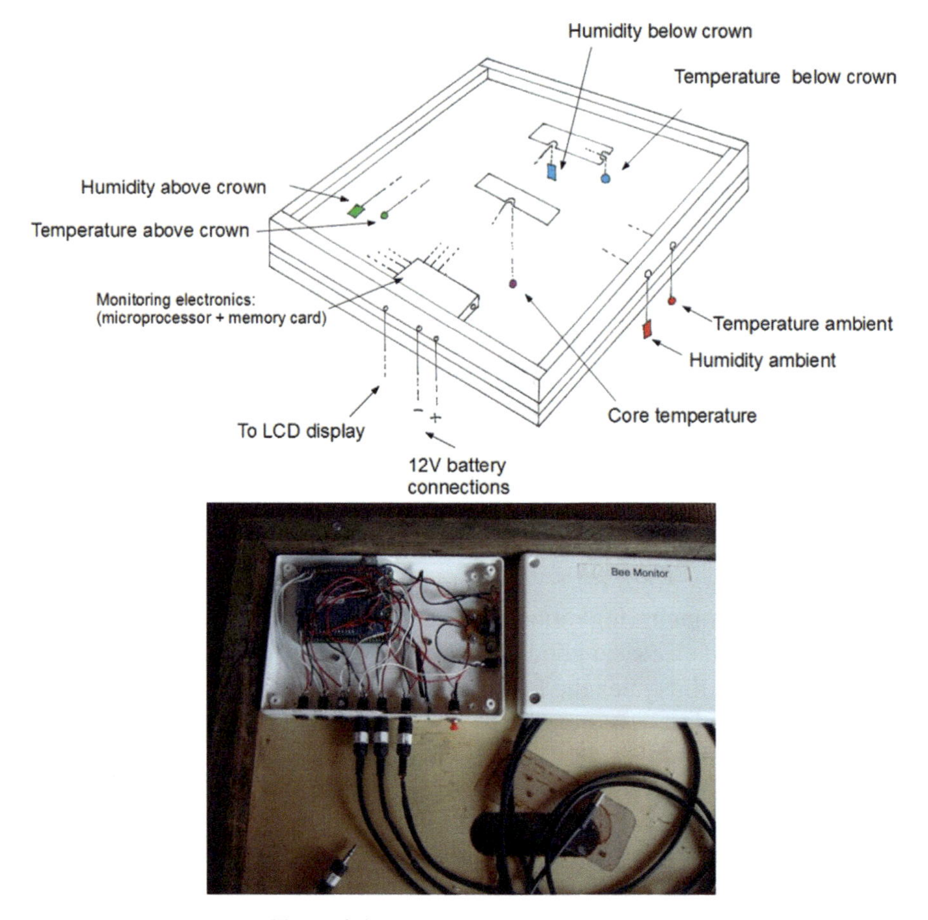

Figure 3.4 National crown board setup.

GoPro cameras modified to have high-quality non-distorting lenses and NIR filters. Figure 3.7 shows a snapshot.

3.5.3 Case study 3: Tomato borer tracking

IoT-dependent Borer insect estimation in Indian tomatoes was recently reported. The researchers employed a robot with a wireless web camera and Azure cloud service. Real-time web camera footage of tomato plants region is sent to Java-based SaaS to detect unripe tomatoes and borders.

The Azure cloud database matches the data with pesticide amalgamation. Intel handles all software activities and orders the robot to spray pesticides on tomato plates for a certain period depending on Azure's advice.

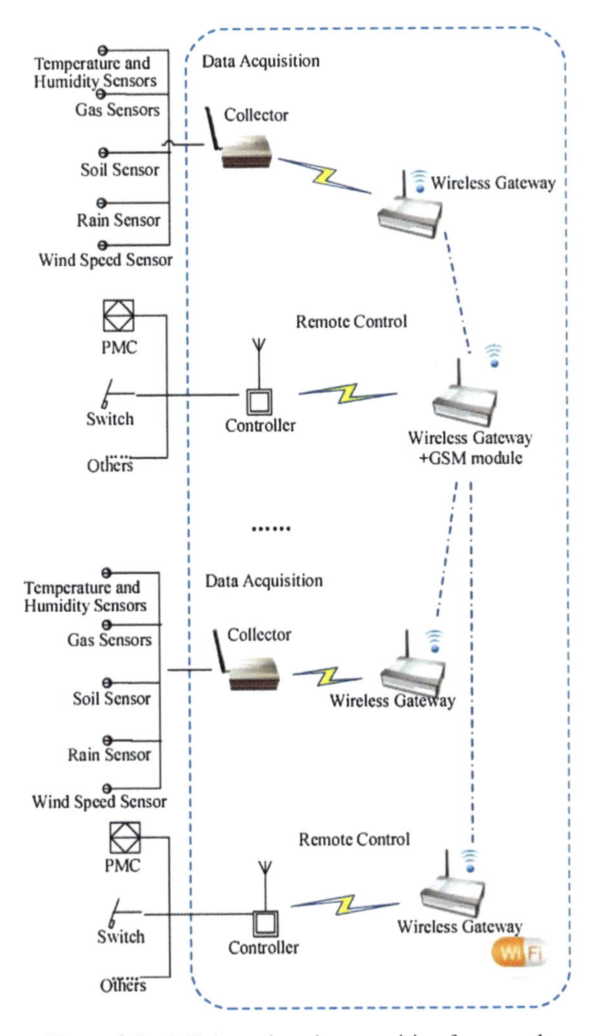

Figure 3.5 IoT-dependent data acquiring framework.

The procedure has two steps. In step 1, cloud end accesses wireless webcam footage in real time, which is converted into gray-scale imagery, image segmentation extracts tomatoes, leaves, and branches, which is processed to remove leaves and branches, dilation retains more tomato images, and masking retrieves RGB images. In step 2, the quantity and kind of tomato pests are recognized and enough pesticide is applied. Figure 3.8 shows the picture algorithm's Borer insect detection phases: (a) unripe tomato image taken 1.5 feet away, (b) Cb and Cr components, (c) Cr and (d) histogram-based segmentation, (e) erosion, (f) dilation, (g) masking, and (h) Borer insect

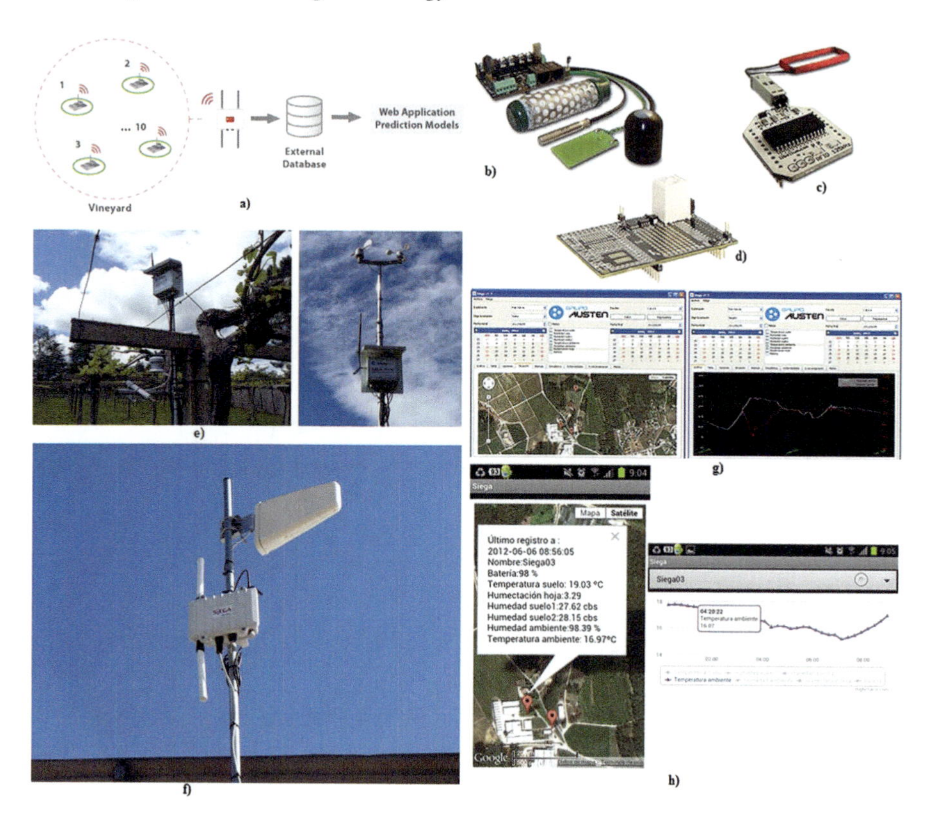

Figure 3.6 IoT vineyard monitoring and visualization.

Figure 3.7 Drone image analysis.

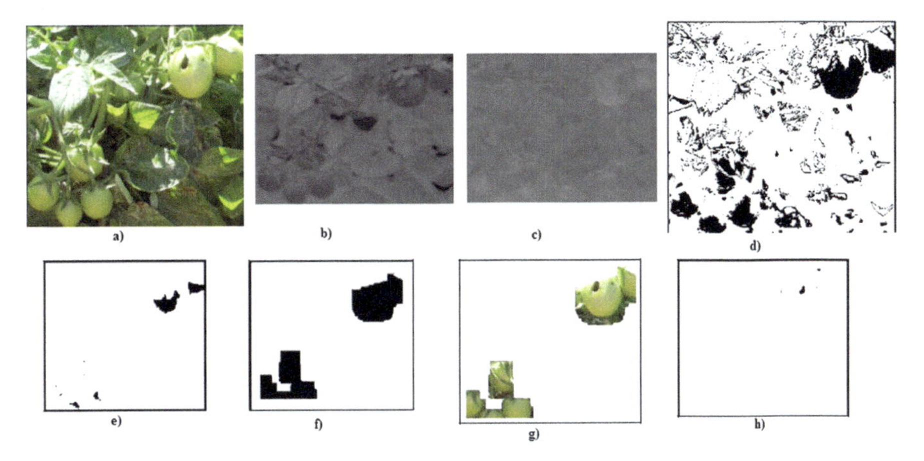

Figure 3.8 Image algorithm analysis for Borer insect detection.

extraction. Pros: Effective image algorithm. Data transfer rate and real-time analysis are disabled.

3.5.4 Case study 4: Smart urban garden

IoT use in urban gardening is rising. A recent effort produced an IoT-based garden framework for house installation. The processing unit has an AT Mega 328 microcontroller connected to light, temperature, soil moisture, and an integrated Wi-Fi module. When positioned on the soil in proximity to the plant being cultivated, the system informs the farmer of the best vegetable to plant in that condition after 24 hours. Otherwise, it sends a notification to the farmer of its needs (light and water) via smartphone messages. The framework is smart and ubiquitous to comprehend agriculturist speech (NLP algorithms), retrieve and represent particular information, and manage that information by generating logical inferences. All system parameters are shown in Table 10. Figure 3.9 shows smart garden implementation: (a) system in plant tubs, (b) hardware system, and (c) web-dependent output. Pros: Simple to use. Cons: No distributed strategy.

3.5.5 Case study 5: Cloud-supported plant industrial

Figure 3.10 shows an overall plant manufacturing management framework. The framework has three parts: farm gate way (FGW), data collecting, storage, and dissemination system, and application unit. The FGW device transmits sensor data, receives cloud commands, and saves data in the data center.

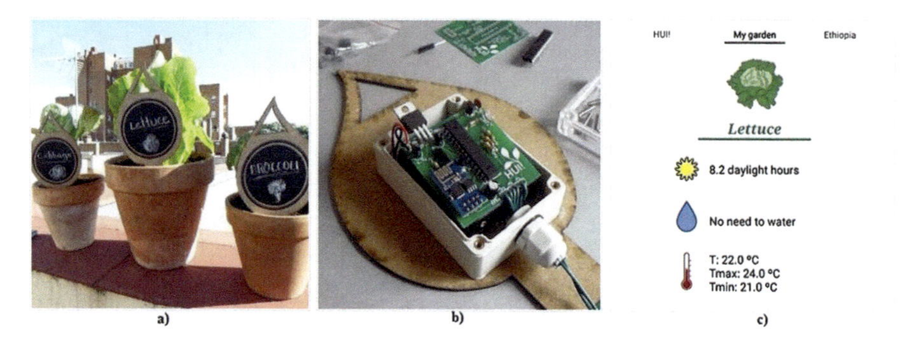

Figure 3.9 Intelligent city garden experimentation and output.

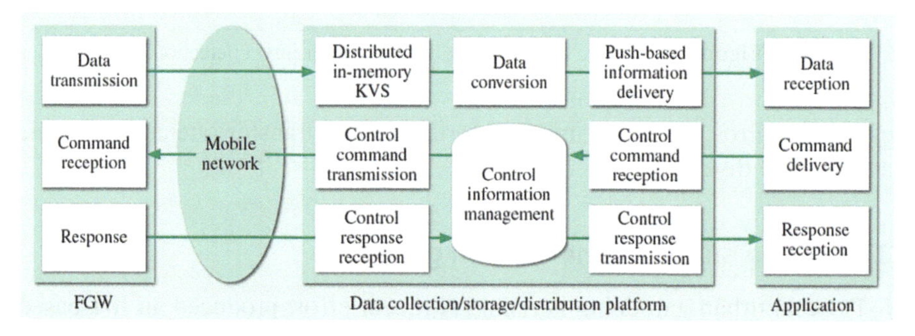

Figure 3.10 A plant manufacturing management system.

It measures sensor data including sun radiation, temperature, humidity, nutritional content, pH, and electrical conductivity. FGW monitors important functional operations including heating and cooling, nutritional solution pumps, shade curtains, and other production equipment. Key-value storage, HTTP protocol-dependent data conversion, and push-dependent data delivery on sensory data are managed by a cloud-dependent data collecting, storage, and dissemination framework. The application unit receives sensor data, sends commands to the FGW, and receives FGW responses. A smart sensor framework is used. Cons: Energy efficiency is poor.

3.6 Future Challenges and Road Map

The present solutions use IoT-based intelligent apps to solve agriculture and agricultural problems. We talk about the possibilities of these implementations to illustrate the present remedies below, and the next parts will explain how to enhance the states point-by-point.

Cost-efficiency: Scientists worldwide are concentrating on reducing IoT H/W and S/W expenditures while increasing framework output. Developed countries want cost-efficient equipment to minimize the additional expense of using foreign imported gadgets to develop systems. Although foreign farms are developing cutting-edge technology, the question remains how to lower costs. Uneconomical works are shown. Thus, such point is intentionally timely.

Standardization: Current efforts do not follow standardized data and process representations. Standardization is another component that may precisely increase IoT. Standardization in IoT lowers early hurdles for service providers and active users, improves interoperability between applications or systems, and improves application-level events. Security, connectivity, and identification standards must change alongside IoT technologies while creating new solutions.

Additionally, colleague academics will establish industry-specific principles and set standards for IoT adoption. IoT should rigorously follow agriculture standards.

Heterogeneity: IoT is a complex heterogeneous network. The agricultural operations indicated cannot interact with heterogeneous units or communication technologies. This increases the complexity of equipment and communication technology, making network rudeness deceptive and delayed. To overcome the depletion of the common platform, IoT must be reviewed to thrive in agriculture.

Context awareness: While millions and millions of sensor-enabled items are linked to the Internet, users cannot manage all the sensor data. Better context-awareness computing methods are needed to determine what data to analyze. Discussions of agri-tasks lack context. This suggests that information validation is a continuous process disruption. Surrounding environment and self-assessment may shift local context to others when producing well-connected self- and periphery-conscious IoT ecosystem.

IoT node identity: The IoT will have many nodes. IPv4 protocol identifies every node with a 4-byte address. All associated devices and data must be retrievable. As IPv4 addresses are fast diminishing and will approach zero in a few years, new addressing strategies will favor IPv6. Communication in presented systems is largely IPv4. However, future networks may be densely populated, making it impossible to provide nodes distinct identities.

Energy management: IoT systems' biggest difficulty is energy management. While storing energy, IoT devices, network antennae, passive modules, and core algorithms should be correctly handled. Testing non-conventional energy harvesting technologies like solar power, wind, biomass, and vibration cloud is very important when creating IoT-based smart farm systems. Already, solar-powered IoT solutions are used. Future researchers may work on different sources.

Fault tolerance: The aforementioned methods lack fault tolerance. To constitute a faultless system, fault tolerance should be great so that technological errors do not stop it. Depleted batteries or other factors might cause hardware module failure. Similar to sensor error and calibration failure, communication may malfunction. Solar power may be a solution to battery-operated modules. Multiple communication protocols may increase power consumption yet provide smooth connection. In this circumstance, activating one protocol at any time may reduce power usage. Before installation, calibrate properly.

Real-time solution: Many given solutions do not account for real-time. However, climate and soil data must be effectively combined with contemporary developments for precision agriculture.

3.6.1 Future road map

We have suggested numerous IoT uses in agriculture and farming. Current implementations include irrigation management, vineyard tracking, intelligent gardening, crop disease estimation, micro-climatic estimation, and aquaculture equipment, cloud solutions, and systems.

3.6.2 Needed improvements

The following IoT factors need future attention:

- Autonomy: Future apps must be totally autonomous to meet specific demands.
- Low-cost IoT solutions are important for expansion and use.
- User control panel: Field workers usually employ IoT-agri-based technologies. Thus, a control panel with a user-friendly interface is best for efficient applications.
- Energy: Green computing concepts should be spread to IoT-based agriculture, where IoT equipment will use less energy and be more reliable and productive.

- IoT devices have the most interoperability concerns. Devices should be able to interact with others from other genres to form a living ecological setting.
- Machine learning and artificial intelligence will be used combined to do predictive and behavioral analyses using enhanced decision support systems and real-time evaluations.
- Maintenance: IoT solutions should decrease maintenance time and cost to a level that ignorant users may tolerate.
- System portability: Current system designs need to be improved to meet agricultural needs.
- Robustness: IoT framework must be robust and fault-tolerant to ensure application sustainability.
- Climate, soil, and water: Building an IoT-dependent agricultural framework is much complex due to global temperature, soil, and water differences. Farmers will have local weather sensors, which may interact with national meteorological centers to warn them of environmental conditions before an unfavorable circumstance. This may boost crop yield and precisely resist agri-product deterioration. Segmented land structure: Many nations struggle with partitioned land farming. Suitable IoT architecture will be developed for this challenge. Before tackling this issue, effective strategy and preparation are needed.
- Low maintenance: Maintenance at once causes major issues. Thus, a low-maintenance system that can optimize processes automatically without human involvement is needed.
- Portability: Portability may make system management harder. SoCs, SiPs, and other portable equipment might supply the final output.

3.6.3 Applications in the future

With sensor-cloud, big data analytics, and ubiquitous computing, new groundbreaking applications are imagined. We briefly explain the principles below.

Sensor-cloud-based computing: A new approach for on-field IoT applications cloud computing and retrieving relevant information from vast volumes of data of many sorts created at high velocity. The sensor-cloud also makes agriculture smarter by improving data administration, data access, and device management. Few applications may be made below.

- A cloud-enabled storage system for monitoring seasonal soil and environmental characteristics.
- A crop health monitoring mobile sensor-cloud service may be created.
- A cloud-based architecture might anticipate agricultural yields in the future. The autonomous sensor-cloud may operate a smart irrigation system for big areas.
- A cloud-controlled greenhouse system can monitor off-season fruit and vegetable output under a predetermined environmental model.
- A remote-controlled field crop planting method is needed to plate fields without farmers. Developing a cloud-based horticultural management system can generate fruits year-round.
- Cloud-enabled smart floriculture settings are needed to keep flowers fresh and aromatic.
- Cloud services can enable real-time livestock monitoring systems to automatically boost farmers' incomes. Fish farming is important when cloud-enabled services can monitor and regulate the process from breeding to sale.
- Cloud-enabled agro-logistics will help farmers make more money by delivering veggies and other agri-products on time.

Big data analytics: Pandey et al. [28] say big data analytics may help identify the meaning of vast amounts of fast-growing, diverse data. Big data analytics may reveal future company trends, undiscovered structural patterns, client preferences, catastrophe prediction, systematic connection between multiple factual components, and more. We list small agricultural applications that can use big data.

- A crop-growth prediction system may be created using farm data.
- Big data analytics may tackle agricultural disease control utilizing statistical models.
- A big-data-enhanced cloud-based web-based information system can help farmers learn about agriculture.

- Big data analytics will provide a centralized agricultural equipment control system for large-scale farming. Big data analytics might improve agricultural output with a decision support system.
- Create a pollination prediction system.
- Big data analytics for cow grazing patterns.
- Big data analytics may enable cost optimization in large-scale agriculture.
- A farmer-friendly climatic information analysis service is needed.
- To analyze agri-product profit margins, supply chain management systems may use big data analytics.
- Analytics may help farmers predict agri-product rot time, reducing waste and increasing value.

The government should value and manage agriculture by using big data analytics at a higher abstraction level. Policymakers should collaborate to improve farm ecosystems using big data.

Ubiquitous computing means "existing everywhere." During operation, ubiquitous computing devices and systems are always connected and informed. Pervasive computing devices can recognize, communicate, and interact with nearby devices. This allows ubiquitous setups to adjust field parameters reliably, real-time, and flexible. These words advocate considering ubiquitous computing as a potential option for agricultural applications with comparable circumstances. Below are some ubiquitous computing-based agricultural applications.

- Development of cost-effective and energy-efficient RFID modules for crop and agricultural component identification.
- Using smartphones to improve farming.
- Remote farm monitoring system development.
- Development of a ubiquitous field pest counting method.
- Design of web-based pesticide spray scheduling system based on crop genre, rate, and time.
- Monitoring water pipe breaches and informing farmers.
- The ubiquitous water flow control system must be created.
- Remote monitoring of field pump power use.

- System development for cattle gaze monitoring utilizing ubiquitous devices.

3.7 Conclusion

By adding additional dimensions, IoT might help the agriculture and farming sectors. I evaluate IoT implementation for innovative agricultural applications in this analytical article. An IoT-based agricultural framework is presented to fully integrate agriculture with IoT. I introduce IoT's idea, description, and characteristics. I then discuss significant IoT applications in agriculture. This study further discusses agricultural hardware platforms on the market. Also featured is agriculture-friendly wireless communication technology. A few IoT cloud service companies are prominent in agriculture. Based on important criteria, I have listed cloud service companies operating in this industry. IoT-based systems implemented in farms and regions worldwide are then surveyed. A few case studies provide deployment details across several applications like beekeeping, vineyard monitoring, precision farming, river water quality monitoring, etc. Lastly, I discuss the current applications' issues. Several potential study and application directions are proposed. Minimal cost, autonomous, energy-efficient, interoperable, standardized, heterogeneous, resilient solutions with AI, decision assistance, and minimal maintenance are needed. Overall, IoT descriptions should make agriculture smart and widespread.

References

[1] Reepu, S. Kumar, M. G. Chaudhary, K. G. Gupta, S. Pramanik and A. Gupta, Information Security and Privacy in IoT, in Handbook of Research in Advancements in AI and IoT Convergence Technologies, Eds, J. Zhao, V. V. Kumar, R. Natarajan and T. R. Mahesh, IGI Global, 2023.

[2] S. Pramanik, "An Effective Secured Privacy-Protecting Data Aggregation Method in IoT", in Achieving Full Realization and Mitigating the Challenges of the Internet of Things, Eds, M. O. Odhiambo and W. Mwashita, IGI Global, 2022, DOI: 10.4018/978-1-7998-9312-7.ch008

[3] S. Choudhary, V. Narayan, M. Faiz and S. Pramanik, "Fuzzy Approach-Based Stable Energy-Efficient AODV Routing Protocol in Mobile Ad hoc Networks", in Software Defined Networking for Ad Hoc Networks, M. M. Ghonge, S. Pramanik and A. D. Potgantwar, Eds, Springer, 2022, https://doi.org/10.1007/978-3-030-91149-2_6

[4] S. Pramanik, An Adaptive Image Steganography Approach depending on Integer Wavelet Transform and Genetic Algorithm, Multimedia Tools and Applications, 2023, https://doi.org/10.1007/s11042-023-14505-y.
[5] R. Jayasingh, J. Kumar R.J.S, D. B. Telagathoti, K. M. Sagayam and S. Pramanik, "Speckle noise removal by SORAMA segmentation in Digital Image Processing to facilitate precise robotic surgery", International Journal of Reliable and Quality E-Healthcare, vol. 11, issue 1, DOI: 10.4018/IJRQEH.295083, 2022
[6] S. Pramanik, M. G. Galety, D. Samanta, N. P Joseph, Data Mining Approaches for Decision Support Systems, 3rd International Conference on Emerging Technologies in Data Mining and Information Security, 2022.
[7] D. Samanta, S. Dutta, M. G. Galety and S. Pramanik, A Novel Approach for Web Mining Taxonomy for High-Performance Computing, The 4th International Conference of Computer Science and Renewable Energies (ICCSRE'2021), 2021, DOI:10.1051/e3sconf/202129701073.
[8] S. Praveenkumar, V. Veeraiah, S. Pramanik, S. M. Basha, A. V. Lira Neto, V. H. C. De Albuquerque and A. Gupta, Prediction of Patients' Incurable Diseases Utilizing Deep Learning Approaches, ICICC 2023, Springer, doi.org/10.1007/978-981-99-3315-0_4
[9] V. Vidya Chellam, V. Veeraiah, A. Khanna, T. H. Sheikh, S. Pramanik and D. Dhabliya, A Machine Vision-based Approach for Tuberculosis Identification in Chest X-Rays Images of Patients, ICICC 2023, Springer, doi.org/10.1007/978-981-99-3315-0_3
[10] P. T. Khanh, T. H. Ngọc, and S. Pramanik, Future of Smart Agriculture Techniques and Applications, in Advanced Technologies and AI-Equipped IoT Applications in High Tech Agriculture, Eds, A. Khang, IGI Global, 2023.
[11] T. H. Ngọc, P. T. Khanh and S. Pramanik, Smart Agriculture using a Soil Monitoring System, in Advanced Technologies and AI-Equipped IoT Applications in High Tech Agriculture, Eds, A. Khang, IGI Global, 2023.
[12] V. Veeraiah, V. Talukdar, Manikandan K., S. B. Talukdar, V. D. Solavande, S. Pramanik, A. Gupta, Machine Learning Frameworks in Carpooling, in Handbook of Research on AI and Machine Learning Applications in Customer Support and Analytics, Eds, Md S. Hossain, R. C. Ho and G. Trajkovski, IGI Global, 2023.
[13] V. Jain, M. Rastogi, J. V. N. Ramesh, A. Chauhan, P. Agarwal, S. Pramanik, A. Gupta, FinTech and Artificial Intelligence in Relationship Banking and Computer Technology, in AI, IoT, and Blockchain

Breakthroughs in E-Governance, Eds, K. Saini, A. Mummoorthy, R. Chandrika and N.S. Gowri Ganesh, IGI Global, 2023.

[14] S. Pramanik and S. Bandyopadhyay, Identifying Disease and Diagnosis in Females using Machine Learning, in Encyclopedia of Data Science and Machine Learning, Eds. John Wang, IGI Global, 2023, DOI: 10.4018/978-1-7998-9220-5.ch187.

[15] S. Pramanik and S. Bandyopadhyay, Analysis of Big Data, in Encyclopedia of Data Science and Machine Learning, Eds. John Wang, IGI Global, 2023, DOI: 10.4018/978-1-7998-9220-5.ch006

[16] R. Bansal, B. Jenipher, V. Nisha, Jain, Makhan R., Dilip, Kumbhkar, S. Pramanik, S. Roy and A. Gupta, "Big Data Architecture for Network Security", in Cyber Security and Network Security, Eds, Wiley, 2022, https://doi.org/10.1002/9781119812555.ch11

[17] K. Dushyant, G. Muskan, A. Gupta and S. Pramanik, "Utilizing Machine Learning and Deep Learning in Cyber security: An Innovative Approach", in Cyber security and Digital Forensics, M. M. Ghonge, S. Pramanik, R. Mangrulkar, D. N. Le, Eds, Wiley, 2022, https://doi.org/10.1002/9781119795667.ch12

[18] A. Gupta, A. Verma and S. Pramanik, Security Aspects in Advanced Image Processing Techniques for COVID-19, in An Interdisciplinary Approach to Modern Network Security, S. Pramanik, A. Sharma, S. Bhatia and D. N. Le, Eds, CRC Press, 2022.

[19] A. Bhattacharya, A. Ghosal, A. J. Obaid, S. Krit, V. K. Shukla, K. Mandal and S. Pramanik, "Unsupervised Summarization Approach with Computational Statistics of Microblog Data", in Methodologies and Applications of Computational Statistics for Machine Learning, D. Samanta, R. R. Althar, S. Pramanik and S. Dutta, Eds, IGI Global, 2021, pp. 23–37, DOI: 10.4018/978-1-7998-7701-1.ch002

[20] Y. Meslie, W. Enbeyle, B. K. Pandey, S. Pramanik, D. Pandey, P. Dadeech, A. Belay, A. Saini, "Machine Intelligence-based Trend Analysis of COVID-19 for Total Daily Confirmed Cases in Asia and Africa", in Methodologies and Applications of Computational Statistics for Machine Learning, D. Samanta, R. R. Althar, S. Pramanik and S. Dutta, Eds, IGI Global, 2021, pp. 164–185, DOI: 10.4018/978-1-7998-7701-1.ch009

[21] A. Gupta, A. Verma, S. Pramanik, "Advanced Security System in Video Surveillance for COVID-19", in An Interdisciplinary Approach to Modern Network Security, S. Pramanik, A. Sharma, S. Bhatia and D. N. Le, CRC Press, 2022.

[22] S. Pramanik, "Carpooling Solutions using Machine Learning Tools", in Handbook of Research on Evolving Designs and Innovation in ICT and Intelligent Systems for Real-World Applications, K. K. Sarma, N. Saikia and M. Sharma, IGI Global, 2022, DOI: 10.4018/978-1-7998-9795-8.ch002.

[23] G. Taviti Naidu, KVB Ganesh, V. Vidya Chellam, S Praveenkumar, D. Dhabliya, S. Pramanik, A. Gupta, Technological Innovation Driven by Big Data, in "Advanced Bioinspiration Methods for Healthcare Standards, Policies, and Reform", Hadj Ahmed Bouarara IGI Global, 2023, DOI: 10.4018/978-1-6684-5656-9

[24] R. R. Chandan, S. Soni, A. Raj, V. Veeraiah, D. Dhabliya, S. Pramanik, Ankur Gupta , Genetic Algorithm and Machine Learning, in "Advanced Bioinspiration Methods for Healthcare Standards, Policies, and Reform", Hadj Ahmed Bouarara, IGI Global, 2023, DOI: 10.4018/978-1-6684-5656-9

[25] D. Mondal, A. Ratnaparkhi, A. Deshpande, V. Deshpande, A. P. Kshirsagar and S. Pramanik, Applications, Modern Trends and Challenges of Multiscale Modelling in Smart Cities, in Data-Driven Mathematical Modeling in Smart Cities, IGI Global, 2023, DOI: 10.4018/978-1-6684-6408-3.ch001

[26] P. K. Mall, S. Pramanik, S. Srivastava, M. Faiz, S. Sriramulu and M. N. Kumar, FuzztNet-Based Modelling Smart Traffic System in Smart Cities Using Deep Learning Models, in Data-Driven Mathematical Modeling in Smart Cities, IGI Global, 2023, DOI: 10.4018/978-1-6684-6408-3.ch005

[27] B. K. Pandey, D. Pandey, V. K. Nassa, A. S. Hameed, A. S. George, P. Dadheech and S. Pramanik, A Review of Various Text Extraction Algorithms for Images, in The Impact of Thrust Technologies on Image Processing, Nova Publishers, 2023.

[28] B. K. Pandey, D. Pandey, V. K. Nassa, A. S. George, S. Pramanik and P. Dadheech, Applications for the Text Extraction Method of Complex Degraded Images, in The Impact of Thrust Technologies on Image Processing, Nova Publishers, 2023.

4

Integrating Artificial Intelligence into Pest Management

Gobi Natesan[1], Priyank Singhal[2], Manoj Kumar Mishra[3], Aarti Kalnawat[4], Abhishekh Benedict[4], Sabyasachi Pramanik[5], and Korhan Cengiz[6]

[1]Department of Computer Science and IT, School of CS and IT, Jain (Deemed to be) University, India
[2]College of Computing Science and Information Technology, Teerthanker Mahaveer University, India
[3]Department of Agriculture, Sanskriti University, India
[4]Symbiosis Law School, Nagpur Campus, Symbiosis International (Deemed University), India
[5]Department of Computer Science & Engineering, Haldia Institute of Technology, India
[6]University of Fujairah, UAE

Email: gobi.n@jainuniversity.ac.in; priyanksinghal1@gmail.com; manoj.ag@sanskriti.edu.in; aartikalnawat@slsnagpur.edu.in; abhishekhbenedict@slsnagpur.edu.in; sabyalnt@gmail.com; korhancengiz@uof.ac.ae

Abstract

Unpredictable insect outbreaks are a problem for farmers due to climate change, a rise in the worldwide trade of infected materials, and pest control issues. In order to solve these issues, integrated pest management (IPM), a sustainable pest control strategy that makes efficient utilization of natural resources, is required. IPM is an ecologically dependent control management method, which takes into account various aspects, including climatic details, economic thresholds, and herb sensitivity and reproduction features. The most important component of IPM is expert employees. The expert is involved in system design, ecological factor monitoring, and decision-making processes. Artificial intelligence may be used to carry out regular tasks for sustainable pest control, like tracking biological and ecological constituents and selecting the best time and approach. This chapter will describe how artificial intelligence is applied in IPM and provide details on the tools, algorithms, and techniques used.

4.1 Introduction

Modern agriculture is now having a lot of problems. Farmers and other players that provide input into agriculture today compete in a worldwide market and must take into account regional and climatic variations as well as global economic and political considerations. Food production must be increased annually to meet the rising global population's nutritional demands; yet arable land is in short supply. While the demand for food worldwide has increased, housing and transportation issues also significantly reduce the amount of land that can be used for agriculture. Recent decades have seen an increment in abrupt precipitation, global warming, dryness, and the prevalence of severe weather conditions due to changes in the global climate. These adverse circumstances put conventional producing regions in peril and increase the dangers and uncertainties facing global agriculture. Climate change is boosting pest outbreaks or making existing pests the primary pest in regions where they were not previously an issue. Unpredictable insect issues experienced by farmers lead to a rise in the worldwide trade of afflicted material (seed, plants, and soil). Along with all of these other worldwide difficulties, insect-related problems including pesticide defiance, subordinate pest outbursts, and host herb resistance breakdowns have exacerbated the difficulty in farming productivity. In order to meet these difficulties, all agricultural production regions must continuously and sustainably boost productivity, and resources such as water, energy, pesticides, and fertilizers must be utilized carefully and effectively.

In order to solve these issues and uncertainties, agriculture needs assistance. New approaches are needed to every facets of farming yield, from improved and expected commodity planning to precision agriculture and budget optimization. IPM [1] is now the most well-liked and environmentally friendly method of pest management. However, this strategy is highly complex and needs a great deal of information, skill, and observation. In actuality, it is required to keep an eye on the pest in the field, identify the pests' most vulnerable life stages, choose the optimal management strategy (pesticide or other alternative ways), and administer it at the ideal moment.

IPM, therefore, needs extensive field observation, skilled personnel, and data mining. It has become clear in this situation that deploying AI approaches is necessary for managing, monitoring, and utilizing these farming inputs at the best period. In this chapter, we will look at how Industry 4.0 has changed contemporary agriculture historically and how IPM advances have affected information technology.

4.2 Background

4.2.1 Agriculture Industry 4.0

According to [2], Industry 4.0 is the period of smart production during which every living and non-living entities having a certain economic value can interact and communicate with one another over the Internet. This period also sees advancements in numerous technological areas, such as artificial intelligence, 3D printing, robotics, and nanotechnology. Developed nations are developing for a newer industrial revolution in practically all areas of the economy, including agricultural production, for boosting the additional value of economic output. Many enterprises have already been incorporated into this newer industrial revolution or are in the process of doing so. Virtual and physical computers are merged throughout this age of production's digitalization, and as a result, the production system's connected items will eventually acquire intelligence.

The factories outfitted with "smart" technologies are one of the key locations where Industry 4.0 has enabled change. Because no one works in these industries, they are known as dark (lights out) factories. Our everyday lives are now impacted by these smart systems, which are widely employed in business (like smart water, energy, fire alarm, etc.). Similar "intelligent factories" might be used in the manufacturing of food. The requirement for water, humidity, heat, and light has therefore been determined using sensor-assisted intervention systems, which have begun to be employed widely. However, based on requirements like water, humidity, light, and ventilation, the system may turn the equipment on or off automatically or can alert the authorized person.

Agriculture is built on massive volumes of output across a wide region, and no amount of effort is ever used to produce enough food to supply the whole planet. This is why the machinery and equipment utilized in farming yield benefit from the technology of today. Therefore, automation in agriculture is considered the first agricultural industry revolution in order to feed the fast expanding global population. Agriculture's "Industry 2.0" refers to the utilization of synthetic manures and insecticides, while "Industry 3.0" refers to the introduction of biotechnology goods.

Integrating AI into production will certainly lead to the realization of agriculture 4.0. It would seem that in order to reap the benefits of Industry 4.0, the idea of "modern agriculture" needs to be implemented. Agriculture's transition from conventional to present-day production increased the utilization of agricultural machinery and contemporary production components including science, technology, and capital. Agriculture, agricultural production, and management activity growth in novel technologies led to a steady specialization and rise in unit area productivity. Technology including computers, satellites, and remote sensing systems were produced in the previous 20 years as a result of computer science and space projects. To put it another way, modern agriculture is the transition from using manual tools to using machines, and subsequently from manual equipment to automated machinery. While the amount of labor done in a unit area has grown significantly due to the use of agricultural equipment, the agricultural yield in a unit area has developed due to the utilization of artificial manures, hormones, and insecticides. The excessive use of insecticides has led to a rise in pest populations and epidemics, reflecting negatively on contemporary agriculture. Unfortunately, the extensive use of these pesticides in agriculture has led to substantial environmental contamination or the poisoning of creatures that were not intended targets. A new idea known as "good agricultural practices" should be applied to modern agriculture in order to solve these modern agriculture challenges and promote the sustainability of farming.

4.2.2 Modern agriculture's better farming implementations

The potential of agricultural output has been significantly increased by the development of high-grade hybrid categories, biotechnology, chemical preservations, and IT. The adoption of the good agricultural practices paradigm has elevated agriculture to a new level in industrialized nations. The introduction of chemicals in modern agriculture was the beginning of a host of exceptional issues, including resource depletion and environmental damage. Globally, nations have recently focused more on the management and protection of the ecological environment in farming growth, taking into account the efficiency and conservation of resources used in production such as soil, manures, water, and pesticides. Here, newer models for sustainable agriculture emerged, including "good agricultural practices."

IPM is a crucial element of effective agricultural operations. Significant advancements in this approach include accurate pest diagnosis, timely pest

Table 4.1 Methods of pest management.

Pest control techniques	→	Cultural operations Physical and mechanical operations Chemical operations Biological operations Biotechnological operations Genetic operations

management, reduced usage of synthetic pesticides, and ecologically acceptable alternative pest control tools, preventing chemical contamination of food and the environment, and workplace security for agricultural employees. It is obvious that all of this cutting-edge knowledge is the result of extensive scientific research and real-world applications. Despite the fact that IPM has been the subject of countless research, IPM experts are still needed for the processing, analysis, and practical implementation of this knowledge. The biggest issue is that just a few experts are able to communicate with agricultural farmers and disseminate knowledge. As a result, it is feasible to convert the IPM's current and developing data into software, evaluate them, and then provide the appropriate guidance to agricultural producers using AI technology. Following this portion of our chapter, we will talk about IPM's definition and how AI is used in it.

4.3 Practices for Pest Control

If no effort is made to combat pests, it is predicted that 60% of the current agricultural yield would be lost. For instance, uncontrolled olive flies may cause damages of up to 15%–30% in typical years and 70% in years of epidemic. One of the primary pests of apples, the codling moth (*Cydia pomonella*), if not managed, may cause damage of up to 60% or even 100%. In many parts of the globe, grain is seen as a source of food security since it is crucial to human nourishment. In Turkey, 50% of the land is used to grow grains. A rise in the incidence of hunger throughout the globe is directly represented by similar loss in wheat production, the basic vital constituents for human nourishment. In this section, we will discuss several pest management techniques as well as how using technology affects the effectiveness of control. Agricultural pests may be controlled using a variety of techniques (Table 4.1). Cultural practices, mechanical, chemical, biological, and genetic controls are all types of control measures. Chemical control (pesticides) is the most popular pest management strategy since it is rapid acting, affordable, very effective, and simple to use. Bugs are eliminated by chemical control using chemicals that are naturally present in them. This is because pesticides often include harmful materials. Therefore, when the chemicals are employed heavily, they pose serious risks to both people and non-target creatures. Additionally, the improper and excessive usage of these compounds results in several ecological issues and environmental harm. Additionally, the chemicals may leave pesticide residues on crops when used at large doses and at the wrong times. However, farmers' health may be seriously compromised by carelessness and ignorance while handling pesticides, as well as by a lack of knowledge and/or training.

Due to these harmful effects of chemical pesticides on creatures other than their intended targets, a tendency is toward alternate control strategies that are more particular. In order to develop plants that are healthy and pest-resistant, general cultivation methods should be employed within the context of cultural practices. Another option is the employment of biological agents tailored to the target species or biotechnological and physical control instruments. To limit insect populations from causing significant economic harm, techniques including biotechnological and biological management are offered. Growing genetically modified plants, releasing genetically sterile organisms, and other biotechnological techniques have all made substantial advancements in pest management in recent years.

Since learning about the dangerous side effects of pesticides, scientists have created newer and innovative pest control techniques. In the most recent decades, a thorough strategy for pest management, specifically

IPM, is created to combat major pests worldwide. IPM is an ecologically dependent control management method that takes into account natural elements, such as climatic conditions and the presence of natural enemies. Programs for integrated pest management (IPM) now in place take into account variables such as insect and natural enemy populations, economic thresholds, herb vulnerability, herb breeding parameters, and pest biology. In IPM techniques, two or more strategies that work well together are coupled, and any possible interactions between them are assessed. IPM techniques should give top emphasis to cultural practices, resistant crops, and biological (parasites and predators) and biotechnological (messenger chemicals) techniques. According to [3], pesticides should only affect pests and have no adverse impact on other species or natural adversaries. As a result, IPM is a type of pest control that prioritizes natural ingredients in the contest against damaging microbes and utilizes a combined method appropriate for economic, and ecological, requests in order to solve agricultural pest problems rather than relying solely on one method. Six key components make up an IPM program: (1) persons (software developer and project manager); (2) knowledge and information; (3) monitoring (of pests);

(4) decision-making levels (pest count needed to initiate pest control check); (5) techniques (combination of control tactics for decreasing pest counts); and (6) implementation.

4.4 Pest Management and Artificial Intelligence are Integrated

IPM expert is the most important component, as was described in the previous section. However, in addition to acting as the system designer, expertise also plays a part in carrying out everyday tasks (such as monitoring and selecting the best strategy and timing). Even if many particular items that may be used to manage agricultural pests have improved, the individual who chooses which way is best plays a crucial role in decision-making. IPM uses a variety of approaches in conjunction to providing successful pest control outcomes. First, the pest must be appropriately identified by using a suitable approach. It has been created using a variety of pest detecting methods thanks to developments in electronics and information technologies. The following section will provide several methods and equipment that may be utilized to establish the diagnosis and pest density. It is essential to gather data quickly and accurately, including information on the degree of insect population as well as immediate changes in plant phenology. For estimating the growth of the microbes utilizing statistical prediction models, climatic data from the ecology in which the herb and pest dwell should also be granted into consideration. The incorporation of AI [4–8] might maximize control success by using these methods in conjunction.

Instantaneous data transmission over a wireless sensor network is a hallmark of this level. All of these technical advantages may be taken into account within the context of IPM, allowing for the implementation of an effective pest management plan. This integration improves pest control efficiency while lowering material and personnel costs. Due to current technological advancements and Industry 4.0's effects on farming, AI approaches are applied, and effective outcomes are attained within the confines of IPM (Table 4.2). The authors will discuss these methods, research works, and applications of artificial intelligence in IPM.

4.4.1 Pest monitoring and detection systems

In recent years, wireless sensors and certain very sensitive motion sensors are used in remote sensing to identify plant pests. There are sound sensors available that can record certain frequencies emitted by bugs. However, it

Table 4.2 Methods of pest control.

Pest control operations	Expenditure	Ecology beneficial	Product standard	Labor reduction	Influence
Chemical operation	Lower	Lower	Lower	Lower	Higher
Biological	Average	Higher	Higher	Average	Average
Genetic	Higher	Average	Higher	Higher	Higher
IPM with artificial intelligence	Average	Higher	Higher	Average	Higher

is difficult to use these sensors to find insect populations over the whole region. At least, it is feasible to identify certain areas of our region where certain insects are present. For direct pest detection, image processing, optical sensor, and pheromone trap technologies are also available. When using an IPM technique, precise pest detection is essential. These days, pest species-specific acoustic, ultrasonic, and optical sensor technologies may be used to identify diagnosis and pest density.

4.4.2 Sound-based sensors

Since the early 20th century, sensitive acoustic technology has been utilized to find insect infestations in soil, trees, and plants. Digital signal processing approaches that enable the detachment of parasite sounds from surrounding noise and serve as the foundation for analysis on newer audible vehicles to determine and track the intrusion of below-the-surface pests have been made possible by the advancement of modern computer technology. The detection of insects in feeding sites has recently been made possible thanks to acoustic technologies. Low noises (0.5–150 kHz) made by insects while they are in flight, foraging, or signaling their other sex may often be used to identify their species. In order to identify insects in stored wheat, Yan et al. [9] employed sound sensors. They discovered that the technique could be used to calculate the population density in the region.

Nagrale et al. [10] claimed that the approach is beneficial for fast pest monitoring in grain repository approaches by using computer software that compares the data from sound sensors and sensors to locate bugs in a wheat heap. In case of the pest *Cephus cinctus* Norton in wheat, Su et al. [11] employed acoustic sensors, claiming that these sensors could detect motion and nourishing actions in the herb. Furthermore, *Sitophilus oryzae* and *Rhynchophorus ferrugineus* Olivier pests were located and fed upon using acoustic sensors [12], and they reported that the solutions provided an exact detection with a rate of 98.9%.

4.4.3 Infrared sensors

Ultrasonic signals released during feeding activities may be used to locate pests that are concealed in seedlings, trunk, and certain different fibrous herb components. The ultrasonic wave emitted from the ultrasonic sensor reverts back as it strikes the pests in a system made up of these parts, sending the signals to the device that can handle digital data. As a result, it is possible to estimate in real time the motion direction and original location of the pests in the region. However, this technology has not yet been able to operate reliably in many situations. It is yet unknown how it will handle the identification of insects of several species and what outcomes it will yield in locations with higher than typical background noise. Such investigations, i.e., further research initiatives, are required to create a smart, mobile ultrasonic instrument for pest tracking in the farm. The fact that there has not been a fresh research since 1990 supports this conclusion.

4.4.4 Laser sensors

This device, which is made utilizing sensors like IR rays [13–15] and paired with an attractant (pheromone, light, etc.), is put at the entrance of pest traps and makes a track of how often the insect goes in for the trap. When the targeted pest reaches the trap housing the appealing things, the optical sensors shoot lights to the trap entry, cutting their route. The sensor's control system experiences voltage fluctuations as a result of this motion. This variation serves as a signal that the pest has entered the trap. In the IPM, determining the time and population compactness at which pests are initially collected in traps is crucial. Each species' unique attractants must be employed for precise detection.

To identify the Scolytus species of bark beetles, Jain et al. [16] created a light ray trap (Figure 4.1). They had created a framework that identifies the time of the insect's admission in minutes, hours, and days and records the information on memory chip by inserting optical sensors inside the entry of the trap. The authors noted that by using this technology, efficient IPM techniques against bark beetles may be established.

If not discovered, the Red Palm Weevil, which sustains on palm trees, may cause severe harm. With the right management measures, harm may be reduced if the pest is identified in time. Pheromone traps are employed to keep an eye on the pest because of this. However, the Mediterranean basin and other difficult-to-access places, such as those that are far away, make hand counting of these pheromone traps extremely challenging.

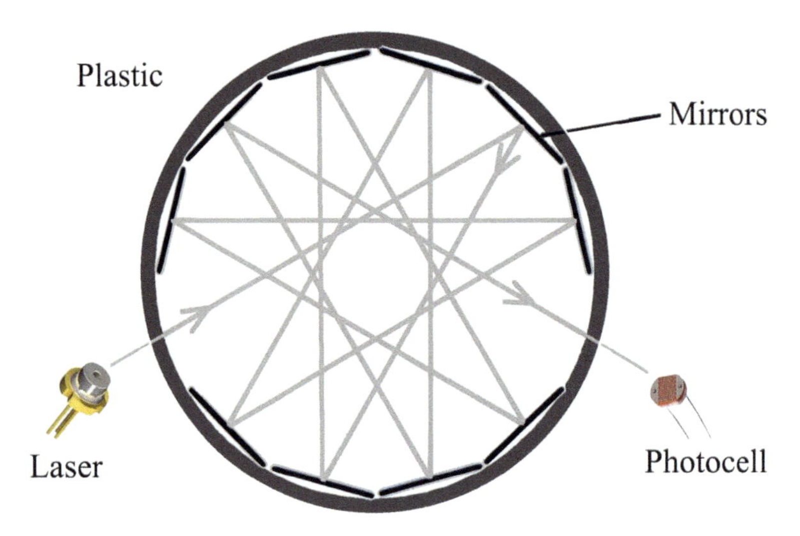

Figure 4.1 Application design for laser scanning.

In order to address this issue, Potamitis et al. explain a novel Red Palm Weevil surveillance technique. To track the insect's descent into the trap, they employed an optical sensor. The electrical circuits attached to the sensor count the bugs and use the GSM network [17–20] to deliver the results as SMS to a specified phone. This technique allowed for the monitoring of the pest population and emergence across broader regions while lowering personnel expenses. When we consider all of this research, optical sensors provide a substantial advantage when it comes to counting pests. However, it has the drawback of operating inefficiently under artificial light and being unable to identify the pest species in the traps when employing no particular species attractant. According to studies, further study is required to accurately identify species.

4.4.5 Systems for processing images of insect pests

Monitoring pest populations enables the efficient use of various IPM approaches. It is possible to learn more about the movement and appearance of insects using the data collected from different pest traps. This information relates to how many pests have been caught in traps and if this amount beats the monetary harm threshold. For instance, the management of the codling moth in apple orchards in the Emilia Romagna area of Italy starts with the recognition of two male moths taken in each trap on two successive days. This condition requires hanging a minimum of two pheromone baits

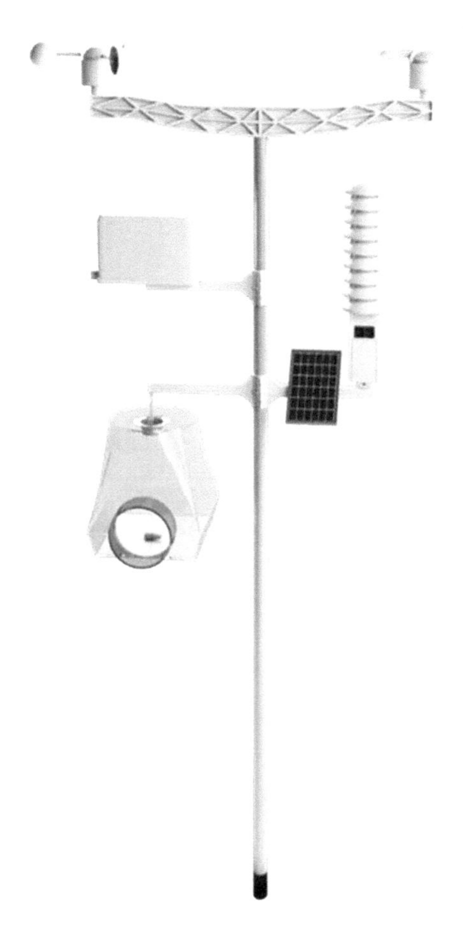

Figure 4.2 Remote monitoring device for pests.

in a fruit orchard measuring one acre, checking them daily until males are caught. The traps must be examined occasionally, the quantity of people caught must be noted, and the processes must be carried out all through the season in order to offer this data. All of these activities waste time and energy and demand heavy work. Similar to the optical sensor frameworks outlined in the preceding section, it is impossible to precisely identify the insect species infiltrating the trap if a species-based attracting insect is not employed in the traps. Additionally, it is not feasible to develop an attractant tailored to each pest species. Automatic pest calculation using a camera installed in the bait and image processing methods in the picture captured by the camera might be used to resolve this issue and provide effective outcomes on distant trap monitoring systems. Figure 4.2 displays a sample system view.

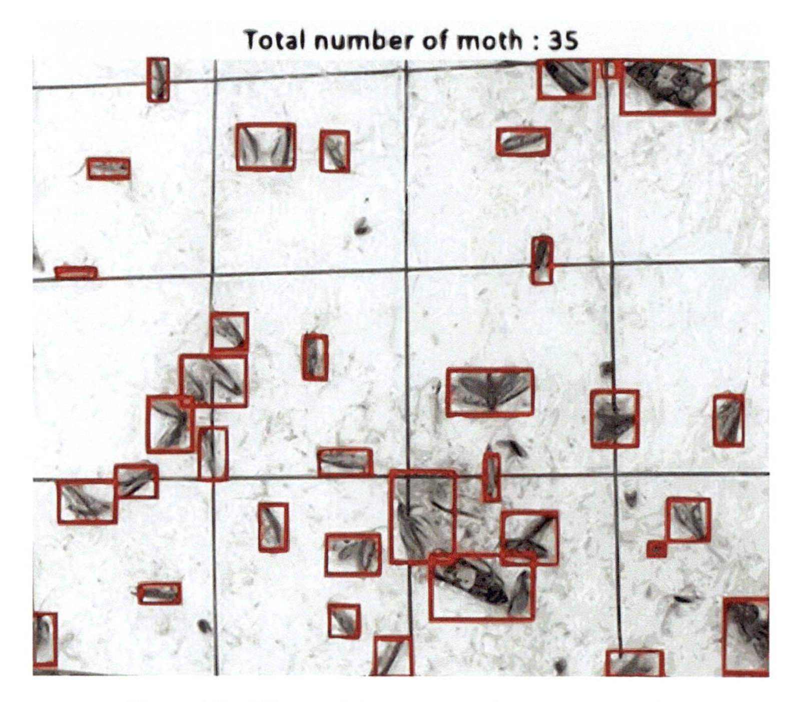

Figure 4.3 Olive moth image processing output example.

According to [21], they employed a pheromone trap together with an image processing method to calculate the diamondback moth individuals. The technique involves utilizing a smartphone to regularly take pictures of the trap and then using a wireless LAN system to send those pictures to a computer. According to them, this method has a 91% success rate. Similar to this, Teixeira et al. [22] reported that they employed image processing to identify adult whitefly pests on rose leaves and underlined that these techniques may be used to integrated management strategies and the early identification of pests. The yellow sticky traps that attract whiteflies may be simply modified to use the technique described by Jose et al. [23]. A technique to automatically count pests in the *Vinca catharantus* plant was created by Todorović et al. [24]. They have devised a framework that takes pictures at certain intervals, distributes them wirelessly to the host, and uses surrounding detachments to count the number of pests. Similar to this, Todorović et al. [24] created a technique to automatize the pest calculation of pheromone traps utilized for rice beetle pests. The photos that the camera took in the pheromone trap were relayed over wireless LAN to a physical server that was set up on the field. They used background-separation-based image processing software on

Table 4.3 Basic procedures for identifying and classifying plant diseases.

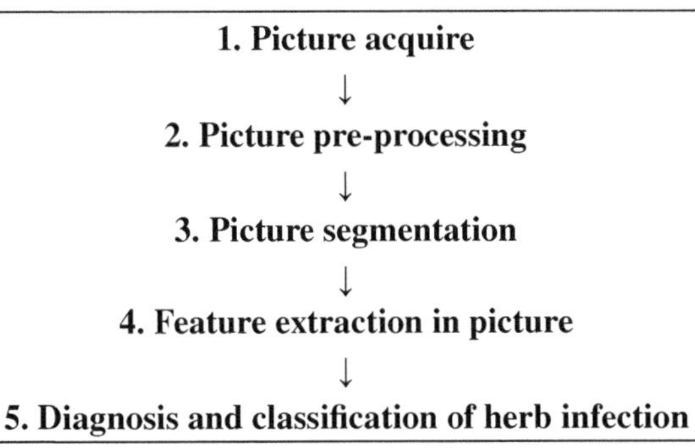

1. Picture acquire

↓

2. Picture pre-processing

↓

3. Picture segmentation

↓

4. Feature extraction in picture

↓

5. Diagnosis and classification of herb infection

the land server to correctly count a trap, with 89.1% success compared to the technique used by hand [25]. By mounting a temperature-humidity sensor and an IP camera on pheromone traps, Sudharson et al. [26] had created a framework that enables data to be recorded on a server at predetermined time-gap to track pests in fruit and vineyards. By transmitting data to the central server through Wi-Fi, they are powered by the solar panel and accessible from any device with internet connectivity. In Table 4.3, an example of an image processing result is shown. These studies make it possible for remote sensing traps to collect information on pest monitoring, pest diagnosis, the quantity of pests caught, and their flight times, all of which are often employed in IPM. However, it only counts as one input for the decision-making process for calculating the control time. Additional biological, physical, and environmental characteristics must be linked with this data.

4.4.6 Plant disease image processing systems

Significant financial losses are brought on by plant diseases in the agriculture sector. An efficient way to stop disease outbreaks is by the early discovery of the pathogen by tracking the herb for the goal of managing contrarily to the illnesses. The condition must be diagnosed using expert knowledge, and this control is given by a person. In addition to the typical practice of observing disease signs, DNA-dependent and serological approaches facilitate the essential resources for a precise estimation of herb disease. The study takes at least 1–2 days, despite the fact that DNA-dependent and serological approaches have restructured the diagnosis of herb disease. Determining the illness diagnosis and the appropriate control plan requires time. The sickness may be detected using image processing on the picture of the sick leaf in order to expedite this procedure and provide a quick diagnosis.

The identification of plant diseases using various image processing techniques is described. An illustration of a sick leaf, i.e. the illness, was identified using an RGB (red, green, and blue) [27–30] picture (Figure 4.4). The green pixels were determined using *K*-means clustering methods, and the variable threshold estimate was subsequently calculated utilizing the Otsu approach. The technique of color identification was utilized to extract features. The RGB picture was transformed into a version with HIS (hue–saturation–intensity). For the purpose of calculating texture statistics, the SGDM matrix was created, and the outcomes were computed utilizing the GLCM function. In order to detect wheat illnesses, Patnala et al. [31] designed newer spectral indices. Powdery mildew, yellow rust, and aphids were three separate plant pests and diseases that they employed in their research on winter wheat. The RELIEF-F method was used to determine the wavelengths that are most and least likely to be detected for various disorders. These novel indicators had high classification accuracy in herbs with powdery mildew, and a robust plant. Image processing methods were used by [32] to identify fruit illnesses. To find the illness, they employed an artificial neural network. They built two distinct databases, one for applying query pictures and the other for training already stored illness images. They focused on shape, color, and three distinct factors. They discovered via their trials that the morphological traits outperformed the other two factors in terms of producing good outcomes.

Khalid et al. [33] employed image processing methods for identifying illness in pepper plant leaf pictures. To make sure that pesticides are only sprayed to sick peppers, this research was conducted. In their research, they classified and recognized images using MATLAB's Fourier filtering, edge detection, and morphological approaches. Phadikar and Hong et al. [34] created a software framework for viral detection in rice herbs and utilized image recognition methods to identify rice illness. To segment the picture of the sick plant, they employed the HIS model. The diseased leaf portion was then located using boundary and point detection. These experiments demonstrated that image processing might be used to correctly identify and categorize plant diseases.

Figure 4.4 shows an illustration of the results of *K*-means clustering for an early scorch disease-infected leaf. (a) The image of the diseased leaf. The pixels from the first, second, third, and fourth clusters are (b), (c), (d), and (e). (e) A single gray-scale picture where the colors of the pixels are determined by their cluster indices.

Many agricultural models need weather data as an input, equally at the separate farm level and at the local or basin level. The information that these models use as input is climatic data, like wind speed, wind direction, temperature, humidity, precipitation, amount of moisture on the leaves, sun radiation,

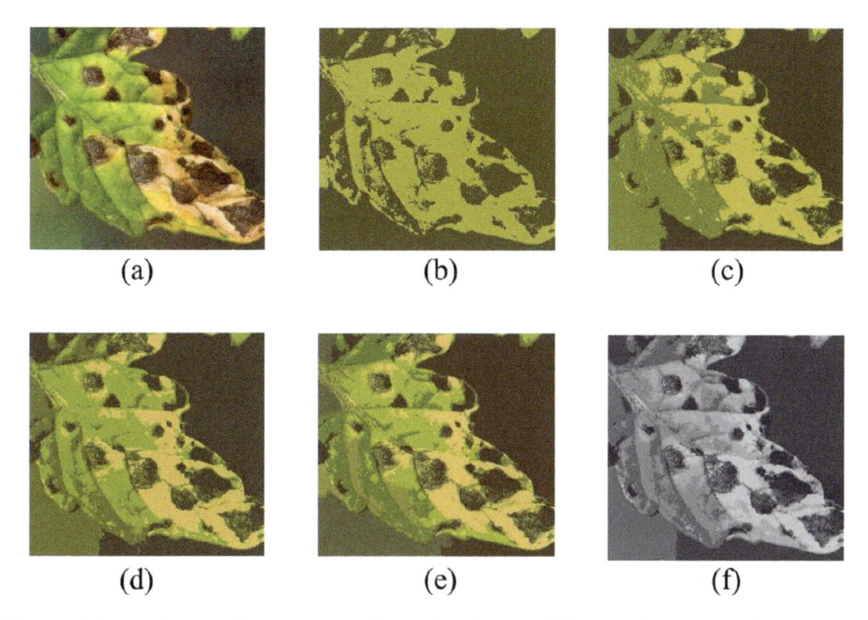

Figure 4.4 An illustration of the result obtained by applying K-Means clustering to a leaf affected by early scorch disease. (a) The image of the leaf that is diseased. The pixels of the first, second, third, and fourth clusters are represented by (b, c, d, e). (e) a monochromatic picture where the pixels are colored according to their cluster index.

and soil moisture. In the early 1970s, weather data was first included in agricultural models. PCs are utilized to generate weather-dependent approaches from the 1980s, thanks to advancements in microcomputers and microprocessors. Following 2006, software assets created or traded by meteorological station builders like ADCON and METOS have assisted agriculturists in gaining access to weather data documented on or near their farms. In this time, crop modeling software called "DSS for Agrotechnology Transfer" was devised. With the growth of the Internet in recent years, it is now feasible to provide consumers with weather-based models. The usage of weather stations has increased significantly as a result of internet technologies. GPRS technology and a common interface, which enables preview this data, facilitate communication between producers and suppliers of meteorological data (Figure 4.5). A smart agriculture model was presented by Yang [35] for real-time monitoring of factors including soil temperature and moisture as well as air quality. They devised a framework combining real-time climatic data and an herb disease detection framework to deploy decision support advising frameworks in a Pest&Disease diagnosing. They said that the framework can manage different field activities from any distance at any time via a web application or a mobile device. Farm side, server side, and client side are

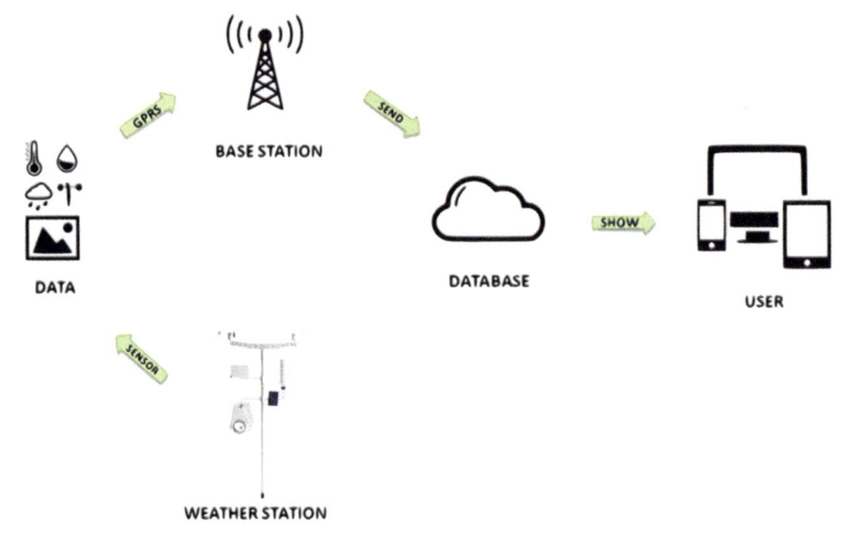

Figure 4.5 The way that weather stations work.

the three elements that make up the suggested system. The research suggests an intelligent agriculture paradigm that integrates information and communication technology (ICT). ICT is important in the field of farming. While villagers' agriculturists have grown "similar" crop for generations, throughout time, weather patterns, soil conditions, insect and disease outbreaks, and other factors changed. Utilizing the suggested method, current information gained enables the farmers to manage and even gain from these modifications. Because agricultural information is extremely specialized and tailored to certain situations, disseminating such knowledge is a difficult endeavor. It is anticipated that having full real-time and historical environmental data would contribute to effective resource management. Their system comprises the following six techniques:

1. Detecting agricultural conditions in the area.
2. Locating the sensor's position and gathering data.
3. Transferring agriculture field data for use in decision-making.
4. Using data analysis, subject-matter expertise, and history developed, decision assistance and early warming are provided.
5. Depending on a choice, action, and control.
6. Crop observation using a camera module.

A distant real-time tracking and control framework for greenhouses was created by Araghian et al. [36]. Additionally, they asserted that the system may employ rational management practices and stop crop epidemics. They created a few sensors to measure greenhouse climatic variables like temperature, pressure, humidity, and CO_2. The authors showed how the framework can regulate the operation of sprinklers to maintain humidity, curtains to block sunlight, and air fans to circulate air. This method enables remote, smart greenhouse tracking and command for herb growth and pest management, which benefits farmers looking to sow crops in a methodical, scientific manner. It is obvious that this design will be popular and especially useful for agriculturists. This methodology may help agriculturists have reliable, cost-efficient access to meteorological data and models. Inputs to DD models may be created using GRPS-based rural meteorological stations to determine insect outputs and plan land-based spraying and fertilization operations. Climate data alone is not sufficient for AI algorithms, which are employed in IPM to determine when to control pests. The O/P of the DD methods and the biological schemes of the pest should be assessed simultaneously.

4.4.7 Models in degree-days

Because they are cold-blooded creatures, agricultural pests need certain temperature ranges to grow. Although organisms begin their physiological functions at this temperature, which is known as the development threshold, other activities need particular temperature accumulations. The total number of degree-days (DD) between higher and lower temperature thresholds in the course of the pest's whole growth cycle may be used to anticipate these physiological activities. The biofix date, which is connected to the organic scheme of the pest, is the day on which DD collection officially starts. The calculation of sprinkling time is essential to preventing extra pesticide usage and minimizing the negative consequences of pesticides, according to Sun et al.'s [37] research on DD approaches and the potential applications in herb shielding. In general, the average daily temperature (DD) may be determined using a straightforward formula derived from the everyday highest and lowest temperatures.

DD is equal to [(daily highest + daily lowest)/2] – baseline temperature.

As an instance, a day with a maximum temperature of 24.6°C and a minimum of 6.3°C will add up to 4 DD utilizing 10°C as the baseline:

Instance 1:

$$DD = (24.6 + 6.3)/2 - 10.$$

Zero DD has collected at temperatures below 10°C. The simplest and least accurate computation approach is this one. Let us look at an example of how this calculating technique might be used to calculate a codling moth pest's biological periods.

The extremely frequent pest of apple trees is the codling moth. The larvae dig holes in the fruit, create galleries there, devour the tissue and the central house, and neglect mud behind. This causes direct injury to the fruit. They therefore either make the fruit to fall to the ground or they reduce the number and quality of fruits that may be left on the tree without spilling. They therefore cause the market value of the apple to decline. Up to 60% or even 100% of the damage might occur in unregulated locations. All of the apple-producing areas in our nation are in conflict. Producers often use chemicals in a hurry. Unfortunately, they mistakenly think that good yield and quality are directly correlated with the quantity of chemical utilized. Calendar-based spraying is often used as a result. Of course, choosing the right time is essential if a chemical control method is to be employed to get rid of a pest. This is a vital step in minimizing the harmful effects of pesticides and preventing their unneeded and inefficient usage. Because of this, several research works on anticipating and cautioning in the codling moth were conducted in an effort to get the best outcomes with the fewest pesticides. To find the codling moth's initial adult flights, pheromone traps are put nearby. To determine if the moths originate from external regions or spend the winter in the vineyard, the traps are suspended close to the orchard's center.

The first moth is captured after the first one or two days of trap checking. If at least one moth is traced on successive days, the date is recorded on the DD model as a biofix date. According to researchers, the codling moth extends at intervals between 12 and 31.1°C. As a result, the following formula is used to calculate DD:

$$DD = \int_{s0}^{s} (S - Sh).ds,$$

where DD is degree-days, s is the date, $S0$ is the biofix date, S is average everyday temperature, and Sh is the threshold for lower development,

As shown in Table 4.4, pesticide control must be used for the first generation in 50–75 degree-days, 100–200 degree-days, and 200–250 degree-days for early and freshly born larvae, respectively. Continue applying insecticides at 1000–1050 degree-days and 1100 degree-days for the second generation. The apples must be inspected at 2160 DD for the purpose of controlling the third generation, and the interval between pesticide treatment and harvest should be taken into account.

Table 4.4 A software for managing codling moths' major events depending on gathered DD.

DD	% Adults appeared	% Egg breed	Management action
120*	1	1	• Position traps in vineyards
145–225	First moth excepted	1	• Examine traps for each of 1–2 days up to when biofix is resolved
		First generation	
0 (biofix)	First consistent catch	1	• Reset DD to 0
65–80	8–12	1	• First eggs are laid • Use insecticides that require to be existing prior to egg-laying
125–245	18–45	1	• Earlier egg-laying duration • Use insecticides that choose earlier egg-laying time
235–265	50–65	1–5	• Start of egg hatch • Use insecticides that target newly hatched larvae
355–655	76–97	18–65	• Critical time for control, higher rate of egg hatch • Essential to keep fruit preserved for this time duration
880	105	89	• Termination of egg hatch for first generation
		Second generation	
1035–1235	6–9	0	• Firstly eggs of second generation are laid • Employ insecticides to select early egg-laying
1260	15	1	• Start of egg hatch • Apply insecticides that target newly hatched larvae
1450–1855	48–98	15–82	• Critical period for control, high rate of egg hatch
2320	125	99	• End of egg hatch for second generation
		Third generation	
2350	2	18	• Start of egg hatch • Remain fruit secured through September • Examine pre-harvest interval of material utilized to check that final spray is not too closer to harvest

Table 4.5 The following is a list of several insects that have temperature thresholds.

Target insect		Lower threshold (C)	Upper threshold (C)
Common name	**Scientific name**		
Alfalfa weevil	*Hypera postica*	−13.8	35.6
Armyworm	*Pseudaletia unipuncta*	−13.6	29.7
Black cutworm	*Agrotis ipsilon*	−13.4	25
Cabbage maggot	*Delia radicum*	5.6	28
Codling moth	*Cydia pomonella*	−15.4	32.2
Corn earworm	*Helicoverpa zea*	13.5	34.5
European pine shoot moth	*Rhyacionia bouliana*	−3.5	–
European red mite	*Panonychus ulmi*	12.7	–
Lilac/ash borer	*Podosesia syringae*	−17.6	–
Obliquebanded leafroller	*Choristoneura rosaceana*	4.6	30.8
Peach twig borer	*Anarsia lineatella*	−14.6	34.6

The effectiveness of managing the codling moth considerably improves using DD models. Studies using the DD model have also been carried out on numerous significant pests, as shown in Table 4.5. Getting the right temperature readings is the most important aspect of using DD. The resultant DDs do not accurately reflect the growth of the target insect if the location of a thermometer or weather station does not accurately reflect the environment in which it happens. Furthermore, temperatures in one area may not be a good indicator of the circumstances in another area a few kilometers away. This shows that different microclimates are influenced by mountains, lakes, and deserts.

In Turkey's Bursa area, researchers used temperature measurements and pheromone traps to analyze the population density of olive moths. With the help of their investigation, they were able to demonstrate that these two relationships might be utilized as a predictor for precise timing when applying insecticides to olive moth larvae. Insecticide exercising opposed to larvae at the proper time can also be utilized as a prediction approach; it was discovered by researchers after they carried out a survey to find out the hatching time of olive psyllid larvae utilizing essential temperatures. In their DD modeling of *Lobesia botrana* (Lepidoptera: Tortricidae), researchers utilized 12-year data and claimed that their model might be a beneficial tool in enhancing the efficacy of IPM tactics. According to the research works that have been done, the IPM method's prediction and early warning studies employ the biofix date and DD models. Adults of the (Lepidoptera: Tortricidae) family deposit their eggs when the night-time temperature is at least 15°C. Because of this, spraying is initiated in estimation and warning surveys when the cumulative essential temperature as of January 2 hits 100 DD and when night-time temperatures reach 17°C or higher. As a consequence of these research works, three sprayings – the first during the first offspring larvae outflow, the second 20 days later, and the third during the second offspring larvae outflow – produce favorable effects. But, depending on the durability, the number of sprayings varies from year to year and from one orchard to another.

In the future, moth population estimates should combine sexual pheromone traps with DD models, and biofix date – instead of January1 – must be utilized as the date of the first biological action identified by pheromone trap. To ascertain the timing of the integrated control, a combination approach including a pheromone trap and DD models should be used. The goal of using DD techniques in forecasting systems is for assessing all variables that might affect the pest population and decide whether or not the pest may cross the economic loss threshold. The technology can estimate the time and alert the makers if all indicators are favorable. Consequently, utilizing timely and well-executed solutions will greatly maintain natural balance and environmental health.

4.4.8 Utilization of AI in IPM where all technological approaches are integrated

Finding the right time for remote tracking of temperature data, remote pheromone trap tracking, and DD models is made easier by the technology and methods employed in IPM. However, these methods need heavy physical effort, which raises labor costs for both businesses that cover a vast geographic region and farmers that employ conventional methods of production. In this situation, a framework that offers a higher degree of control timing together with an AI technique that makes use of the information received from these approaches is required.

In the proposed framework:

1. The framework must be able to diagnose any biological stage of the pest in the field (utilizing information like color, pattern, width-length, etc.) and identify it right away.
2. The same image processing technique should be used to regularly check the quantity of pests.
3. The framework must sensitively gather crucial climatic data from the environment where the insect dwells, such as temperature and humidity that have an impact on the biology of the pest.

4. The framework must include a mathematical model that allows it to assess current climatic information in a certain way.
5. The framework must be capable of interpreting the data from the image processing framework and sensors in accordance with the model, assessing the likelihood of an outbreak or outflow of pests instantly, and conveying the information to the manufacturer or expert through web-dependent instant messaging.
6. Smartphones and tablets running iOS or Android-based applications should be able to watch or interrupt the system (Figure 4.6).
7. According to the pest's crucial biological period as determined by a current official source, this program must be capable of providing the suitable licensed (pesticides or biotechnology) product.

This selected system has undergone infrastructure investigations, and a project dubbed "GRSEN-TAM: Intelligent Agricultural Control System Supported with Image Processing and Sensor" was created (Figure 4.6). With its image processing and AI techniques, the system uses all the technological elements and features listed above and fall within the purview of IPM. GRSENTAM was successfully utilized to combat olive psyllid in an olive garden in Japan.

4.5 Future Directions for Research

IPM is a wide and amazing field where AI algorithms are used. There should be more development and study on this subject. These days, people are increasingly interested in issues like phytosanitary procedures, excellent agricultural practices, and sustainable agriculture. In certain nations, entrepreneurial initiatives on these issues are encouraged. Entrepreneurs should act in the following manner:

1. choosing a pest that affects agriculture to focus on;
2. figuring out this pest's development thresholds, such as a minimum of 7.5 °C and a maximum of 44.5 °C;
3. creating the mechanism to gather field data;
4. creation of DD models relevant to pests;
5. creation of equipment to monitor biological activity in the field (such as a camera-equipped pheromone trap system);

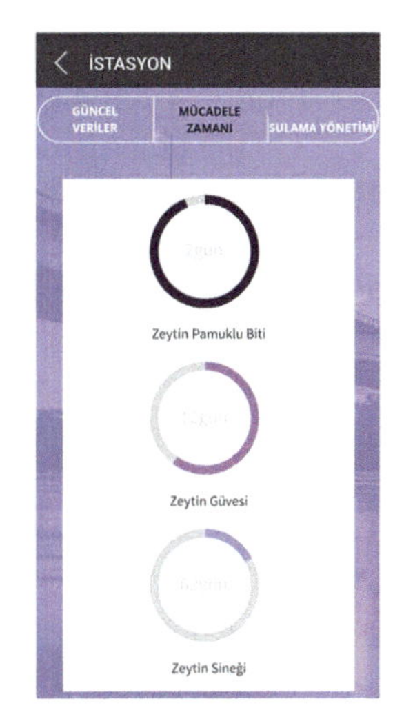

Figure 4.6 Images from the GORSENTAM smartphone app.

6. software creation that automatically tracks pest identification, count, and biological activity (using an image processing technique, for example);
7. creation of an AI program that analyzes all of these facts and renders reasoned judgments;
8. creation of a mobile application or online interface to allow for real-time data monitoring;
9. system verification;
10. addition of newer farming pests to the framework to continue research and development efforts.

4.6 Conclusion

AI algorithms for IPM can provide successful outcomes, but better R&D efforts are required in this area. As farming pests are cold-blooded creatures, the climatic data in their surroundings affects how they behave biologically. The locations of these pests might refer to the inside of a fruit as well as the soil and plant surfaces. For bugs that reside in fruit, the climatic data in their habitat would have a consequence on the pest's development. The appropriateness and sensitivity of the sensors used to gather climate data from the pest's surroundings for the DD modeling are crucial in this situation. To improve the precision of the data that will produce input to AI algorithms on this subject, specific agricultural sensors should be created.

On the other hand, for AI algorithms to be applicable for all pests, hundreds of distinct species must be created. Let us say, for illustration purposes, that Turkey has 150 primary pests. The development thresholds of these pests must be identified, and methods for pest detection and counting using DD modeling and image processing must be developed. In this context, an interdisciplinary research including computer science and agricultural science is required. The development of pest monitoring tools and systems that track pest populations in the field is necessary to perform effective IPM. These tools and systems provide supporting data for decision-making.

Although there are numerous pesticides that are now approved, their successful usage requires specialist expertise. Depending on the pest and its phenology, different insecticides have different fatal effects. Expert knowledge is necessary to choose the appropriate pesticide at the appropriate moment. In this situation, the approaches' reporting of the appropriate pesticide advice must moreover increase the degree of benefit for the authority in addition to calculating the period of control. The algorithm must check the country's registered pesticide list before recommending it to the user by choosing pesticides that are appropriate for the phenology of the pest. Utilizing a mobile app or interface, the field data and the algorithm IPM's outcomes should be monitored. The Internet must be used to access this mobile app or interface in order to be used.

References

[1] A. C. Teixeira, J. Ribeiro, R. Morais, J. J. Sousa, and A. Cunha, "A Systematic Review on Automatic Insect Detection Using Deep Learning," *Agriculture*, vol. 13, no. 3, p. 713, Mar. 2023, doi: 10.3390/agriculture13030713.

[2] K. Haricha, A. Khiat, Y. Issaoui, A. Bahnasse and H. Ouajji, "Recent technological progress to empower Smart Manufacturing: Review and Potential Guidelines," in IEEE Access, doi: 10.1109/ACCESS.2023.3246029.

[3] S. Azfar, A. Nadeem, K. Ahsan, A. Mehmood, H. Almoamari, and S. S. Alqahtany, "IoT-Based Cotton Plant Pest Detection and Smart-Response System," *Applied Sciences*, vol. 13, no. 3, p. 1851, Jan. 2023, doi: 10.3390/app13031851.

[4] L. -B. Chen, X. -R. Huang, W. -H. Chen, W. -Y. Pai, G. -Z. Huang and W. -C. Wang, "Design and Implementation of an Artificial Intelligence

of Things-Based Autonomous Mobile Robot System for Cleaning Garbage," in IEEE Sensors Journal, vol. 23, no. 8, pp. 8909–8922, 15 April15, 2023, doi: 10.1109/JSEN.2023.3254902.

[5] Y. Meslie, W. Enbeyle, B. K. Pandey, S. Pramanik, D. Pandey, P. Dadeech, A. Belay, A. Saini, "Machine Intelligence-based Trend Analysis of COVID-19 for Total Daily Confirmed Cases in Asia and Africa", in Methodologies and Applications of Computational Statistics for Machine Learning, D. Samanta, R. R. Althar, S. Pramanik and S. Dutta, Eds, IGI Global, 2021, pp. 164–185, DOI: 10.4018/978-1-7998-7701-1.ch009

[6] S. Pramanik, "Carpooling Solutions using Machine Learning Tools", in Handbook of Research on Evolving Designs and Innovation in ICT and Intelligent Systems for Real-World Applications, K. K. Sarma, N. Saikia and M. Sharma, IGI Global, 2022, DOI: 10.4018/978-1-7998-9795-8.ch002.

[7] A. Gupta, A. Verma, S. Pramanik, "Advanced Security System in Video Surveillance for COVID-19", in An Interdisciplinary Approach to Modern Network Security, S. Pramanik, A. Sharma, S. Bhatia and D. N. Le, CRC Press, 2022.

[8] R. R. Chandan, S. Soni, A. Raj, V. Veeraiah, D. Dhabliya, S. Pramanik, Ankur Gupta , Genetic Algorithm and Machine Learning, in "Advanced Bioinspiration Methods for Healthcare Standards, Policies, and Reform", Hadj Ahmed Bouarara, IGI Global, 2023, DOI: 10.4018/978-1-6684-5656-9

[9] D. Mondal, A. Ratnaparkhi, A. Deshpande, V. Deshpande, A. P. Kshirsagar and S. Pramanik, Applications, Modern Trends and Challenges of Multiscale Modelling in Smart Cities, in Data-Driven Mathematical Modeling in Smart Cities, IGI Global, 2023, DOI: 10.4018/978-1-6684-6408-3.ch001

[10] J. Yan, M. Yi, X. Yang, X. Luo and X. Guan, "Broad Learning-Based Localization for Underwater Sensor Networks with Stratification Compensation," in IEEE Internet of Things Journal, doi: 10.1109/JIOT.2023.3260192.

[11] N. K. Nagrale, V. N. Nagrale and A. Deshmukh, "Iot Based Smart Food Grain Warehouse," 2023 2nd International Conference on Paradigm Shifts in Communications Embedded Systems, Machine Learning and Signal Processing (PCEMS), Nagpur, India, 2023, pp. 1-5, doi: 10.1109/PCEMS58491.2023.10136049.

[12] Y. Su, Y. Xu, Z. Pang, Y. Kang and R. Fan, "HCAR: A Hybrid-Coding-Aware Routing Protocol for Underwater Acoustic Sensor Networks," in

IEEE Internet of Things Journal, vol. 10, no. 12, pp. 10790-10801, 15 June15, 2023, doi: 10.1109/JIOT.2023.3240827.

[13] W. Tian et al., "A Centralized Control-Based Clustering Scheme for Energy Efficiency in Underwater Acoustic Sensor Networks," in IEEE Transactions on Green Communications and Networking, vol. 7, no. 2, pp. 668–679, June 2023, doi: 10.1109/TGCN.2023.3249208.

[14] P. K. Mall, S. Pramanik, S. Srivastava, M. Faiz, S. Sriramulu and M. N. Kumar, FuzztNet-Based Modelling Smart Traffic System in Smart Cities Using Deep Learning Models, in Data-Driven Mathematical Modeling in Smart Cities, IGI Global, 2023, DOI: 10.4018/978-1-6684-6408-3.ch005

[15] Reepu, S. Kumar, M. G. Chaudhary, K. G. Gupta, S. Pramanik and A. Gupta, Information Security and Privacy in IoT, in Handbook of Research in Advancements in AI and IoT Convergence Technologies, Eds, J. Zhao, V. V. Kumar, R. Natarajan and T. R. Mahesh, IGI Global, 2023.

[16] V. Veeraiah, V. Talukdar, Manikandan K., S. B. Talukdar, V. D. Solavande, S. Pramanik, A. Gupta, Machine Learning Frameworks in Carpooling, in Handbook of Research on AI and Machine Learning Applications in Customer Support and Analytics, Eds, Md S. Hossain, R. C. Ho and G. Trajkovski, IGI Global, 2023.

[17] V. Jain, M. Rastogi, J. V. N. Ramesh, A. Chauhan, P. Agarwal, S. Pramanik, A. Gupta, FinTech and Artificial Intelligence in Relationship Banking and Computer Technology, in AI, IoT, and Blockchain Breakthroughs in E-Governance, Eds, K. Saini, A. Mummoorthy, R. Chandrika and N.S. Gowri Ganesh, IGI Global, 2023.

[18] S. Pramanik, "An Effective Secured Privacy-Protecting Data Aggregation Method in IoT", in Achieving Full Realization and Mitigating the Challenges of the Internet of Things, Eds, M. O. Odhiambo and W. Mwashita, IGI Global, 2022, DOI: 10.4018/978-1-7998-9312-7.ch008

[19] R. Bansal, B. Jenipher, V. Nisha, Jain, Makhan R., Dilip, Kumbhkar, S. Pramanik, S. Roy and A. Gupta, "Big Data Architecture for Network Security", in Cyber Security and Network Security, Eds, Wiley, 2022, https://doi.org/10.1002/9781119812555.ch11

[20] D. Pradhan, P. K.Sahu, N. S. Goje, H. Myo, M. M.Ghonge, M., Tun, R, Rajeswari and S. Pramanik, "Security, Privacy, Risk, and Safety Toward 5G Green Network (5G-GN)", in Cyber Security and Network Security,Eds, Wiley, 2022, DOI: 10.1002/9781119812555.ch9

[21] S. Choudhary, V. Narayan, M. Faiz and S. Pramanik, "Fuzzy Approach-Based Stable Energy-Efficient AODV Routing Protocol in Mobile Ad

hoc Networks", in Software Defined Networking for Ad Hoc Networks, M. M. Ghonge, S. Pramanik and A. D. Potgantwar, Eds, Springer, 2022, https://doi.org/10.1007/978-3-030-91149-2_6

[22] H. O. Choe, M. Hun Lee and H. Yoe, "Automatic Pest Image Acquisition System Hardware Design," 2023 International Conference on Artificial Intelligence in Information and Communication (ICAIIC), Bali, Indonesia, 2023, pp. 069–071, doi: 10.1109/ICAIIC57133.2023.10067130.

[23] A. C. Teixeira, J. Ribeiro, R. Morais, J. J. Sousa, and A. Cunha, "A Systematic Review on Automatic Insect Detection Using Deep Learning," *Agriculture*, vol. 13, no. 3, p. 713, Mar. 2023, doi: 10.3390/agriculture13030713.

[24] J. Jose and M. S. U P, "Development of a Multipurpose Solar Powered Insect Pests Trap: KRISHI VIKAS," 2023 International Conference on Power, Instrumentation, Control and Computing (PICC), Thrissur, India, 2023, pp. 1–6, doi: 10.1109/PICC57976.2023.10142478.

[25] Todorović, D., Perić-Mataruga, V., Mirčić, D. *et al.* Estimation of changes in fitness components and antioxidant defense of *Drosophila subobscura* (Insecta, Diptera) after exposure to 2.4 T strong static magnetic field. *Environ Sci Pollut Res* 22, 5305–5314 (2015). https://doi.org/10.1007/s11356-014-3910-8

[26] Albattah, W., Masood, M., Javed, A. *et al.* Custom CornerNet: a drone-based improved deep learning technique for large-scale multiclass pest localization and classification. *Complex Intell. Syst.* 9, 1299–1316 (2023). https://doi.org/10.1007/s40747-022-00847-x

[27] K. Sudharson, B. Alekhya, G. Abinaya, C. Rohini, S. Arthi and D. Dhinakaran, "Efficient Soil Condition Monitoring with IoT Enabled Intelligent Farming Solution," 2023 IEEE International Students' Conference on Electrical, Electronics and Computer Science (SCEECS), Bhopal, India, 2023, pp. 1–6, doi: 10.1109/SCEECS57921.2023.10063050.

[28] B. K. Pandey, D. Pandey, V. K. Nassa, A. S. Hameed, A. S. George, P. Dadheech and S. Pramanik, A Review of Various Text Extraction Algorithms for Images, in The Impact of Thrust Technologies on Image Processing, Nova Publishers, 2023.

[29] B. K. Pandey, D. Pandey, V. K. Nassa, A. S. George, S. Pramanik and P. Dadheech, Applications for the Text Extraction Method of Complex Degraded Images, in The Impact of Thrust Technologies on Image Processing, Nova Publishers, 2023.

[30] R. Jayasingh, J. Kumar R.J.S, D. B. Telagathoti, K. M. Sagayam and S. Pramanik, "Speckle noise removal by SORAMA segmentation

in Digital Image Processing to facilitate precise robotic surgery", International Journal of Reliable and Quality E-Healthcare, vol. 11, issue 1, DOI: 10.4018/IJRQEH.295083, 2022.

[31] S. Pramanik and S. Suresh Raja, "A Secured Image Steganography using Genetic Algorithm", Advances in Mathematics: Scientific Journal, vol. 9, issue 7, pp. 4533–4541, 2020. DOI: https://doi.org/10.37418/amsj.9.7.22

[32] A. Patnala, S. Stadtler, M. G. Schultz and J. Gall, "Generating Views Using Atmospheric Correction for Contrastive Self-Supervised Learning of Multispectral Images," in IEEE Geoscience and Remote Sensing Letters, vol. 20, pp. 1–5, 2023, Art no. 2502305, doi: 10.1109/LGRS.2023.3274493.

[33] D. Banerjee, V. Kukreja, S. Hariharan and V. Sharma, "Fast and Accurate Multi-Classification of Kiwi Fruit Disease in Leaves using deep learning Approach," 2023 International Conference on Innovative Data Communication Technologies and Application (ICIDCA), Uttarakhand, India, 2023, pp. 131–137, doi: 10.1109/ICIDCA56705.2023.10099755.

[34] A. Khalid, S. Akbar, S. A. Hassan, S. Firdous and S. Gull, "Detection of Tomato Leaf Disease Using Deep Convolutional Neural Networks," 2023 4th International Conference on Advancements in Computational Sciences (ICACS), Lahore, Pakistan, 2023, pp. 1–6, doi: 10.1109/ICACS55311.2023.10089689.

[35] Q. Hong et al., "A Distance Transformation Deep Forest Framework With Hybrid-Feature Fusion for CXR Image Classification," in IEEE Transactions on Neural Networks and Learning Systems, doi: 10.1109/TNNLS.2023.3280646.

[36] J. Yang, Z. Pan, Z. Wang, B. Lei and Y. Hu, "SiamMDM: An Adaptive Fusion Network With Dynamic Template for Real-Time Satellite Video Single Object Tracking," in IEEE Transactions on Geoscience and Remote Sensing, vol. 61, pp. 1–19, 2023, Art no. 5608619, doi: 10.1109/TGRS.2023.3271645.

[37] M. H. Araghian, M. Rahimiyan and M. Zamen, "Robust Integrated Energy Management of a Smart Home Considering Discomfort Degree-Day," in IEEE Transactions on Industrial Informatics, doi: 10.1109/TII.2023.3234083.

[38] Y. Sun, Y. Hu, H. Zhang, H. Chen and F. -Y. Wang, "A Parallel Emission Regulatory Framework for Intelligent Transportation Systems and Smart Cities," in IEEE Transactions on Intelligent Vehicles, vol. 8, no. 2, pp. 1017–1020, Feb. 2023, doi: 10.1109/TIV.2023.3246045.

5

Practices of Deep Learning in Farming: What Deep Learning Can Do in Intelligent Agriculture

J. Bhuvana[1], Rajendra P. Pandey[2], Umesh Kumar Tripathi[3], Nuzhat Fatima Rizvi[4], Sachin Tripathi[4], and Ahmed J. Obaid[5]

[1]Department of Computer Science and IT, School of CS and IT, Jain (Deemed to be) University, India
[2]College of Computing Science and Information Technology, Teerthanker Mahaveer University, India
[3]Department of Agriculture, Sanskriti University, India
[4]Symbiosis Law School, Nagpur Campus, Symbiosis International (Deemed University), India
[5]University of Kufa, Iraq

Email: j.bhuvana@jainuniversity.ac.in; panday_004@yahoo.co.uk; umesh.ag@sanskriti.edu.in; nuzhatrizvi@slsnagpur.edu.in; sachintripathi@slsnagpur.edu.in; ahmedj.aljanaby@uokufa.edu.iq

Abstract

Deep learning, a subset of machine learning, is a cutting-edge method for processing pictures and scrutinizing large amounts of data. DL techniques are effectively utilized in many industries for improving performance. Given that present agricultural practices cannot keep up with the population's fast increase, the agriculture industry is one that might profit from DL methods. The most present day developments in the agriculture sector's use of deep learning methods will be covered in this chapter, along with an overview of related research projects. Additionally, the implemented models will be examined and contrasted with other models already in use.

5.1 Introduction

In addition to giving people a means of subsistence, agriculture makes a major contribution to national revenue and, in turn, to national development. When processed and exported, agricultural goods are a particularly important source of foreign currency. The money produced in this way greatly aids in a nation's growth, guarantees the stability of the currency, and provides the government with a practical instrument for imports. Since the dawn of time, the primary source of human nutrition has been agricultural goods. It is impossible to overstate the importance of agriculture in world civilization since few people can go many days without eating. However, the globe will soon confront a grave calamity brought on by a food crisis and unheard-of levels of hunger due to the fast and ongoing growth in human population. The worldwide population is anticipated to surpass 10 billion individuals by the year 2035, according to some analysts. To fulfill the population's increasing needs, food production must be doubled in this scenario. Until then, obstacles like urbanization and climate change will make it difficult to increase food production. Planning becomes more challenging due to weather and season volatility as a result of global warming drying up once productive land. Abysmally less land is now accessible for agriculture as a result of urbanization, which has taken over agricultural fields and turned them into cities. This makes commercial farming challenging and significantly lowers overall agricultural production.

Furthermore, the current agricultural systems must be drastically altered due to the combined impacts of climate change, energy shortages, and water limitations. In addition to polluting lands and destroying crops and other plants, industrial remains, unburned carbon, and oil spills have aggregated to taint the seas and eliminate the availability of underwater farming goods worldwide. Therefore, it is important to address each of these issues while simultaneously producing enough goods to fulfill the population's ever increasing food demands. In order to increase the manufacturing rate, machine learning (ML) may be really helpful in this situation. Agriculture machine learning will usher in and support existing attempts to develop smart agriculture. In order to increase the amount and standard of farming commodities, yield effectiveness, and resource optimization, intelligent agriculture uses the most recent technologies, including the GPS [1], soil inspection, Internet of Things, data processing, and management.

To address issues with harvest yield like crop diseases, sustainability, food quality, and environmental influence, smart agriculture is essential. Modern research has shown that deep learning algorithms' new ideas are quite accurate. Deep learning [2] algorithms analyze photographs and images from interesting phenomena to draw conclusions for usage in the future. It delves deeply into such phenomena, examining their traits all the way down to their genetic composition. Smart agriculture is enabled by these deep learning algorithms. These deep learning algorithms have several uses in agriculture, including the categorization of leaves, the detection of weeds, yield estimation, and the prediction of soil moisture and plant diseases. This chapter will address these applications by contrasting and evaluating the deep learning processes with the already-in-use methodologies.

5.2 Background

DL is a subset of ML, which, by progressively extracting higher-level characteristics from several datasets, may provide the desired output utilizing multiple layers. A sophisticated depiction of the input data is created by deep learning at every level. With the help of experience and data, computer systems can perform better thanks to a deep learning algorithm. The ability of machines to solve complicated problems increases every day. They study these tasks until they become experts at them and can do them repeatedly without becoming stressed out. These are activities that, for the most part, involve people. By striving to think exactly like a person doing that action, deep learning robots, using carefully crafted algorithms, understand how people accomplish these tasks. They eventually develop artificial intelligence (AI), or knowledge of how humans accomplish this task, and they keep learning until they have all the information necessary to carry out the task in the same manner as a person. Ivakhnenko et al. reported the first deep learning algorithm that was ever a success in 1967. Since then, several algorithms have been created, and more are continually being created, to provide trustworthy answers for many problems that people are now facing. In order to make life simpler for people, save very important time and rare resources, and save human energy that is constantly draining, it is necessary to have robots do tasks that would otherwise be performed by people.

In machine learning, there are three distinct kinds of learning models: supervised learning, semi-supervised learning, and unsupervised learning. Unsupervised learning systems are left to master self-sufficiently beyond being fed any training data, supervised learning machines are fed training data that are correctly labeled, and semi-supervised learning machines are fed some but not all of the necessary training data. In order to provide solutions, these learning approaches make use of certain neural network topologies. Few most popular NN topologies in DL are the ones listed below:

- DNN
- RNN
- CNN

5.2.1 DNN

DNN are a subtype of ANN [3] that uses mathematical functions to link different layers between the I/P layer and the O/P layer. The network moves from one layer to the other by figuring out the likelihood of every possible result. DNNs may be taught to analyze something and draw inferences regarding anything else by calculating the likelihood that something will be the same as the item it was trained to evaluate. It executes this by thoroughly examining the item of interest, comprehending every characteristic, and drawing conclusion that it may mostly utilize to distinguish the specific entity from a field of many. A DNN that has been trained to recognize cassava stems is one example. When presented with a picture, DNN analyzes it and compares it to that it has determined to be the characteristics of a cassava stem. Moreover, it indicates the likelihood that the picture is of a cassava stem. The network is certain that the picture is of a cassava stem if the probability is 1.

If the closest thing to a cassava stem is zero, the picture is most definitely not of a cassava stem. These networks strip whatever it is they are researching down to the atomic level. Artificial neurons and their connections are given fractional numerical values at the atomic level. When the network is used, these values or weights produce probability.

5.2.2 Recurrent neural network

Another area of ANNs where data may flow in either direction is recurrent neural networks (RNN). RNNs are composed of a network of nodes that resemble neurons that are all linked to one another in a single direction and arranged into consecutive layers. Each node contains values and has the ability to accept data from an external network, produce results, or alter data as it moves from the input to the output. There are two kinds of recurrent networks. These recurrent networks have both infinite and finite impulses. A directed graph in a finite impulse RNN cannot be unrolled because there are no directed cycles. On the other hand, an infinite impulse recurrent network possesses a directed graph and unrollable directed cycles. The storage states of the two varieties of RNNs are managed by the neural network. The LSTMs [4] of RNNs is made up of these storage states. RNNs may handle variable-length data input because they can process the data in their internal memory. As a result, these RNNs are mostly used in voice recognition and language modeling.

5.2.3 Convolutional neural network

CNNs, a subset of DNN, are multilayer perceptrons where each layer's neurons are connected to each other on a layer-by-layer basis. For image processing, these structures are often used. CNNs [5–10] can categorize pictures without the need of any additional human effort or prior knowledge. It relatively slowly becomes familiar with the characteristics of photos before processing them to get the intended outcome. By using filters, convolutional neural networks are able to capture all of the spatial and temporal relationships in a picture. There are input, output, and multiple hidden layers in a CNN. There are weights associated with every input neuron. All of the weighted input is subjected to a mathematical procedure to produce the result. Weights may be reused and parameters are minimized as much as feasible. This makes it simpler to teach the network how to comprehend even complex pictures.

If an image I/P is received by a convolutional neural network (CNN), the system creates 3D maps of the picture, breaking it up into layers of the object. The picture is then split into additional stacks of separate layers, and all of the volumes of neurons that make up the image are connected locally. The I/P quantity eventually changes into the O/P quantity. The CNN architectures RESNET [11], DENSENET [12], Inception, and SQUEEZENET [13] are only a few examples. Any of these may assess how well datasets perform and how well management systems work.

5.3 Deep-Learning-based Crop Monitoring

Plant diseases pose a serious danger to global food security and have catastrophic consequences for small-sized agriculturists, where the sole source of earnings is the products they grow. It is estimated that small-scale farmers in this developing globe provide close to 80% of the overall farming output. Nevertheless, it might be far greater because what is generated is seldom sufficient. It has been determined that every year, more than 50% of the overall crop is lost. The same research determined that pests and illnesses were the usual reasons for the failure of more than 50% of the crop. This kind of number ought to alarm everyone. Such enormous losses have a negative impact on the globe's food supply and farmers who lose billions of dollars yearly. It becomes essential to find solutions. Until far, agricultural researchers have been successful in developing treatments for these illnesses. But since they were created to treat certain ailments, these medications are disease-specific. Farmers may not always be aware of the specific illness that is harming their crops or the appropriate medication to treat a condition. It becomes essential to find alternatives. Precision farming has become feasible because of deep learning.

The term "satellite farming" also applies to precision agriculture. By creating a DSS [14–17] for optimizing agriculture production when maximizing the utilization of agricultural resources, it is an idea in farming management that monitors, computes, and reacts to field variability in harvest. Precision agriculture's [18] ideals are realized through real-time sensor arrays on GPS devices that help pinpoint a specific object's location on a farm and enable the generation of maps that display a variety of phenomena including organic matter content and moisture levels. These sensor arrays not only provide these maps but also assess things like plant hydration status and chlorophyll levels, which, when paired with satellite pictures, are incredibly helpful to planters, sprayers, and farmers. These devices, commonly referred to as variable rate technology, convey data and allocate resources in the most efficient way possible without the need for human intervention.

With the use of precision agriculture, crop diseases may be quickly diagnosed and treated, fruits may be tallied, and crop types may be simply categorized, consisting of which one among others will produce the most. Deep learning crop monitoring often uses mobile cameras to collect datasets. The object is a portable camera with the ability to send and collect data. This portable camera may be a camera for the Internet of Things [19]. Other reliable technologies for data collection include mobile cameras, cell phones, autonomous pinking robots, and autonomous spray robots.

The application of deep learning algorithms to resolve the challenges stated will be discussed in the following areas.

5.3.1 Identification of plant ailments and pests

The majority of plant diseases and pests that jeopardize food security impact the crops. Farmers sometimes do not become aware of these pests and illnesses until it is too late, at which point they are not under control. Instead of recognizing it for what it is, farmers can think that a plant's unusual appearance is just a result of the plant going through some of its typical development stages, such as being more green than usual or having some spots. It may have taken over the whole farm before they know it is a sickness and significantly decreased their farm's output for that reason. It is crucial to assist farmers in recognizing illnesses early enough to know how to combat them, particularly those without specialized agricultural knowledge.

Prior to now, local plant clinics and organizations dedicated to agriculture helped the diagnosis of the illness in plants. The primary approach used in practice for the diagnosis of illness in plants is the trained professional's naked eye inspection since the majority of diseases show some kind of visual symptom. But sometimes, this might fail as a result of weak observational abilities. In addition to having limits, human skills may also be overstretched by conditions like eye difficulties or stress. In these situations, the viewer should not be accomplished to draw appropriate conclusions from what he is observing. Therefore, a variety of image-dependent DL approaches may be utilized to avoid these issues and streamline the illness diagnosis process. Certain traits of all potential illnesses must be recognized and identified in order for this to be achievable. The following is a list of some of the traits that may be used to analyze certain illnesses or pests.

- Infection stage: Throughout its life cycle, a plant exhibits a variety of patterns with varying levels of infection. This may help in the technique of diagnosing the illness. The specific pattern that an herb exhibits at a certain phase of infection must be recognized and appropriately labeled. It must be carried out at every step so that it is clear that when a certain pattern occurs, the infection of a plant is at that particular stage.
- Shape and color: Depending on the illness, a plant may display a variety of colors and forms of infection. A table of illnesses and the corresponding shapes and colors may be created in the algorithm so that the disease can be recognized from the detected shapes and colors.
- Position of the infected neighborhood: The infected area is not only restricted to the leaf area; it may also exhibit a variety of symptoms on branches or fruits. Various diseases affect herbs at various stages. Getting the position of the infection on a herb makes it simpler to diagnose the condition, bringing the search for the ideal treatment one step closer.
- Kind of fungus: Knowing the kind of fungus an illness is caused by is a simple way to grasp the variations between diseases. Knowing the features of every recognized fungus, such as the color and form they take on afflicted plants and the portion of the herb they impact, would be important to determine the kind of fungus attacking the plant.

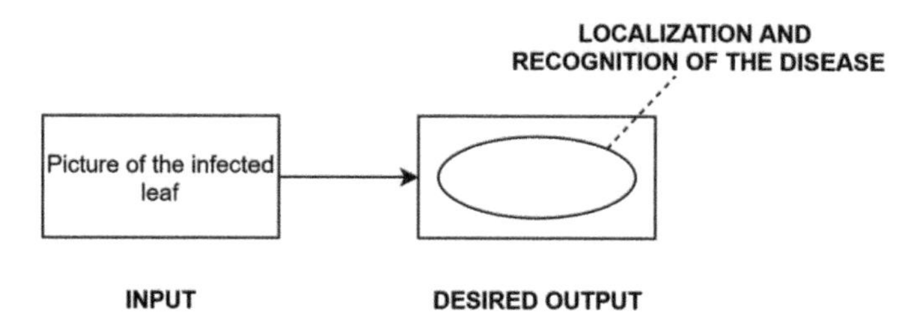

Figure 5.1 The algorithm used to identify diseases in its overall process.

- Changes in leaf pattern: Relying on the illness, changes in leaf pattern might also be seen. It will be simple to determine if a plant has been afflicted by the illness in question if the disease in question is documented to quit a definite pattern on the leaves of damaged herbs.

While some of these traits may be recognized visually, doing so is very taxing on anybody and is likely to lead to mistakes. The aforementioned traits as well as other significant ones will be included into the network thanks to deep learning. The network would then get a large number of photos to analyze. Different plants, some healthy and others harmful, would be shown in the photographs. The network would be able to examine and process everything when the characteristics were mapped to the proper illnesses.

The formulation of the issue, or how to recognize and detect the illness on the leaf, is shown in Figure 5.1. According to [20], NNs offer the mapping from the I/P (a picture of the sick herb) to the O/P (spotting the illness). Each node in the NN is a mathematical function, in which the inward linkages are utilized to gather I/Ps and give results as an outward link in the neural network.

Convolutional neural networks (CNNs), which have potent approaches to represent composite processes and detect patterns from pictures, are the deep learning technology that is most often employed in this application. With the use of numerous datasets that include pictures of both healthy and diseased plants, these CNN architectures are trained and evaluated. Therefore, when these deep learning architectures are given a leaf picture, they evaluate it using the aforementioned criteria to identify the diseased area and show the kind of disease or pest that has affected the herb. An appropriate treatment for the plant may also be offered to the user based on this information.

5.3.2 Recognition of plant phenology

The review of a herb's life cycle and how it would be impacted by climatic variations and environmental factors is known as phenology of agricultural plants. It is an essential knowledge for precision farming. Studies on plant phenology take into account the day an herb was sowed, the day its leaves initially emerged, the day it began to have flowers and draw insects, the date its leaves began changing color and falling, if it is a deciduous tree, and the time the fruits start to form, became ripe, and were finally picked. When examining how changes in weather and temperature, and habitat elements like elevation and soil topology, can have impacted the herb, all of these are taken into consideration.

Plant yields may be severely impacted by even little climatic differences, like alterations in temperature and humidity. For making future agricultural choices, past phenological data from a specific farming area might be helpful. These records may also provide higher temporal resolution for the survey of weather change and global warming by serving as a calculation appropriate for utilization as an indication of the estimates of temperature in historical climatology.

Applying deep learning algorithms would improve outcomes such as precise harvest timing, insect management, production prediction, farm monitoring, etc., by precisely identifying plant phenological state changes. Various nations have been developing agricultural monitoring network systems throughout the years. However, the majority of them just create a phenology monitoring metric based on the colors shown in the photographs. However, they are inaccurate since they primarily rely on color analysis, which is susceptible to change for a number of causes, including jitter, and unexpected alterations in camera settings. Few herbs lack greenish hues altogether. Within a few months of sowing, plants like maize and rice reach maturity and alter color even in the rainy season. How can one determine the phenology of the herbs using just the colors in their image? Since utilizing plant colors to determine phenology has its limitations, deep learning algorithms may greatly aid in this process.

To quickly determine a picture's properties and provide results, a pre-trained CNN like AlexNet [21], VGG_Net [22], GoogleNet [23], etc., may be utilized. Researchers apply a pre-trained CNN architecture that is dependent on the range of the dataset and the similarity of the data. A dataset on Kaggle [24] may be utilized to test the technique.

The lowest layer of the pre-trained model should be adjusted when the dataset to be utilized has lesser size and resemblance. The first k layers of the predetermined model must be frozen, and the remaining layers must be trained, in order to calibrate the bottom layer. As a result, the top layers are adjusted to the new set of data. Images must be correctly collected using appropriate technologies in order to prevent the aforementioned issues. These photos are trustworthy and may be used to identify plant phenological phases precisely. To better comprehend the phenological phases of these herbs and to better understand how the plants develop, these photographs must be taken at certain intervals of time at the same location. Additionally, a sizable collection of photographs must be taken for increasing the accuracy of the findings; these images should not be impacted by any disruptions created during the capturing of the images. Throughout a plant's lifespan, this capture should be done as often as feasible. With the use of this technique, farmers may get a deeper knowledge of the plant phenological states, which are essential to precision farming.

5.3.3 Irrigation of crops

In terms of agriculture, irrigation systems are one of the most crucial components. Watering the land to maintain plant growth is the process of irrigation. It is crucial for farmers in regions with irregular or seasonal rainfall, little to no annual precipitation, or perhaps none at all, as is sometimes the case in desert regions. Farmers can grow crops where they otherwise could not by watering the soil. They may also boost productivity and provide the plants with a healthy environment so that they can develop into harvestable crops. However, it takes a lot of work and time to water the plants by hand. Irrigation often involves several steps, creating it to be more costly in commercial farms and much complicated for subsistence farmers. Additionally, irrigation may cause soils to become waterlogged and raise the salt content to a degree where plants may be harmed. Here, a good irrigation system built on DL principles would be highly beneficial for herb development in addition to lowering labor requirements.

Farmers now use a variety of irrigation techniques, whether for commercial or subsistence purposes. The most common irrigation systems are sprinklers and drip systems. These systems are created such that when water is needed to be supplied to the plants, the valves will often open. However, these methods do have certain drawbacks, which hurt the farmer and his property in various ways. Implementing these systems has a number of drawbacks, one of which is improper irrigation scheduling. Inadequate planning causes the plants to either be overwatered or underwatered. Overwatering prevents plants from absorbing enough oxygen, which may lead to plant death. Conversely, underwatering prevents plants from absorbing vital nutrients from the soil. In addition, the program cannot be regularly changed because of weather variations. The owner cannot pre-program the irrigation design to fit the weather at any time since he cannot forecast the seasonal

Table 5.1 The estimated amount of water needed for particular seasonal crops.

Sl. number	Crop	Crop water need (mm/total growing span of time)
1.	Tomato	550–760
2.	Pepper	550–850
3.	Onion	400–600
4.	Rice	600–650
5.	Melon	550–625
6.	Banana	1350–2520
7.	Wheat	580–780
8.	Cotton	625–1250
9.	Peanut	650–750
10.	Potato	650–680
11.	Sugarcane	1680–2650

variations that take place every year. The soil moisture constituent differs from herb to herb, as indicated in Table 5.1. No matter the crop type, giving plants the same quantity of water has negative effects on their ability to grow.

These problems may be solved by using an IoT-based system built on deep learning principles. As they may recognize the nonlinear ties present in the herb data, these deep learning principles are used. It is necessary to snap many pictures of the plant and submit them to a DL software platform. Utilizing pictures, deep learning algorithms are utilized to categorize plants [25]. The program determines the likelihood that a picture corresponds to a given group of plants after the photographs have been submitted. The suitable moisture content of the plant will be acquired from the database when it has been identified. The current amount of moisture and humidity is then computed and sent to the microprocessor with the aid of the temperature sensor, humidity sensor, and soil moisture sensors. The program then makes use of the information gathered by the CPU and database and subsequently modifies the irrigation system's schedule in accordance with the specifications, as illustrated in Figure 5.2.

5.4 Yield Prediction and Harvesting using Deep Learning

For the global production of food, crop yield forecast is crucial. The yield of novel hybrid seeds that demonstrate how the seeds behave in various situations must be predicted by seed producers. To decide whether to import or export the items, dealers and policymakers need accurate projections.

According to [26], forecasting yield helps farmers make financial choices and decide which crop to sow and where. Lack of such information

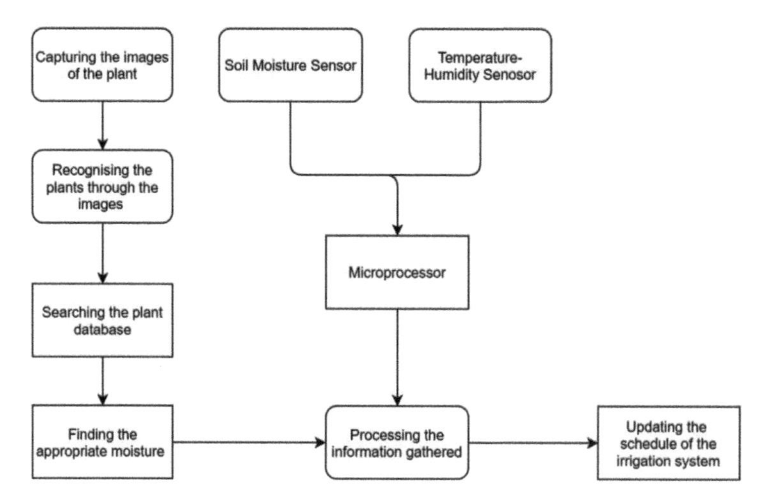

Figure 5.2 The fundamentals of an IoT-dependent smart irrigation framework.

might cost the farmer money and cause all agricultural investors to lose money. Just before the land is cleared, it is critical to understand if the investment may be recovered from the anticipated return. Because of the numerous complex factors, estimating crop output is a difficult procedure. Estimating yield production cannot be done simply by knowing the precise amount of fruits, flowers, etc. The current practice of having staff hand count fruits is more time-consuming and costly. A workable answer to this problem is provided by automatic yield estimate in robotic agriculture.

5.5 Selection of Crop Variety

The process of selecting which crops to plant is known as crop selection. In order to anticipate production, crop choice is crucial. There are several aspects to take into account when choosing crops, including the soil, current agricultural conditions, the season, equipment, advertising and profitability, security, expenditure of planting supplies, and the least support cost in the particular year. It may be challenging to predict that such variables should have the most positive or negative impact on a given crop's production. When there are hundreds of different crop kinds available, it gets increasingly harder. Crop variety selection method (CVSM), an algorithm, would take into account all potential elements including seasonal and economical considerations in order to solve this issue. This approach would be utilized to choose the crop type that will contribute to the highest yield.

The basic goal of CVSM is to forecast the yield rate in the particular crop. Three sections of this algorithm are crucial to attaining the desired result. The divisions include choosing the crop, taking the market price into consideration, and choosing the type of crops. ANNs are used by the CVSM approach to determine the optimal variety. ANNs are utilized as they excel at pattern recognition and must be the finest at forecasting yield rate. There should be input and output layers in the networks. Environmental factors including temperature, rainfall, humidity, and soil elements like nitrogen, potassium, and phosphorous are all included in the neurons of the input layer. Following the processing of the input layers, one output layer with a single neuron that carries the anticipated yield rate will be produced. And for all other crops as well, this procedure will be repeated. The rate yield of every category is anticipated to be called at the conclusion, and it will be obvious which variety is superior from this point on.

The crop picker will choose the appropriate crop types based on the specified soil type. Knowing which crops will not thrive in some soil types is crucial, as is picking just research crops that are appropriate for a given soil category. Although without accomplishing so, if the soil type is taken into account as one of the criteria, the low rate at the conclusion of processing would be sufficient to deter farmers from growing crops in inappropriate soil types. To determine the profit for each crop that is chosen, the yield rate is multiplied by the MSP for the relevant year. Following that, the crop picker selects the crop that yields the most profit.

Naturally, a farmer would want to determine how many fruits his plant has produced. Fruits are counted to help with planning and to determine if security measures have been violated in any way. Farmers may sometimes measure their harvest by snapping pictures of their plants and then calculating the quantity of fruits they can see in the picture. For numerous purposes, along with variation in appearance owing to lighting, clarity, and occlusion due to nearby leaves and fruits, detecting the quantity of fruits depending on the images is a challenging and difficult process. The current fruit counting algorithms are depending on computer vision techniques that call for particular characteristics that make use of the form, color, and texture of a range of fruits. However, these techniques have drawbacks; they perform best in certain circumstances. These are often fruit-specific and incapable of being applied to other fruits or even fruits of a different tree that are similar. Again, the environment must be carefully managed to get a good enough outcome for even that one fruit. Without a doubt, this is time-consuming and takes away from time that might have been better spent doing anything else.

Deep learning outperforms computer approaches that are more complicated when dealing with these unstructured situations. One of the most efficient ways to count fruits is to run DL approaches in a pipelining architecture. Firstly, a team of human labelers gathers truth labels to create the

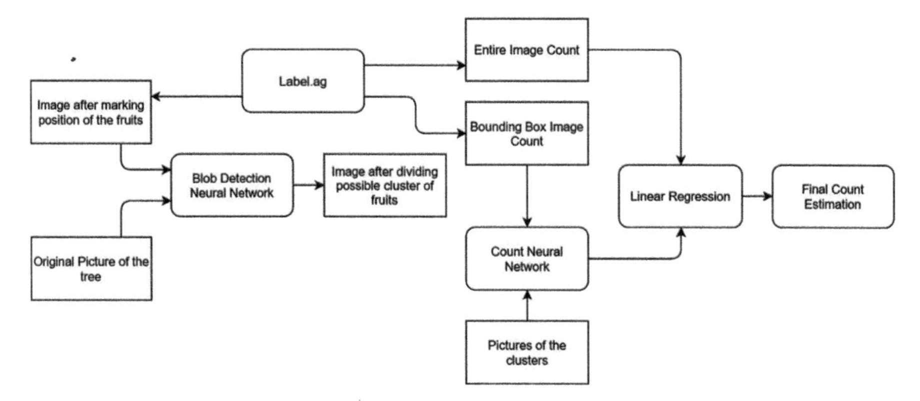

Figure 5.3 The method for counting fruits step-by-step using deep learning.

ground-truth labels using a crowd-sourcing labeling platform. These are kept in the web-based labeling framework Label.ag. By circling the photos in this method, a user creates SVG 11 (scalable vector graphics) data from the fruits.

They are then kept for further analysis. The main method for utilizing deep learning to count fruits from photos is shown in Figure 5.3.

A fully convolutional network is then used to accomplish blob detection in order to separate potential fruit clusters from the background. The benefit of having this kind of network is that it can accommodate images of any size. It is trainable on one side of the picture and then used on the other. The following phase uses a neural network to estimate the number of fruits. Each identified blob may sometimes include many fruits, which might lead to an inaccurate calculation of the number of fruits. Convolutional neural networks, which take into account the bounding boxes around every blob as I/P and output the quantity of fruits that the box holds, may thus be used to solve this issue. And each training picture is subjected to these procedures in order to get the estimated count. Additionally, the neural network is counted. The count estimate is then linearly regressed against the ground truth count. This kind of DL technique may then be used to estimate the fruit count.

5.6 Harvesting of Fruits

Because of a shortage of farming workers and the higher cost of fruit harvesting, the agriculture industry has faced several issues in recent years. It hurts that after investing so much money in planting crops, fruit picking itself turns out to be a major hassle. Farmers start to add up their costs from the moment they cleared their field to the amount they would now pay the very few workers who are still available, whose salaries are rocketing through the ceiling on a regular basis. Every farmer will breathe a sigh of relief upon learning that there is a less expensive option available. One of the better solutions to this issue is to create an autonomous fruit-harvesting robot. Three tasks that are carried out in the phases indicated below are involved.

Step 1: Find the fruit's location on the tree.

Step 2: Place the robotic arm where the fruit is.

Step 3: Lastly, gather the fruit without harming either the tree or the fruit.

A stereo camera is attached to the harvesting robot, which is set up having a robotic hand. The pre-written algorithm has three phases. Finding the fruit's location in 2D [27] is the initial stage. The gadget was used by the researchers to collect apple fruits for the study. The stereo camera provides an image that may be used to determine the location of the apple fruit when it detects it. This technique locates the position using an SSD. A method known as SSD was created using the CNN idea. It recognizes many items from a single picture by using a single DNN and high-speed and accurate object detection techniques.

Choose a fruit that is close to the robotic arm from the ones that SSD has recognized. Then, using a stereo camera, three-dimensional reformation is required to get a 3D position after obtaining a point cloud from the stereo camera and the pixel-selected 2D location. By using the parallax between the images on the right and left, this is accomplished. Then calculate the separation between the stereo camera [28] and the apple. Finding the distance will enable the robotic hand's motion to be more precisely coordinated as it attempts to gather the fruit.

The bot finishes its work by doing harvesting. The robot will initially travel a few millimeters in the direction of the desired fruit before situating the robotic arm underneath it. This will facilitate the robot arm's access to the fruit. The robots are cautiously constructed to avoid bruising the fruits while selecting them. Once the robot has gotten near enough to the desired fruit, the robotic hand – which is equipped with suction grippers to prevent fruit spoilage – grabs the fruit and twists it off the stalk supporting it by spinning four times. According to the experimental findings, it took the apple tree 2 seconds to determine the apple's location and 14 seconds to collect the fruit. Rethinking these ideas will boost the pace. The nearby species of apples may be harvested using the suggested method.

5.6.1 Future works

There are various opportunities in deep learning approaches in smart agriculture. The methods may be used to address a number of issues in agriculture, including irrigation, measuring water erosion, and monitoring greenhouses.

Unmanned aerial vehicle images may be utilized to build a database that will improve the effectiveness of the present image processing technologies. These DL methods may be added to mobile programs, giving farmers easier access to them. With the use of satellite pictures and smartphone apps driven by deep learning, farmers may be alerted to unforeseen weather conditions and insect assaults that might harm their crop.

5.7 Conclusion

This chapter reviews deep learning's uses in agriculture. Here, we demonstrated how deep learning techniques may be used to simplify some agricultural tasks for farmers. Deep learning techniques have been widely applied in a variety of agricultural fields. The processes showed consists of the estimation of plant illnesses, the recognition of weeds and pests, the calculating fruits, the forecasting of yields, the harvesting of fruits, and crop quality. According to the analysis, while DL research is well received in various industries, like healthcare, responses to findings in the field of agriculture are very dissimilar, despite being far and few in number. This is mostly due to the fact that most discoveries have only found applicability in theoretical frameworks. But over time, key actors in the business would get engaged, and these discoveries would completely transform the agricultural industry. Other image processing methods, in addition to deep learning, might improve agriculture, but DL performs superior to the various common methods. These methods enable numerous agricultural tasks to be completed with less labor and thus at a lower cost. DL also assures to boost the overall production as well as the effectiveness of agricultural processes. For meeting the needs of the expanding human population, DL applications will enable a greater rate of food production. It would maximize resources, improve the efficiency of the whole process, and reduce the amount of food that is now wasted to pests and illnesses. There is little question that this will boost yields, revenue, and the food supply.

References

[1] X. Shao, S. Li, J. Zhang, F. Zhang, W. Zhang and Q. Zhang, "GPS-free Collaborative Elliptical Circumnavigation Control for Multiple Nonholonomic Vehicles," in IEEE Transactions on Intelligent Vehicles, doi: 10.1109/TIV.2023.3240855.

[2] N. Shlezinger, J. Whang, Y. C. Eldar and A. G. Dimakis, "Model-Based Deep Learning," in Proceedings of the IEEE, vol. 111, no. 5, pp. 465–499, May 2023, doi: 10.1109/JPROC.2023.3247480.

[3] S. Punitha, T. Stephan, R. Kannan, M. Mahmud, M. S. Kaiser and S. B. Belhaouari, "Detecting COVID-19 From Lung Computed Tomography Images: A Swarm Optimized Artificial Neural Network Approach," in IEEE Access, vol. 11, pp. 12378–12393, 2023, doi: 10.1109/ACCESS.2023.3236812.

[4] Y. Wei, J. Jang-Jaccard, W. Xu, F. Sabrina, S. Camtepe and M. Boulic, "LSTM-Autoencoder-Based Anomaly Detection for Indoor Air Quality Time-Series Data," in IEEE Sensors Journal, vol. 23, no. 4, pp. 3787–3800, 15 Feb.15, 2023, doi: 10.1109/JSEN.2022.3230361.

[5] R. Jayasingh, J. Kumar R.J.S, D. B. Telagathoti, K. M. Sagayam and S. Pramanik, "Speckle noise removal by SORAMA segmentation in Digital Image Processing to facilitate precise robotic surgery", International Journal of Reliable and Quality E-Healthcare, vol. 11, issue 1, DOI: 10.4018/IJRQEH.295083, 2022.

[6] S. Pramanik and S. Suresh Raja, "A Secured Image Steganography using Genetic Algorithm", Advances in Mathematics: Scientific Journal, vol. 9, issue 7, pp. 4533–4541, 2020. DOI: https://doi.org/10.37418/amsj.9.7.22

[7] Reepu, S. Kumar, M. G. Chaudhary, K. G. Gupta, S. Pramanik and A. Gupta, Information Security and Privacy in IoT, in Handbook of Research in Advancements in AI and IoT Convergence Technologies, Eds, J. Zhao, V. V. Kumar, R. Natarajan and T. R. Mahesh, IGI Global, 2023.

[8] V. Veeraiah, V. Talukdar, Manikandan K., S. B. Talukdar, V. D. Solavande, S. Pramanik, A. Gupta, Machine Learning Frameworks in Carpooling, in Handbook of Research on AI and Machine Learning Applications in Customer Support and Analytics, Eds, Md S. Hossain, R. C. Ho and G. Trajkovski, IGI Global, 2023.

[9] V. Jain, M. Rastogi, J. V. N. Ramesh, A. Chauhan, P. Agarwal, S. Pramanik, A. Gupta, FinTech and Artificial Intelligence in Relationship Banking and Computer Technology, in AI, IoT, and Blockchain Breakthroughs in E-Governance, Eds, K. Saini, A. Mummoorthy, R. Chandrika and N.S. Gowri Ganesh, IGI Global, 2023.

[10] S. Pramanik, "An Effective Secured Privacy-Protecting Data Aggregation Method in IoT", in Achieving Full Realization and Mitigating the Challenges of the Internet of Things, Eds, M. O. Odhiambo and W. Mwashita, IGI Global, 2022, DOI: 10.4018/978-1-7998-9312-7.ch008

[11] Q. Wang, J. Du, H. -X. Wu, J. Pan, F. Ma and C. -H. Lee, "A Four-Stage Data Augmentation Approach to ResNet-Conformer Based Acoustic Modeling for Sound Event Localization and Detection," in IEEE/ACM

Transactions on Audio, Speech, and Language Processing, vol. 31, pp. 1251–1264, 2023, doi: 10.1109/TASLP.2023.3256088.

[12] V. T. Q. Huy and C. -M. Lin, “An Improved Densenet Deep Neural Network Model for Tuberculosis Detection Using Chest X-Ray Images,” in IEEE Access, vol. 11, pp. 42839–42849, 2023, doi: 10.1109/ACCESS.2023.3270774.

[13] Z. Wei et al., “Density-Based Affinity Propagation Tensor Clustering for Intelligent Fault Diagnosis of Train Bogie Bearing,” in IEEE Transactions on Intelligent Transportation Systems, vol. 24, no. 6, pp. 6053–6064, June 2023, doi: 10.1109/TITS.2023.3253087.

[14] R. Bansal, B. Jenipher, V. Nisha, Jain, Makhan R., Dilip, Kumbhkar, S. Pramanik, S. Roy and A. Gupta, “Big Data Architecture for Network Security”, in Cyber Security and Network Security, Eds, Wiley, 2022, https://doi.org/10.1002/9781119812555.ch11

[15] D. Pradhan, P. K.Sahu, N. S. Goje, H. Myo, M. M.Ghonge, M., Tun, R, Rajeswari and S. Pramanik, “Security, Privacy, Risk, and Safety Toward 5G Green Network (5G-GN)”, in Cyber Security and Network Security, Eds, Wiley, 2022, DOI: 10.1002/9781119812555.ch9

[16] S. Choudhary, V. Narayan, M. Faiz and S. Pramanik, “Fuzzy Approach-Based Stable Energy-Efficient AODV Routing Protocol in Mobile Ad hoc Networks”, in Software Defined Networking for Ad Hoc Networks, M. M. Ghonge, S. Pramanik and A. D. Potgantwar, Eds, Springer, 2022, https://doi.org/10.1007/978-3-030-91149-2_6

[17] Y. Meslie, W. Enbeyle, B. K. Pandey, S. Pramanik, D. Pandey, P. Dadeech, A. Belay, A. Saini, “Machine Intelligence-based Trend Analysis of COVID-19 for Total Daily Confirmed Cases in Asia and Africa”, in Methodologies and Applications of Computational Statistics for Machine Learning, D. Samanta, R. R. Althar, S. Pramanik and S. Dutta, Eds, IGI Global, 2021, pp. 164–185, DOI: 10.4018/978-1-7998-7701-1.ch009

[18] A. Gupta, A. Verma, S. Pramanik, “Advanced Security System in Video Surveillance for COVID-19”, in An Interdisciplinary Approach to Modern Network Security, S. Pramanik, A. Sharma, S. Bhatia and D. N. Le, CRC Press, 2022.

[19] S. Pramanik, An Adaptive Image Steganography Approach depending on Integer Wavelet Transform and Genetic Algorithm, Multimedia Tools and Applications, 2023, https://doi.org/10.1007/s11042-023-14505-y.

[20] Z. Liu et al., B-Spline Wavelet Neural Network-Based Adaptive Control for Linear Motor-Driven Systems Via a Novel Gradient Descent

Algorithm, in IEEE Transactions on Industrial Electronics, doi: 10.1109/TIE.2023.3260318.

[21] T. H. Ngọc, P. T. Khanh and S. Pramanik, Smart Agriculture using a Soil Monitoring System, in Advanced Technologies and AI-Equipped IoT Applications in High Tech Agriculture, Eds, A. Khang, IGI Global, 2023.

[22] P. T. Khanh, T. H. Ngọc, and S. Pramanik, Future of Smart Agriculture Techniques and Applications, in Advanced Technologies and AI-Equipped IoT Applications in High Tech Agriculture, Eds, A. Khang, IGI Global, 2023.

[23] R. R. Chandan, S. Soni, A. Raj, V. Veeraiah, D. Dhabliya, S. Pramanik, Ankur Gupta , Genetic Algorithm and Machine Learning, in "Advanced Bioinspiration Methods for Healthcare Standards, Policies, and Reform", Hadj Ahmed Bouarara, IGI Global, 2023, DOI: 10.4018/978-1-6684-5656-9

[24] D. Mondal, A. Ratnaparkhi, A. Deshpande, V. Deshpande, A. P. Kshirsagar and S. Pramanik, Applications, Modern Trends and Challenges of Multiscale Modelling in Smart Cities, in Data-Driven Mathematical Modeling in Smart Cities, IGI Global, 2023, DOI: 10.4018/978-1-6684-6408-3.ch001

[25] A. Reyana, S. Kautish, P. M. S. Karthik, I. A. Al-Baltah, M. B. Jasser and A. W. Mohamed, "Accelerating Crop Yield: Multisensor Data Fusion and Machine Learning for Agriculture Text Classification," in IEEE Access, vol. 11, pp. 20795–20805, 2023, doi: 10.1109/ACCESS.2023.3249205.

[26] A. Mandal, S. Dutta, S. Pramanik, "Machine Intelligence of Pi from Geometrical Figures with Variable Parameters using SCILab", in Methodologies and Applications of Computational Statistics for Machine Learning, D. Samanta, R. R. Althar, S. Pramanik and S. Dutta, Eds, IGI Global, 2021, pp. 38–63, DOI: 10.4018/978-1-7998-7701-1.ch003

[27] S. Pramanik, "Carpooling Solutions using Machine Learning Tools", in Handbook of Research on Evolving Designs and Innovation in ICT and Intelligent Systems for Real-World Applications, K. K. Sarma, N. Saikia and M. Sharma, IGI Global, 2022, DOI: 10.4018/978-1-7998-9795-8.ch002.

[28] R. R. Chandan, S. Soni, A. Raj, V. Veeraiah, D. Dhabliya, S. Pramanik, Ankur Gupta , Genetic Algorithm and Machine Learning, in "Advanced Bioinspiration Methods for Healthcare Standards,

Policies, and Reform", Hadj Ahmed Bouarara, IGI Global, 2023, DOI: 10.4018/978-1-6684-5656-9

[29] B. K. Pandey, D. Pandey, V. K. Nassa, A. S. Hameed, A. S. George, P. Dadheech and S. Pramanik, A Review of Various Text Extraction Algorithms for Images, in The Impact of Thrust Technologies on Image Processing, Nova Publishers, 2023.

6

Building a Solar-powered Greenhouse having SMS and a Web Information Framework

N. Beemkumar[1], Rupal Gupta[2], Vinaya Kumar Yadav[3], Deepti Khubalkar[4], Vaidehi Pareek[4], and Farid Nait-Abdesselam[5]

[1]Department of Mechanical Engineering, Faculty of Engineering and Technology, JAIN (Deemed to be University), India
[2]College of Computing Science and Information Technology, Teerthanker Mahaveer University, India
[3]Department of Agriculture, Sanskriti University, India
[4]Symbiosis Law School, Nagpur Campus, Symbiosis International (Deemed University), India
[5]School of Science & Engineering, University of Missouri Kansas City, USA

Email: n.beemkumar@jainuniversity.ac.in; r4rupal@yahoo.com; vinay.ag@sanskriti.edu.in; deeptik@slsnagpur.edu.in; vaidehipareek@slsnagpur.edu.in; naf@umkc.edu

Abstract

One of the largest expenses in growing crops in greenhouses is energy for heating and cooling. This has caused people to reevaluate energy-saving tactics, inclusive of the need of solar energy as a practical, sustainable, and renewable option for greenhouse farming. This article describes the creation of a solar-powered framework that makes use of GSM and IoT technologies to sense, regulate, and support ideal weather conditions within a greenhouse. The suggested framework is intended to calculate and track alterations in temperature, humidity, soil moisture, and light intensity automatically. Regardless of the user's location, the technique used in the design framework gives them access to information about the regular parameters online and by

SMS. The chapter moreover includes a technique to self-adjust the greenhouse's climate to a favorable level for plant development. A system like this may enhance the yield and grade of crops cultivated in greenhouses. Tests on the system demonstrate its efficacy in light of the design criteria.

6.1 Introduction

According to [1], a greenhouse is a building mostly made of transparent materials like plastic, glass, and fiberglass where controlled climatic conditions ideal for plant development are maintained. According to [2], one of the global industries with the quickest growth rates is the greenhouse sector. The greenhouse offers protection for the crop from the impact of the elements. Thus, it makes it possible to grow crops in regions with erratic seasons or environmental conditions. The greenhouse's enclosure makes it possible to control the crop environment for better, more plant-friendly growing. This causes output to increase, improve, and use fewer pesticides, and last longer.

One of the largest expenses in growing crops in greenhouses is energy for heating and cooling. As a result, energy-saving tactics have been reconsidered, and solar energy is now being sought after as a sustainable option for greenhouse agriculture. The need to produce enough food, as well as factors like rising global citizens, higher energy utilization, uncertain weather patterns, deficient water supplies management, and the requirement to support the agricultural sector, are favoring greenhouses as viable options.

As an example, in Japan, there are around 2,000,000 smallholder or family agriculturists, compared to 40,000 commercial agriculturists. Many of these farmers rely primarily on the land to support their family, with the goal of having some surplus to trade or sell. However, the often unpredictable weather is affecting farming. For instance, dry spells, floods, heatwaves, or strong winds significantly reduce agricultural productivity. Additionally, it makes small-scale agriculturists increasingly helpless. These harsh weather conditions also cause soils to progressively deteriorate, which reduce these areas' ability to support grazing animals and lower crop yields influences millions of people's access to food in the long run. The concept of "smart farming," which encompasses the greenhouse sector, is increasingly being utilized to illustrate how modernization may be applied to raise crop yield and enhance crop quality.

For herbs to survive these unstable, harsh environmental conditions, frequent monitoring is necessary. Farmers with greenhouse settings must be on-site at all times to keep an eye on the plants. However, it is quite difficult for a greenhouse farmer to maintain constant watch on and control over the greenhouse's environment in order to ensure the development of high-quality

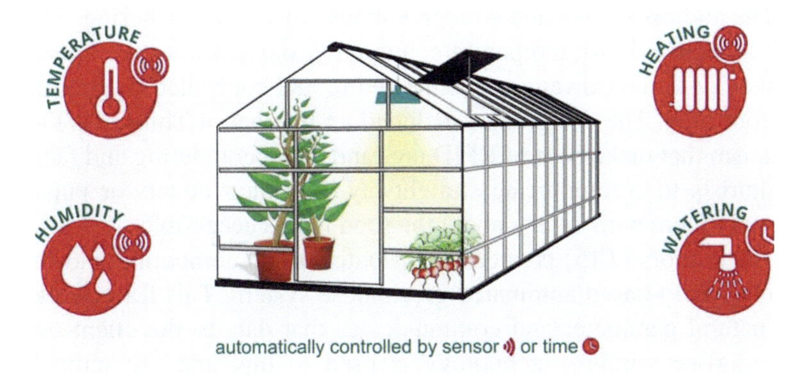

Figure 6.1 A smart greenhouse as an example.

plants. More so, issues with the supply of electricity, particularly in emerging nations, worsen the problem. As a result, the goal of this article is to develop and deploy a solar-powered greenhouse system that makes utilization of GSM and the IoT to remotely monitor greenhouse climatic conditions. Temperature, relative humidity, soil moisture, and lightning would all be replicated in a greenhouse-like environment by this system, which would also transmit frequent updates through the Internet utilizing GSM technology. A typical greenhouse is shown in Figure 6.1 and is automatically regulated by sensors in real time.

6.2 Literature Review

Various studies on greenhouse frameworks are created within a larger precision agricultural area, utilizing technology like IoT, cloud computing, and sensors.

A survey of various relevant studies in solar-powered Internet-of-Things-based greenhouse systems is provided in this section.

The authors of [3] built an automated watering system allowing the greenhouse's environmental changes to be tracked and managed. The prototype system makes use of a client-server web service and a wireless sensor network. Low power usage and remote monitoring and control were the two main design factors for the project. The researchers suggested a range of tracking techniques to account for the various ecological needs of various herbs. In their suggested design, five sensors – humidity, soil moisture, light, temperature, and CO_2 sensors – were utilized. The characteristics of the greenhouse sensor system were also monitored and controlled using an Android app that was also incorporated.

The authors of [4] created a greenhouse using drip watering. The sensors used include LDR, temperature, humidity, and soil wetness. The sensors and other system hardware are connected to and controlled by the Arduino microcontroller. The design also included an Internet of Things (IoT)-based mechanism that makes use of RFID tags and cloud computing and facilitates agriculturists to communicate straightway with their clients or purchasers and supply them with details about the food that is accessible.

The authors of [5] created a distributed cloud computing and Internet of Things (IoT)-based automated greenhouse system. This framework finds every natural parameter and communicates that data to the client over the cloud. ZigBee wireless technology is used to link and use temperature, humidity, soil moisture, and carbon dioxide sensors. When improper changes in the greenhouse take place, the framework is moreover capable of controlling the sensing settings.

In [6], the researchers presented a tracking and controlling framework for greenhouses that would link its many sensors and actuators using an Arduino microcontroller. Moreover, to sprinkler and air vent controls, sensors for humidity, temperature, light, and soil moisture are used. Additionally, the system includes an Arduino Wi-Fi shield for wireless sensor parameter transmission.

According to the examined works, several of the suggested designs include technical flaws such as unnecessary complexity, a failure to take into account renewable energy sources, and a lack of a perfectly integrated IoT-dependent framework to enable simple tracking and management of the greenhouse. Consequently, the research suggests that a cheap greenhouse powered by solar energy is combined with SMS and online notification systems to make it simple to monitor, manage, and regulate the greenhouse environment.

6.3 Systems Design

6.3.1 H/W design

Figure 6.2 presents a block schematic of the greenhouse design. The hardware is linked to the Arduino [7], which serves as the system's main processing unit. The system's detailed circuit diagram is shown in Figure 6.3.

6.3.2 Unit of power supply

Figure 6.4's representation of the greenhouse power supply unit shows a solar panel, a battery, and a charge controller. Through the charge controller,

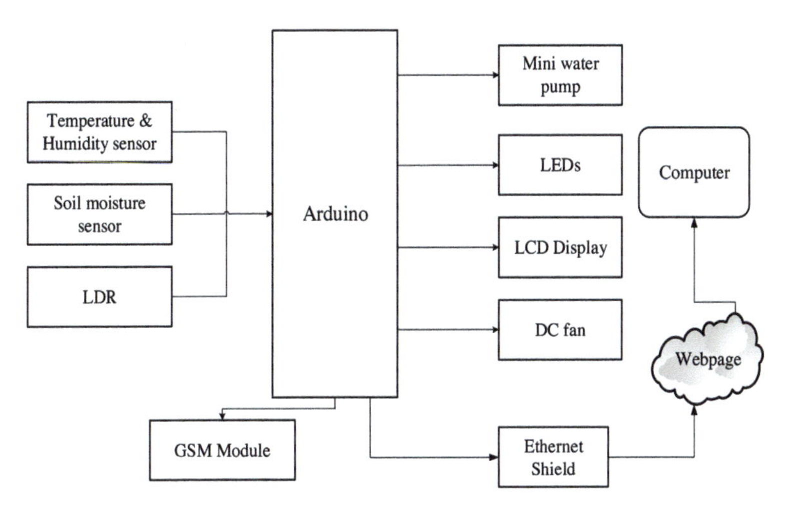

Figure 6.2 The system's block diagram.

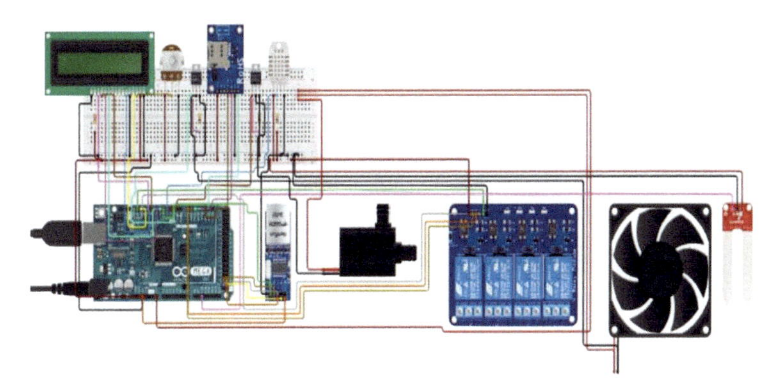

Figure 6.3 A thorough circuit schematic of the device.

the solar panel charges the 12-V battery, which in turn powers all of the greenhouse's components.

6.3.3 Program design

The complete flowchart of the framework in Figure 6.5, which is made up of five sub-systems – the IoT [8–12], notification, server, client, and control systems – depicts how the system performs its purpose.

The sensors are initialized upon system boot. After all of the analog values have been gathered, they are saved, utilized as system parameters, and then cleared. After that, calculations are performed and the current numbers

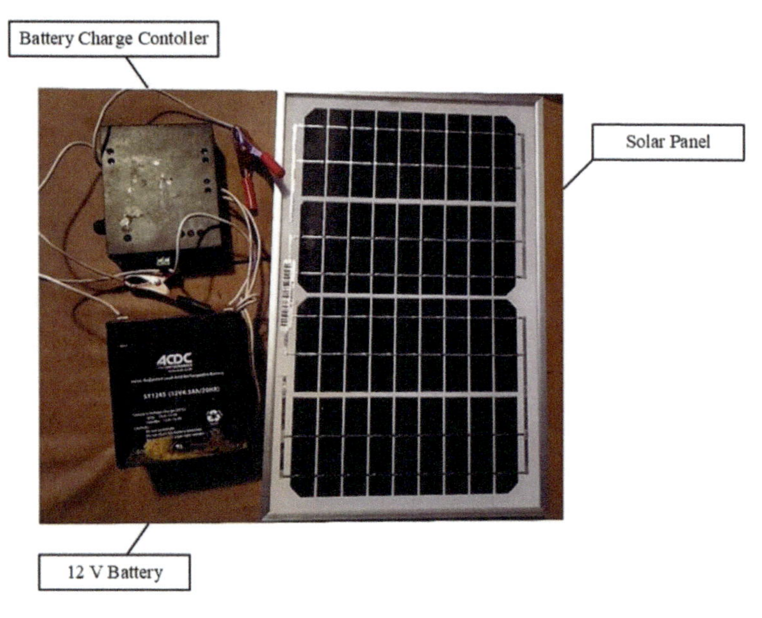

Figure 6.4 The solar power supply unit.

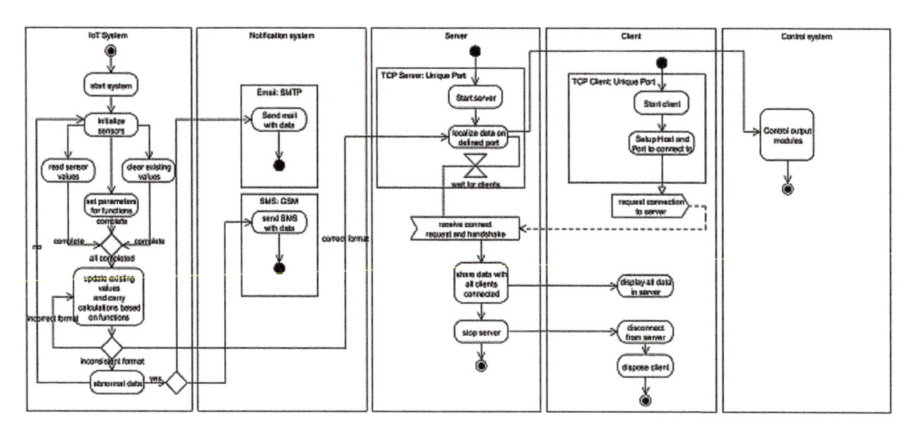

Figure 6.5 Diagram of a flowchart.

are updated. For recalculation, output setup for the erroneous format is undone. Data that has been detected abnormally is reinitialized and gathered again. Transfer control protocol (TCP) [13–18], which is operating on a particular port and continually watching for clients to connect, is then used to send regular data to the server. A handshake is conducted to transfer sensor

Table 6.1 Testing DHT11 sensor values.

Iteration(s)	Temperature, °C (Measured by DHT11) Test 1	Test 2	Test 3	Temperature, °C (measured by Samsung air conditioner)
1	21	20	20	21
2	20	23	20	22
3	24	23	21	23
4	22	23	22	24
5	23	22	22	25
6	26	25	26	26
7	25	25	27	27

data after a client connection has been established. Data is transformed into information in this process, and the information is displayed via a GUI [19]. The session ends when the client disconnects.

6.3.4 Outcomes of the framework's testing

The temperature measurement from the temperature sensor is contrasted with the ambient temperature in Table 6.1. Since the findings are comparable, the temperature sensor is set up properly.

The brief message provided by the GSM [20] element informing the user of the greenhouse parameters is shown in Figure 6.6. Figure 6.7 depicts the website that the user accesses to track the greenhouse's settings. Temperature, humidity, and soil moisture are among the primary metrics shown on the web server.

6.4 Summary and Future Work

Designing and implementing a prototype solar-powered greenhouse utilizing the IoT idea was the main goal of this endeavor. The utilizer of the greenhouse is able to remotely track the greenhouse through SMS including a website interface thanks to the utilization of both internet connectivity in the form of a web server and the GSM element. Based on the intended result, the work's goals and requirements were satisfied.

Incorporating AI like machine learning on the collected sensor data, as a method of forecasting or predicting, might help the greenhouse monitoring system in the future events inside the system. Additional sensor parameters, including pH and carbon dioxide management, and cameras for visual tracking and analysis of plant development, might also be included.

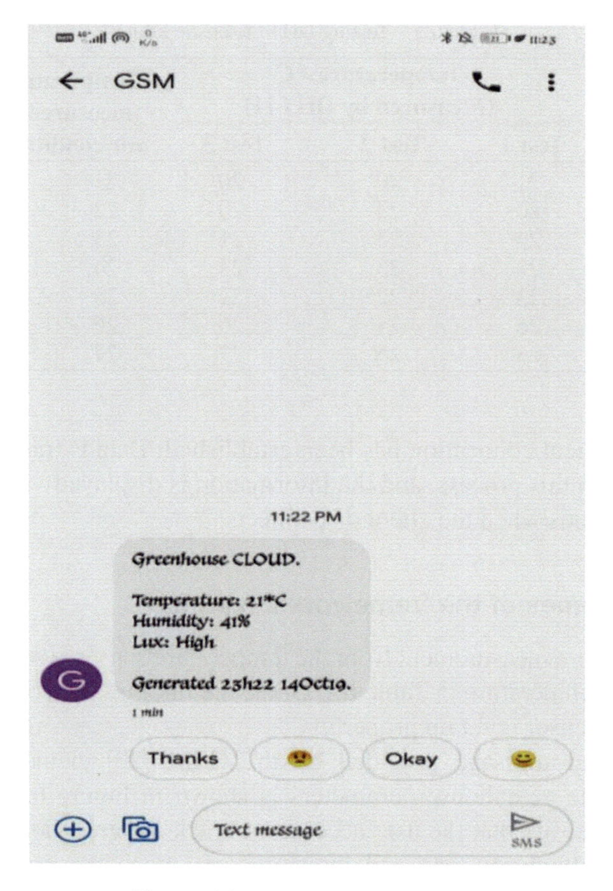

Figure 6.6 A smartphone message.

Figure 6.7 The sensor readings interface on a website.

References

[1] A. Saha, V. Simic, S. Dabic-Miletic, T. Senapati, R. R. Yager and M. Deveci, "Evaluation of Propulsion Technologies for Sustainable Road Freight Distribution Using a Dual Probabilistic Linguistic Group Decision-Making Approach," in IEEE Transactions on Engineering Management, doi: 10.1109/TEM.2023.3253300.

[2] M. U. Ahmed, M. A. Karim, M. S. Tahsin, Y. Rahman and F. Tafannum, "Analyzing Co2 Emission in Developing Countries Using Auto Regression and Auto Regression Walk Forward: A Time Series Approach," 2022 IEEE Delhi Section Conference (DELCON), New Delhi, India, 2022, pp. 1–5, doi: 10.1109/DELCON54057.2022.9752807.

[3] Y. Gamal, A. Soltan, L. A. Said, A. H. Madian and A. G. Radwan, "Smart Irrigation Systems: Overview," in IEEE Access, doi: 10.1109/ACCESS.2023.3251655.

[4] G. B. Cáceres, A. Ferramosca, P. M. Gata and M. P. Martín, "Model Predictive Control Structures for Periodic ON–OFF Irrigation," in IEEE Access, vol. 11, pp. 51985-51996, 2023, doi: 10.1109/ACCESS.2023.3277618.

[5] G. Sathishkumar., P. Siva., D. Anto Maria Joseph., R. Santhosh and R. Reshma, "Study on Distributed Multi-Level Security Scheme For Cloud Computing," 2023 International Conference on Inventive Computation Technologies (ICICT), Lalitpur, Nepal, 2023, pp. 1052–1057, doi: 10.1109/ICICT57646.2023.10134363.

[6] C. Qiu, Z. Wu, J. Wang, M. Tan and J. Yu, "Multiagent-Reinforcement-Learning-Based Stable Path Tracking Control for a Bionic Robotic Fish With Reaction Wheel," in IEEE Transactions on Industrial Electronics, vol. 70, no. 12, pp. 12670–12679, Dec. 2023, doi: 10.1109/TIE.2023.3239937.

[7] S. Pramanik, An Adaptive Image Steganography Approach depending on Integer Wavelet Transform and Genetic Algorithm, Multimedia Tools and Applications, 2023, https://doi.org/10.1007/s11042-023-14505-y.

[8] R. Jayasingh, J. Kumar R.J.S, D. B. Telagathoti, K. M. Sagayam and S. Pramanik, "Speckle noise removal by SORAMA segmentation in Digital Image Processing to facilitate precise robotic surgery", International Journal of Reliable and Quality E-Healthcare, vol. 11, issue 1, DOI: 10.4018/IJRQEH.295083, 2022

[9] S. Pramanik and S. Suresh Raja, "A Secured Image Steganography using Genetic Algorithm", Advances in Mathematics: Scientific Journal, vol. 9, issue 7, pp. 4533–4541, 2020. DOI: https://doi.org/10.37418/amsj.9.7.22

[10] S. Pramanik, K. M. Sagayam and O. P. Jena, Machine Learning Frameworks in Cancer Detection, ICCSRE 2021, Morocco, 2021.

[11] S. Pramanik, M. G. Galety, D. Samanta, N. P Joseph, Data Mining Approaches for Decision Support Systems, 3rd International Conference on Emerging Technologies in Data Mining and Information Security, 2022.

[12] D. Samanta, S. Dutta, M. G. Galety and S. Pramanik, A Novel Approach for Web Mining Taxonomy for High-Performance Computing, The 4th International Conference of Computer Science and Renewable Energies (ICCSRE'2021), 2021, DOI:10.1051/e3sconf/202129701073.

[13] R. Bansal, A. J. Obaid, A. Gupta, R. Singh, S. Pramanik "Impact of Big Data on Digital Transformation in 5G Era", 2nd International Conference on Physics and Applied Sciences (ICPAS 2021), doi:10.1088/1742-6596/1963/1/012170, 2021.

[14] S. Pramanik, S Joardar, OP Jena, AJ Obaid, "An Analysis of the Operations and Confrontations of Using Green IT in Sustainable Farming", Al-Kadhum 2nd International Conference on Modern Applications of Information and Communication Technology, *AIP Conference Proceedings* 2591, 040020 (2023). https://doi.org/10.1063/5.0119513

[15] S. Pramanik, Niranjanamurthy M. and S. N. Panda, Using Green Energy Prediction in Data Centers for Scheduling Service Jobs, ICRITCSA 2022, Bengaluru, 2022.

[16] V. Vidya Chellam, V. Veeraiah, A. Khanna, T. H. Sheikh, S. Pramanik and D. Dhabliya, A Machine Vision-based Approach for Tuberculosis Identification in Chest X-Rays Images of Patients, ICICC 2023, Springer

[17] S. Praveenkumar, V. Veeraiah, S. Pramanik, S. M. Basha, A. V. Lira Neto, V. H. C. De Albuquerque and A. Gupta, Prediction of Patients' Incurable Diseases Utilizing Deep Learning Approaches, ICICC 2023, Springer

[18] P. T. Khanh, T. H. Ngọc, and S. Pramanik, Future of Smart Agriculture Techniques and Applications, in Advanced Technologies and AI-Equipped IoT Applications in High Tech Agriculture, Eds, A. Khang, IGI Global, 2023.

[19] T. H. Ngọc, P. T. Khanh and S. Pramanik, Smart Agriculture using a Soil Monitoring System, in Advanced Technologies and AI-Equipped IoT Applications in High Tech Agriculture, Eds, A. Khang, IGI Global, 2023.

[20] S. Pramanik and S. Bandyopadhyay, Identifying Disease and Diagnosis in Females using Machine Learning, in Encyclopedia of Data Science and Machine Learning, Eds. John Wang, IGI Global, 2023, DOI: 10.4018/978-1-7998-9220-5.ch187.

7

Agriculture using Digital Technologies

Ramkumar Krishnamoorthy[1], Shakuli Saxena[2], Kaushal Kishor[3], Aditee Godbole[4], Himanshi Bhatia[4], Ankur Gupta[5], and Mohammad Zubair Khan[6]

[1]Department of Computer Science and IT, School of CS and IT, Jain (Deemed to be) University, India
[2]College of Agriculture Science, Teerthanker Mahaveer University, India
[3]Department of Agriculture, Sanskriti University, India
[4]Symbiosis Law School, Nagpur Campus, Symbiosis International (Deemed University), India
[5]Department of Computer Science and Engineering, Vaish College of Engineering, India
[6]Department of Computer Science and Information, Taibah University Medina Saudi Arabia

Email: ramkumar.k@jainuniversity.ac.in; shakuli2803@gmail.com; kaushal.ag@sanskriti.edu.in; aditeegodbole@slsnagpur.edu.in; himanshibhatia@slsnagpur.edu.in; ankurdujana@gmail.com; mkhanb@taibahu.edu.sa

Abstract

The reader will get an introduction to the cutting-edge new technologies employed in agriculture in this chapter, including cloud computing, ML, and AI. The whole crop cycle is then briefly discussed, consisting of seven sections: crop choice, soil production, seed choice, seed plantation, irrigation, crop growth, fertilization, and harvesting. This chapter also explains how digital technology might be used for the agricultural cycle. The remainder of the chapter will describe how farming will function in the future and how contemporary digital technology will be combined with the cycle of agricultural crops.

7.1 Introduction

The global population is projected to reach 9.6 billion by 2050. It will be necessary to generate twice as much food as is now being produced. This will put to the test the agricultural advances that are being developed to assist us in achieving this aim. By introducing cutting-edge technology to the agricultural sector and putting them together on the same platform utilizing contemporary digital technologies, it will be necessary to transform a new agricultural age in order to satisfy human needs. Over the last century, the sector has witnessed significant advances.

To put food on 76 million people's tables in 1990, it took 37% of America's total labor force, 5.7 million farms, and 5.7 million workers; by 2005; however, only 2.5% of the country's population is needed to be fed (World Resources Report 2013–14). Humans may now succeed in a variety of other areas of life in addition to farming, where they formerly had to spend their days and nights to meet their most fundamental need for sustenance. The world is moving toward a time of "smart farming," in which robots will cultivate food for people, and that too without the need for human labor. Small sensors and cameras set in the farm will be used to track crop growth. The data gathered from the sensors can be possessed and processed by the processors, and it will then be supplied to the agriculturists on their interface devices or mobile phones in the mode of warnings and survey documents. Drones [1] and robots [2] may be utilized to detect weeds and blast them with laser beams or chemical streaks to deal with issues like weeds. Comparing this method to the traditional blanket spraying method will result in a reduction of up to 90% in the amount of pesticides used on the crops. The degree of current developments and inventions has led to the development of autonomous strawberry and a few other vegetable pickers and gatherers, although they are distinct for various crops. The development of picking robots that can switch between various crops by choosing the correct software will be the next advancement. The huge realm of agriculture has not been left unaffected by humankind's constant drive for advancement and improvement in all aspects of existence. Just a few years ago, the general public first had access to GPS [3] technology, and electronic control systems first entered the huge world of farming. Various sensors were placed in the farm to monitor various farming parameters and inform agriculturists and scientists about crop growth and associated issues, such as uneven soil chemical composition, pest and weed presence, improper or excessive irrigation, and poor crop quality in a particular area. Even though the farmer had access to a lot of data, the primary task was still gathering it, keeping track of it, compiling it

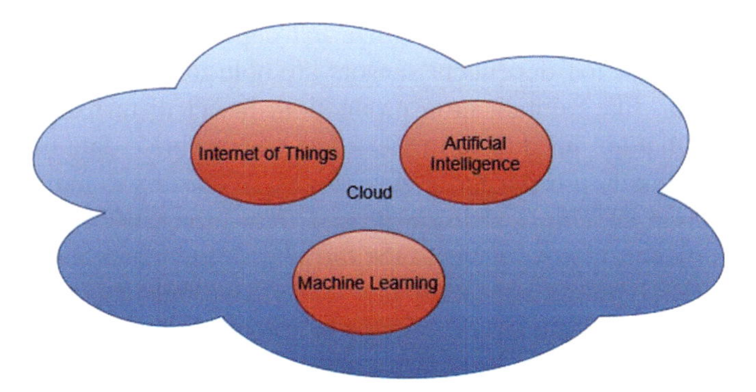

Figure 7.1 Agriculture and digital technologies.

(based on various crops, geographical regions, weathers, and various categories of seeds, fertilizers, and pesticides utilized), and ultimately stringing all the detail beads into an array to utilize them effectively.

7.2 Modernization of Digital Technologies in Farming

Although it is incapable yet to be said that precision farming has been universally accepted in crop yield, it is making considerable progress in that direction by using cutting-edge digital technology. Figure 7.1 illustrates the many technologies being adopted for agricultural developments.

1. IoT
2. AI
3. ML

The paragraphs that follow will provide further information on these technologies.

7.3 Agricultural IoT

A digital technology utilized to increase farming yield while lowering technical obstacles is the Internet of Things (IoT). Now that farmers have access to IoT and are fully informed about new technological advancements, IoT is capable of transforming the agricultural sectors and inspiring them to meet technological challenges.

By continuously monitoring the soil's acidity level, temperature, weather, and many other parameters with IoT, farmers may increase their output. Additionally, farmers may utilize IoT-dependent gadgets to have an eye on their cows. IoT-dependent sensors are able to provide agriculturists with vital data, like rainfall, crop production, insect disturbance, and soil nourishment that are invaluable to yield and give accurate data that may be utilized to gradually prosper farming practices. According to Sangaiah et al. [4], the Internet of Things delivers precise, real-time, and shared features having significant improvements to the farming supply chain and offering a complex tool in building a seamless flow of farming logistics. The following are the main benefits of Internet of Things for agricultural advancement:

1. Water management can be usefully carried out with Internet of Things sensors.
2. Internet of Things makes it possible to continuously monitor the land so that any necessary preventative actions can only be taken in advance.
3. It aids farmers in minimizing manual labor, resulting in more effective and swifter agricultural operations.
4. IoT makes it simple to manage soil properties like pH and moisture level, which aids farmers in planting seeds at the correct depth in the soil.
5. The crucial instruments utilized in the detection of plant and agricultural diseases are RFID chips and sensors. Information is read from RFID [5] tags and passed to the correct person through the Internet. The particular agriculturist or researcher may receive this information and make the necessary defense from distant locations, preserving the crops from the prevalent illnesses.
6. Because farmers may now readily connect with the worldwide market from any location, crop sales will rise on the global market.

7.3.1 IoT's potential in agriculture

Implementing Internet of Things in farming has changed it into smart agriculture, removing the requirements for agriculturists to do bodily labor, and enhanced production overall. The usage of IoT in agriculture has completely changed how farming is done since it allows for real-time monitoring of fields using sensors and their connection, which keeps an eye on key metrics like humidity soil fertility, temperature, etc. Along with this, there has been a decrease in the consumption of expensive resources like water and electricity.

- **Climate:** Climate is a highly important factor in farming. Additionally, having inadequate climatic knowledge seriously degrades the quantity and standard of farming yield. Still, Internet of Things technologies shows one the present weather situations. The farming lands have sensors installed both within and outside the farms. They gather ecological knowledge, which is used to select the finest crops for the specific climatic situations. The entire IoT environment consists of sensors that can reliably measure real-time weather variables including humidity, rainfall, and temperature. Various sensors are present to estimate every constituent and can be set up for meeting one's requirement for smart agriculture. The sensors monitor the crops' wellness and the environmental weather. An alarm is sent when any disturbing weather is found. The need for physical appearance in adverse weather situations is reduced, which ultimately enriches yield and facilitates agriculturists to get greater farming advantages.
- **Precision farming:** Another popular application of IoT in agriculture is precision agriculture (PA). By using smart agricultural applications including animal tracking, vehicle monitoring, farm monitoring, and inventory tracking, it enhances the farming practice's accuracy and command. PA [6–10] tends to assess data generated by sensors and acknowledge properly. By the utilization of sensors, PA provides farmers to collect data, evaluate it, and have rapid, instructed selections. Various precision farming techniques, along with livestock management, vehicle tracking, and irrigation management, mostly provide importantly to lifting yield and usefulness. One can evaluate soil conditions and various compatible specifications with the usage of PA to enhance functioning advantage. Additionally, one may also see how the linked devices are operating in real time to check the water and nutrient levels.
- **Intelligent greenhouse:** Internet of Things has provided for weather stations to naturally alter the ecology related to a particular collection of instructions, making our greenhouses more intelligent. The use of IoT in greenhouses has eliminated the requirement for human interaction that minimizes expenditures and enhances proficiency all through the approach. For example, making modern, cost-effective greenhouses using Internet of Things sensors driven by solar energy, these sensors collect and transmit real-time data that is utilized to correctly monitor the ranking of the greenhouse in real time. The sensors permit for the tracking of greenhouse circumstances and water utilization via emails

or SMS warnings. Internet of Things is utilized for automatic and smart irrigation. Data about the pressure, humidity, temperature, and light levels is supplied in chunk by these sensors.

- **Farming drones:** The advent of farming drones is the current outburst in technology, which has nearly completely altered agricultural operations. Drones are utilized in farm analysis, planting, crop sprinkling, crop tracking, and crop health estimation. Drone technology has changed the agricultural arena by providing it a higher enhancement and a transpose with sufficient plan and arranging depending on real-time data. Drones provided with thermal or multispectral sensors find the positions that require irrigation adaptation. Sensors decide the vegetation index after the crops have started to grow and display their condition of health. Finally, the ecological consequence was reduced by smart drones. The final result is a major minimization in the quantity of chemical penetrating groundwater.

The aforementioned use of IoT has equipped agriculture with cutting-edge technologies which has closed the manufacturing, quality, and/or quantity deficit. The many sensors linked through the Internet of Things collect real-time data that may be utilized to respond quickly and lessen crop damage. Internet of Things has enhanced the agricultural industry's business model by enabling quicker product processing and retail delivery.

7.4 Agriculture and AI

The agricultural industry has embraced AI into its practices to increase efficiency while minimizing negative environmental implications. The production of our food is changing as a result of AI, and the agriculture sector's emissions have fallen by 20%. Farmers can handle any unwelcome natural situation with the aid of AI. Most agriculture-based firms are now concentrating on AI-based production solutions to increase efficiency. The market for AI in agriculture is anticipated to grow to 2678 million US dollars by the end of 2030, according to a market assessment report. By using AI-powered methods, illnesses or climatic changes might be detected earlier and intelligent responses could be made.

7.4.1 Benefits of using AI in agriculture

Farmers can now comprehend the meaning behind statistics such as sun radiation, temperature, precipitation, and wind speed thanks to artificial

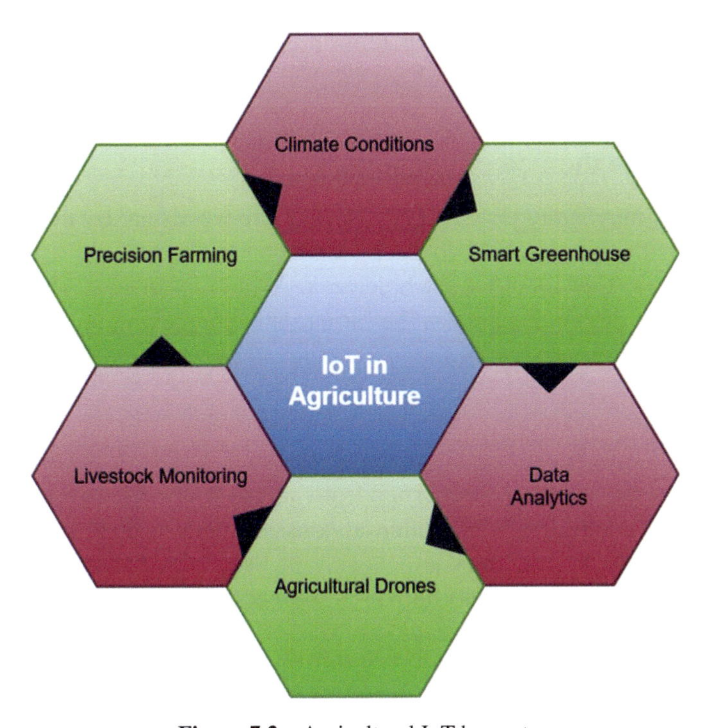

Figure 7.2 Agricultural IoT harvest.

intelligence (AI) [11–15]. The nicest thing about AI's use in agriculture is that, while removing employment, it will aid farmers in advancing their businesses. Figure 7.2 illustrates how AI can grow and market necessary crops more accurately and with more efficiency.

7.4.2 Future weather information

Farmers employ artificial intelligence (AI), a rather cutting-edge technology, to stay informed about the weather so they can reduce crop risk. By comprehending and using AI to learn, the study of the data produced aids the farmer in exercising prudence. Making wise decisions on time is made possible by putting this discipline into practice.

7.4.3 Crop and soil health monitoring

Through AI's picture identification process, farmers may find potential flaws and nutritional deficits in the soil. Numerous applications for assessing the flora patterns in agriculture have been created thanks to the deep learning feature of Al. These AI-powered tools are helpful in comprehending soil imperfections, plant pests, and illnesses.

7.4.4 Reduce the use of pesticides

The farmers may spray pesticides just where the weeds are by using the data collected with the aid of AI to keep an eye on the weeds. As a result, less chemical was used to spray a whole area. As a consequence, AI decreases the amount of herbicides used in the farm relative to the amount of chemicals improperly sprinkled.

7.4.5 AI farming robots

Farmers can now defend their crops from weeds more effectively thanks to AI-enabled agricultural bots, which also assist them to deal with the man-power shortage. AI machines are capable of harvesting crops more quickly and in greater quantities than human employees. Utilizing computer vision enables the monitoring and spraying of weeds.

7.5 Machine Learning's Role in Agriculture

Making robots learn like humans is the goal of machine learning. Like any child, they must acquire knowledge by experience. With machine learning, algorithms are created by analyzing tens of thousands of samples. If the algorithm succeeds in achieving its objective, it is then modified. The software really improves with time. In this way, computers like IBM's Watson are able to do tasks like cancer diagnosis, classical symphony composition, and Ken Jennings' Jeopardy defeat. Little software moreover replicates the neural networks that aid in problem solving in both people and robots.

Prior to continuing, it is important for us to see the procedures used in agriculture. The following eight key tasks that a farmer completes are depicted in Figure 7.3.

1. Crop picking
2. Preparing the soil
3. Selection of seeds
4. Seed planting
5. Irrigation

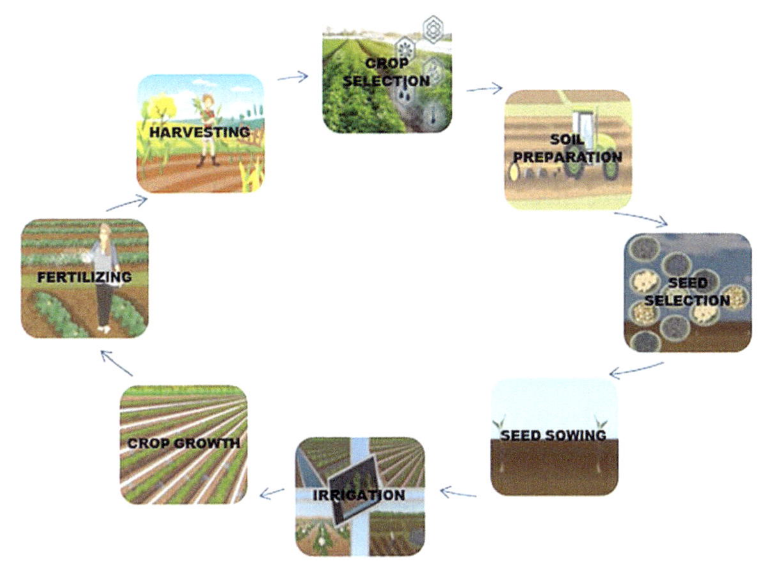

Figure 7.3 Utilizing digital technologies in agriculture.

6. Plant growth
7. Fertilizing
8. Harvesting

7.5.1 Preparing the soil

For the production of crops, soil preparation is necessary. Typically, soil is discovered as hardened, dense pieces of soil material. It is necessary to loosen and soften this dirt. Before cultivating, the soil must be loosening since this will enable more air to reach the plant roots, which will help the plant, develop and result in high crop output. The ability to retain water is another benefit of loosening the soil, which encourages the development of beneficial bacteria that contribute to the production of humus in the soil.

Humus is comprised material that is nutrient-rich. This is the source of nutrients for plants. Additionally, loosening the soil makes it easier for the plant to reach the lower soil by turning it up with a plow. The relevance of moisture, soil fertility, N and P content, and soil texture are now recognized concepts, but the significance of pH value is still unclear. A substance's pH may be used to identify whether it is acidic or basic in property. The scale is based on positive values ranging from 1 to 15, with 7 denoting neutrality, or not being either acidic or basic. While a substance's pH value below 7 indicates that it is acidic and above 7 that it is basic. It is thought that soil with a pH value of 70.5 is best for agricultural use. The pH value shows how many harmful compounds are present.

The BoniRob collects soil specimens, liquidizes them, and finally analyzes the pH, phosphorus, and nitrogen contents in real time. Along with it, several sensors placed in the field will measure the soil's moisture content and offer information on how soft the soil is. Through a suitable application, all of these data are gathered and built accessible to the agriculturist on his cell phone.

7.5.2 Selection of seeds

The choice of seeds is a crucial part of agriculture; high-quality seeds will always result in superior crop yield. The following criteria affect seed selection for sowing:

1. Whole seeds are preferable; broken or crushed seeds are to be avoided.
2. Seeds must have a better sowing quality.
3. They need to have a better capacity to germinate.
4. Viruses must not be located in the seeds.
5. The seeds should not be mixed with different seeds or weed seeds.
6. Disease resistance is a necessity for seeds.
7. A reputable seed agency should provide the seeds.

The NK Seed Selection Tool from Syngenta [16] is a seed analyzer that helps the agriculturist to put goods that are optimal for a farm, not only depending on the soil category but also the weather and takes into consideration the impacts of the yearly change on production. It ensures that only those seed hybrids and types are chosen that will thrive for the next several years in that local environment by receiving the weather prediction for the following years.

7.5.3 Seed planting

The seeds may be sown using one of two techniques.

7.5.4 Broadcasting

This technique is used to seed wheat, maize, and paddy. This technique involves randomly scattering or throwing seeds throughout the farmed land. In locations with poor farming practices, this approach is typical.

7.5.5 Irrigation

A technique for delivering water to the land is irrigation. A framework of pipelines, log channeled ditches, and garden hoses may be used to bring water to a dry location, or it may be as ordinary as utilizing a garden hose to water plants in the garden. Wherever rainfall is either inconsistent or unable to provide enough natural water to support agricultural development, irrigation is employed. The top three irrigation techniques are as follows.

7.5.6 Furrow irrigation or flooding

When water is pumped from a water source and then let to flow on the earth's surface, it can be subterranean water, a river, or a lake.

7.5.7 Watering via drip

Here, hoses with holes are used to provide water. Less water is lost to evaporation owing to less exposure to the environment, which also results in a 25% reduction in water use when compared to flood irrigation.

7.5.8 Irrigation via spray

High-pressure water is pushed via the hoses and attached nozzles in this to spray the water. Due to the high water waste and need for power to pump the water, it is losing favor.

Although irrigation is a crucial step in the agricultural process, there are also drawbacks. It accounts for a sizable portion of a nation's overall water use and represents a significant financial burden for farmers. Now, runoff from irrigation, or too much water flowing across the fields, pollutes the nearby water sources by introducing dissolved fertilizer and pesticide residues.

The widespread usage of AI technology today has been a blessing since it has significantly reduced costs associated with training labor, water waste, and environmental degradation. Let us now look at the case of AI in action – an irrigation system in a greenhouse that, with small adjustments, might be extended to an open farm The following sensors have been used.

7.5.9 Sensors for temperature

In order to maintain the ideal temperature and moisture in the air for plant development, temperature sensors are utilized to turn on and off the fans that blast hot air out of the greenhouse.

7.5.10 Sensor for soil moisture

In order to preserve sufficient moisture in the soil and environment to support plant development, soil moisture sensors are utilized to turn on and off irrigation water pumps as well as misting systems.

7.5.11 Light sensor

When the amount of sunshine inside the greenhouse is insufficient, artificial lights that have been put inside are turned on by light detectors. The demand may be met by 5–8 sensors [17–21] placed on an acre of land. Each of these sensors costs between 20 and 60 rupees, making it affordable.

A microprocessor drives all of the sensors, and the temperature, moisture, and light levels are manually chosen according to the crop and soil type. The Internet of Things will take care of the rest, and the farmer will continue to get information about his greenhouse farm as depicted in Figure 7.4. Additionally, the farmer is given the choice to manually manage temperature and moisture in place of the desired method.

7.5.12 Plant growth

An essential component of agricultural operations is keeping an eye on crop development and taking precautions to minimize crop loss or a reduction in production.

The issue that now emerges is, "How may one find the vegetative greenness of plants utilizing remote sensing?" Relying on the quantity of chlorophyll in the plant, vegetation absorbs, reflects, or reemits solar light. As seen in Figure 7.5, satellites gather the energy that the vegetation reflects or emits.

NDVI, or normalized difference vegetation index, is a remote sensing technique that shows the relative biomass of vegetation as well as its degree of greenness. NIR and red wavelength bands' reflectance is compared

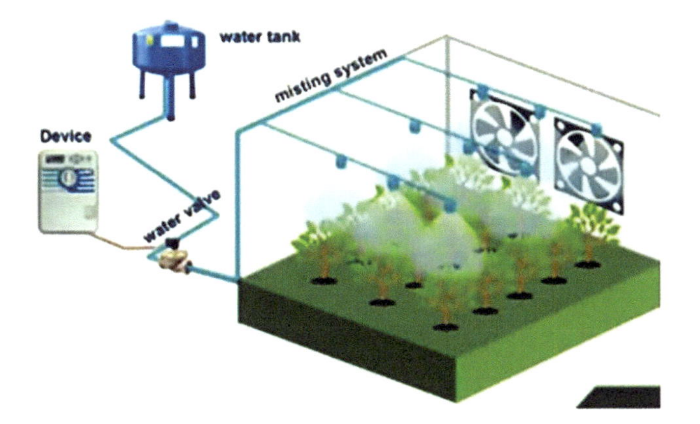

Figure 7.4 Detectors of light in agriculture.

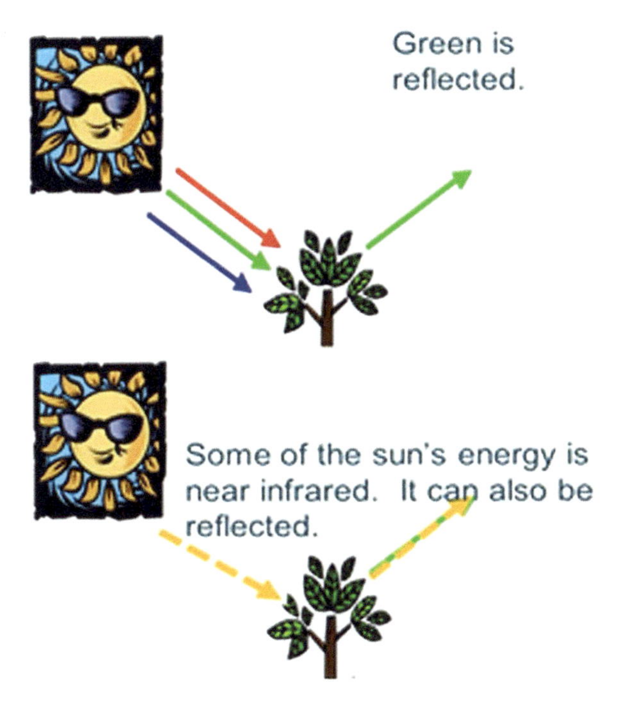

Figure 7.5 Growth stages of crops.

via NDVI. A farmer may use NDVI to get a close-up look of his crop and use remote sensing to track its progress on a daily basis. Prior to harvest, NDVI may identify problems that are harming the health of the plant. Even though few of the issues can be unfixable for the present crop year or cycle, modifications may be made. Higher correlation yield maps are utilized in this.

NDVI may be used to verify the following at higher resolution:

1. Deficiencies in nutrients (tissue analysis/soil samples)
2. Issues with the equipment
3. Find tiles (after a spring rainfall, tiles may briefly become apparent as the dirt begins to drain).

7.5.13 Pressure from insects and diseases

For a good comprehending of how remote sensing captures images and how compiling these images aids in monitoring crop development and associated issues to guide agriculturists in taking necessary preventative actions, please refer to the following satellite imagery. There are several satellite options to monitor agricultural development, including:

- AVHRR, or advanced very high resolution radiometer
- Spatial: The Earth's surface in its entirety
- Temporal: On a consistent basis (often) to recognize seasonal and yearly variations
- 1-km resolution
- 8. Landsat
- NIR + Red/NDVI (Near IR – Red)
- Space: Smaller Area

7.5.14 Fertilizing

Animal dung and artificial fertilizers, which supply crops with the N and P they need to develop and generate food for human consumption, are applied by farmers to their fields as nutrients. Unused nitrogen and phosphorus from agricultural leftovers, however, may overflow from farms and harm the air and water quality of nearby water bodies by gushing with rains or by leaching via the layers of soil when too much irrigation water is used. Hypoxia (also known as "dead zones") may result from contamination of water bodies, which can kill fish and reduce aquatic life. When too many nutrients enter freshwater systems, harmful algae blooms (HABs) develop more quickly disrupting the eco-cycle and producing toxins that are detrimental to people.

AI and IoT have shown to be invaluable in dealing with such dangerous behaviors that pose a major harm to the environment. The next generation of intelligent agricultural machinery is being created by Blue River Technology in the USA. They have created "See & Spray Technology," which makes it possible for every plant to matter in the future. This technique uses a lot of cameras to capture photographs of the plants growing in the farm and distinguish between desired crops and undesirable weeds. Herbicides will be sprayed on the undesirable weeds using various sprayers built into the machinery, or the weeds may be burned using lasers. Meanwhile, just around 20 plants are treated with fertilizers, and the ill or weak desired crop plants are recognized utilizing images of the color, size, and texture of their leaves. By applying pesticides and herbicides just where and as necessary, this system uses 90% less pesticides and herbicides than previous methods.

7.5.15 Harvesting

When agricultural activities are carried out utilizing AI and the IoT, the final step, harvesting, is not limited to harvesting alone but includes a broad variety of processes, including the following:

Harvesting: It is the act of harvesting and cutting ripe crops with a sickle.

Winnowing: The method of detaching grains from chaff.

Threshing: The process of separating grain from hay.

Robotic pickers will determine if a fruit or flower is mature enough to be picked or not in small fruit, vegetable, and flower farms based on the size and color of the produce. They will then select and gather the produce at the appropriate time. Some examples of these robotic pickers are small-scale pickers for potatoes, capsicums, and cotton.

When it comes to huge agricultural fields, however, the photographic data obtained by remote sensing having the aid of drones and satellites may supply farmers the knowledge on the ideal time to harvest the crop. Carrot harvester and separator, maize harvester, sugar cane harvester, combine (grain) harvester, etc., are a few of the machines utilized in vast agricultural fields.

7.6 Conclusion

Numerous businesses have teamed together over the last 20 years to support the expanding agriculture industry. The R&D division of Agro Agency in the USA is planning toward planting and harvesting barley on a 1-hectare farmland with no physical assistance from humans in order to test fully autonomous agriculture technologies. Machine learning is a constant process that enhances a system's capacity to distinguish between a variety of crops and the weeds that pose a danger to them. For instance, the US company Agribotix has created commercial software that analyzes infrared images taken by drones to detect poor crop and unwelcome weed growth. Light aircraft equipped with multispectral cameras are used by the US-based Mavrx firm to collect data over vast agriculture areas dispersed around the nation. In order to have a more comprehensive perspective of the landscape to help with crop monitoring, PlanetLabs runs a fleet of CubeSats that snap photos of huge areas of land from space on a weekly basis. Farm-management system S/W and related H/W are being developed by several businesses across the world, enabling producers of all sizes to succeed in their industry. The agricultural network compiles information from a vast number of fields into a single pool to facilitate its users the macro- and micro-level knowledge for the most productive agriculture with the fewest resources.

References

[1] N. Souli, P. Kolios and G. Ellinas, "Multi-Agent System for Rogue Drone Interception," in IEEE Robotics and Automation Letters, vol. 8, no. 4, pp. 2221–2228, April 2023, doi: 10.1109/LRA.2023.3245412.

[2] C. Della Santina, C. Duriez and D. Rus, "Model-Based Control of Soft Robots: A Survey of the State of the Art and Open Challenges," in IEEE Control Systems Magazine, vol. 43, no. 3, pp. 30–65, June 2023, doi: 10.1109/MCS.2023.3253419.

[3] X. Shao, S. Li, J. Zhang, F. Zhang, W. Zhang and Q. Zhang, "GPS-free Collaborative Elliptical Circumnavigation Control for Multiple Nonholonomic Vehicles," in IEEE Transactions on Intelligent Vehicles, doi: 10.1109/TIV.2023.3240855.

[4] A. K. Sangaiah, A. Javadpour, F. Ja'fari, H. Zavieh and S. M. Khaniabadi, "SALA-IoT: Self-Reduced Internet of Things with Learning Automaton Sleep Scheduling Algorithm," in IEEE Sensors Journal, doi: 10.1109/JSEN.2023.3242759.

[5] G. Shi, X. Shen, L. Gu, S. Weng and Y. He, "Multipath Interference Analysis for Low-power RFID-Sensor under Metal Medium Environment," in IEEE Sensors Journal, doi: 10.1109/JSEN.2023.3253571.

[6] P. T. Khanh, T. H. Ngọc, and S. Pramanik, Future of Smart Agriculture Techniques and Applications, in Advanced Technologies and

AI-Equipped IoT Applications in High Tech Agriculture, Eds, A. Khang, IGI Global, 2023.

[7] T. H. Ngọc, P. T. Khanh and S. Pramanik, Smart Agriculture using a Soil Monitoring System, in Advanced Technologies and AI-Equipped IoT Applications in High Tech Agriculture, Eds, A. Khang, IGI Global, 2023

[8] V. Veeraiah, V. Talukdar, Manikandan K., S. B. Talukdar, V. D. Solavande, S. Pramanik, A. Gupta, Machine Learning Frameworks in Carpooling, in Handbook of Research on AI and Machine Learning Applications in Customer Support and Analytics, Eds, Md S. Hossain, R. C. Ho and G. Trajkovski, IGI Global, 2023.

[9] V. Jain, M. Rastogi, J. V. N. Ramesh, A. Chauhan, P. Agarwal, S. Pramanik, A. Gupta, FinTech and Artificial Intelligence in Relationship Banking and Computer Technology, in AI, IoT, and Blockchain Breakthroughs in E-Governance, Eds, K. Saini, A. Mummoorthy, R. Chandrika and N.S. Gowri Ganesh, IGI Global, 2023.

[10] S. Pramanik, "An Effective Secured Privacy-Protecting Data Aggregation Method in IoT", in Achieving Full Realization and Mitigating the Challenges of the Internet of Things, Eds, M. O. Odhiambo and W. Mwashita, IGI Global, 2022, DOI: 10.4018/978-1-7998-9312-7.ch008

[11] S. Pramanik and S. Bandyopadhyay, Identifying Disease and Diagnosis in Females using Machine Learning, in Encyclopedia of Data Science and Machine Learning, Eds. John Wang, IGI Global, 2023, DOI: 10.4018/978-1-7998-9220-5.ch187.

[12] S. Pramanik and S. Bandyopadhyay, Analysis of Big Data, in Encyclopedia of Data Science and Machine Learning, Eds. John Wang, IGI Global, 2023, DOI: 10.4018/978-1-7998-9220-5.ch006

[13] R. Anand, J. Singh, D. Pandey, B. K.Pandey, V. K.Nassa and S. Pramanik, "Modern Technique for Interactive Communication in LEACH-Based Ad Hoc Wireless Sensor Network", in Software Defined Networking for Ad Hoc Networks, M. M. Ghonge, S. Pramanik and A. D. Potgantwar, Eds, Springer, 2022, https://doi.org/10.1007/978-3-030-91149-2_3

[14] K. Dushyant, G. Muskan, A. Gupta and S. Pramanik, "Utilizing Machine Learning and Deep Learning in Cyber security: An Innovative Approach", in Cyber security and Digital Forensics, M. M. Ghonge, S. Pramanik, R. Mangrulkar, D. N. Le, Eds, Wiley, 2022, https://doi.org/10.1002/9781119795667.ch12

[15] M. Sinha, E. Chacko, P. Makhija and S. Pramanik, Energy Efficient Smart Cities with Green IoT, in Green Technological Innovation for

Sustainable Smart Societies: Post Pandemic Era, C. Chakrabarty, Eds, Springer, 2021, DOI: 10.1007/978-3-030-73295-0_16

[16] T. A. Sathi, M. A. Hasan and M. J. Alam, "SunNet: A Deep Learning Approach to Detect Sunflower Disease," 2023 7th International Conference on Trends in Electronics and Informatics (ICOEI), Tirunelveli, India, 2023, pp. 1210–1216, doi: 10.1109/ICOEI56765.2023.10125676.

[17] A. Bhattacharya, A. Ghosal, A. J. Obaid, S. Krit, V. K. Shukla, K. Mandal and S. Pramanik, "Unsupervised Summarization Approach with Computational Statistics of Microblog Data", in Methodologies and Applications of Computational Statistics for Machine Learning, D. Samanta, R. R. Althar, S. Pramanik and S. Dutta, Eds, IGI Global, 2021, pp. 23–37, DOI: 10.4018/978-1-7998-7701-1.ch002

[18] A. Mandal, S. Dutta, S. Pramanik, "Machine Intelligence of Pi from Geometrical Figures with Variable Parameters using SCILab", in Methodologies and Applications of Computational Statistics for Machine Learning, D. Samanta, R. R. Althar, S. Pramanik and S. Dutta, Eds, IGI Global, 2021, pp. 38–63, DOI: 10.4018/978-1-7998-7701-1.ch003

[19] Y. Meslie, W. Enbeyle, B. K. Pandey, S. Pramanik, D. Pandey, P. Dadeech, A. Belay, A. Saini, "Machine Intelligence-based Trend Analysis of COVID-19 for Total Daily Confirmed Cases in Asia and Africa", in Methodologies and Applications of Computational Statistics for Machine Learning, D. Samanta, R. R. Althar, S. Pramanik and S. Dutta, Eds, IGI Global, 2021, pp. 164–185, DOI: 10.4018/978-1-7998-7701-1.ch009

[20] S. Pramanik, "Carpooling Solutions using Machine Learning Tools", in Handbook of Research on Evolving Designs and Innovation in ICT and Intelligent Systems for Real-World Applications, K. K. Sarma, N. Saikia and M. Sharma, IGI Global, 2022, DOI: 10.4018/978-1-7998-9795-8.ch002.

[21] G. Taviti Naidu, KVB Ganesh, V. Vidya Chellam, S Praveenkumar, D. Dhabliya, S. Pramanik, A. Gupta, Technological Innovation Driven by Big Data, in "Advanced Bioinspiration Methods for Healthcare Standards, Policies, and Reform", Hadj Ahmed Bouarara IGI Global, 2023, DOI: 10.4018/978-1-6684-5656-9

8

Agriculture Digitization: Perspectives on the Networked World

Thiruvenkdam[1], Akanchha Singh[2], Umesh Kumar Mishra[3], Nikhil Polke[4], Gabriela Michael[4], Dharmesh Dhabliya[5], and Ayman Noor[6]

[1]Department of Computer Science and IT, School of CS and IT, Jain (Deemed to be) University, India
[2]College of Agriculture Science, Teerthanker Mahaveer University, India
[3]Department of Agriculture, Sanskriti University, India
[4]Symbiosis Law School, Nagpur Campus, Symbiosis International (Deemed University), India
[5]Department of Information Technology, Vishwakarma Institute of Information Technology, India
[6]College of Computer Science and Engineering, Taibah University, Saudi Arabia

Email: thiruvenkadam.t@jainuniversity.ac.in; singhakanchha3@gmail.com; umeshm.ag@sanskriti.edu.in; nikhilpolke@slsnagpur.edu.in; dharmesh.dhabliya@viit.ac.in; anoor@taibahu.edu.sa

Abstract

The beginning of the chapter focuses on the existing state of digitization in the agricultural sector. It identifies the necessity and explains the rise of fresh AgTech businesses that focus on cutting-edge digital technology. The chapter seeks to cover the advantages of digitization in the agriculture sector as well as the post-COVID ramifications. In addition, it covers a variety of topics related to the digitization of agriculture in general, such as telematics, precision agriculture, blockchain, AI, etc. Finally, few of the major issues were explored, including connection, interoperability, portability, and the requirement of collaboration between the public and commercial sectors.

8.1 Introduction

8.1.1 Present situation

One of the main factors affecting the food and farming markets of current globe is the high need for food, restricted availability of natural resources, and uncertainty in agricultural production. The UN Department of Economic and Social Affairs forecasts that by 2035, the globe's population will raise from 8.2 billion to 10.4 billion people. It is predicted that this extreme increment in population would influence on entire food consumption. The predicted 15% rise in the globe's city population between 2023 and 2035 is a major constituent for the estimated expansion in food stock and need. In accordance with estimations, the globe will see a 35% reduction in water supply and a 35% reduction in agricultural land by 2045. By 2050, the output of cereals must grow by 3 billion tons in order to meet global demand. The need for cattle is predicted to climb by a similar margin by 2030, rising to a level of 200% by 2050.

About 600 million cultivated fields are situated in various regions of the earth, and the agricultural and food industry employs 30% of all population over the globe. Because these details and numbers are understandable, the agricultural business is one that will need continuous technological progress in order to feed the growing population for time to come. The paper predicts that over 821 million people still experience hunger as a result of a shortage of food supply, despite the fact that it has been shown that traditional technologies are enough to fulfill the expanding food demand. By itself, this figure raises the important issue of how to feed 10 billion populations by 2045 in a sustainable and equitable way.

Because of these factors, the globe has to quickly implement the newer, scaled-up, and transformed farming environment shown in Figure 8.1. Industry 4.0 [1] is in charge of transforming numerous industries by introducing innovations via these disruptive digital technologies. Moreover, the agriculture domain is not an exclusion which is free from this digital transformation technique. Way back, this transition did not appear practical because of the shortage of knowledge on the basic requirements and major drawbacks, like the shortage of access to I/Ps like seed, fertilizer, loans, or affordable costs for smallholder farmers.

Smallholder agriculturists are getting accustomed with the current growth in mobile communication via cell phones and various associated technologies because of the modern technology uprising. The growth of the new electronically handled agri-food frameworks is helped by all of these.

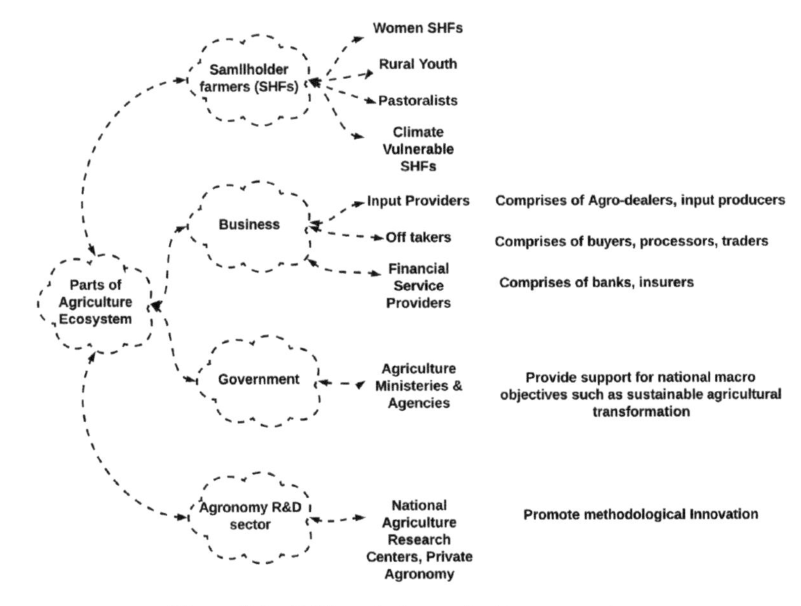

Figure 8.1 Differentiating agriculture ecosystems.

If these capabilities are ramped up, the agricultural sector might include a change that is only feasible via digitization.

The bulk of the people in rural areas, where agricultural work occupies the majority of daily activities, have been the main origin of the next generation of mobile connections. According to the World Bank Group (2016), the poorest 20% of people in developing nations have access to almost 70% of these mobile devices.

Worldwide, 35% of population has access to the net connection, and there exist continuous digital activities to link the remaining people, mainly those in rural parts of developed nations. The agri-food market system is anticipated to have significant alterations during the upcoming 10 years because of Industry 4.0. This is only feasible because of cutting-edge digital technologies like artificial intelligence.

The foundation of e-commerce and its consequence on the worldly agricultural-food business, alterations in the climate, and various other important components that can be to blame for the transfer in the framework are among the other factors that have contributed to this change. Through open and robust agri-food systems, the Food and Agricultural Committee is having major projects to boost food production, productivity, and sustainability. The major goal of these inventions is to fulfill the 2030 global hunger eradication goal set out by the US's Sustainable Development Goals. Only by implementing the shift using these digital agriculture technologies can these accomplishments be realized.

We could seek to ensure that digital agriculture is successfully implemented, which would then aid in recognizing potential possibilities and averting potential risks like the "digital divide" to the world's agricultural sector. This can only be done by imagining distinct future use cases and situations that primarily emphasize the various problems to varying degrees. Currently, the challenges posed by the digital divide do not exist in impoverished or rural regions. Enlarging the gap between various industries and economies is mostly due to digitization [2]. Early adopters vs. hesitant parties, gender, and levels of urbanization all play a role in this discrepancy. The continent of Africa is said to have had the fastest rise in terms of internet use, with usage rising from 3.3% in 2010 to 28.5% in 2022.

In spite of all the changes the globe is going through, there are still some significant problems that are to blame for the digital divide. The farming practices do not really reflect the benefits of modern technology. A few of these significant obstacles are the objectives of rising economies, a weak technology infrastructure, a lack of awareness of digital literacy and skills, restricted access to the Internet, and other services. However, these difficult circumstances in the agricultural industry demand that digital technology be strongly integrated into the sector.

Through some creative thinking processes, it is essential to figure out how to manage this new environment in the agricultural industry.

To find a solution, corporate executives, policymakers, and international organizations must think creatively rather than following the tired maxim of "business as usual." When the whole globe, which is being driven by the young population, is heading toward extension and effective digitalization to make up the huge technologies and inventions, the agri-food business has never been faced with such challenges. Bringing digitization to the agri-food industry remains difficult. In a number of disciplines, we must make specific adjustments. Changes to the agricultural system, rural economies, and communities, coupled with improved resource management, can demonstrate to be successful techniques to aid the transition extend its full ability.

8.1.2 Explore the definition

Through recent technology advancements, modern agriculture has seen efficiency, yield, and significantly higher profits than before. The first agricultural revolution, which occurred in about 10,000 BC and allowed humans to settle, allowed for the creation of the first civilizations and cultures. Mechanization and other developments were brought about by the revolutions between 1900 and 1930 in the food industry.

Additional revolutions, such as the green revolution [3] of the 1960s, encouraged the use of agro-chemicals to boost agricultural production levels while also creating new crop kinds that were more resistant than before. The establishment of newer technology to change the genetics of crop production occurred between 1990 and 2005. The human race has the potential to survive and flourish in the future with the aid of the electronic farming revolution. Industry 4.0 makes more data-intensive computing tools more widely available.

Simply said, digitalization is the act of using digital technologies that operate on digital data to automate different corporate operations and procedures. For instance, digital papers might be uploaded to the cloud and made available from anywhere at any time. Another example is using Google Sheets as a collaboration platform instead of the time-consuming process of sending an excel file to another individual when necessary. Another example of digitization is the ability of business intelligence systems to generate reports with a few clicks.

The last concept is called "digital transformation" [4], which is the use of digital technology to enhance productivity [5], consumer events, and business operations. Instead of conducting initiatives separately, it may be looked on as establishing a mentality to accomplish the customer-driven strategic business. It usually automates manual business processes and provides business with agility, a customer-focused approach, and data-driven choices.

8.1.3 Development of digital agriculture

If we examine the digital revolution across several sectors, the agri-food [6] industry is by far the most disruptive and transformative. Not only will this transition alter current agricultural practices and farmer behavior, but it will also have an effect on every link in the value chain for the industry. Along with suppliers, retail and processing firms may also make substantial modifications to how the items are processed, marketed, priced, and sold. The shift in resource management brought about by digitization in the agriculture sector will greatly improve it. The interconnectedness of the sector's subcomponents and the enormous reliance on the data provided would allow for far more intelligent and personalized use of this entire digital infrastructure in real time. Additionally, it is important to watch out for and handle each animal separately.

We may anticipate some traceability and teamwork in the agri-food sector's value chains at the most basic level of granularity. Systems with high levels of production are secure, proactive, and well-adapted to changing environmental conditions. In an ideal world, these kinds of systems would result from the changes that digitization would bring to the agricultural industry. Some of the expectations from these cutting-edge technologies include an improvement in food security, profit levels, and sustainability. According to certain market projections and forecasts, the agricultural business and the food sector will undoubtedly alter in the next 10 years because of the development of digital technology. Distinct technologies will play distinct roles and have varying effects on the various components of the agri-food industry's value chain.

The management of the consolidation of these technologies for each individual component of the agri-food value chain is often determined by two major aspects. The first is the degree to which that specific component is difficult, and the second is the degree to which that component of the chain – into which the technology will be integrated – is mature. Digital technologies may be categorized using the following framework that is dependent on the level of complexity of each separate technology and the maturity of the specific agri-food chain segment that the technology will be used in. According to a research study, the use of digitization tools and technology may enhance income creation and yield productivity to considerably greater levels.

According to predictions, by 2030, these changes may together add 14% to the increase of India's GDP, which at the present pace would amount to 15 trillion US dollars. Currently worth $7.8 trillion USD, the agricultural sector has seen significant change as a result of technology, which has also had an influence on other sectors. For the purpose of feeding the globe, the agri-food industry employs 40% of all people worldwide.

As a result, it is important to deal with some of the difficulties that the digital transformation process is posing. Some of them include insufficient levels of standardized digital solutions for issues with data, such as erroneous format settings. The characteristics of the data, as well as its use and access control procedures, are unknown. In this strange situation, large worldwide corporations are using digital transformation in agriculture to conduct their agri-food business. Firms like the government and public sector, which are a crucial component of this technique, are being impacted by unemployment difficulties as they work to address the socioeconomic challenges and issues surrounding the transformation of rural life. Making prudent use of these disruptive technologies is a significant challenge since they may have an influence on the social and economic environment capable of gaining an edge.

8.1.4 Reasons for why the digital revolution in agriculture is needed

By the year 2050, the population of the globe is likely to expand by a factor of 40%, reaching a staggering six billion. According to the FAO (UN Food and Agriculture Organization), there would be a need for 70% more food production to meet demand as a result of the growing world population.

While using the same amount of agricultural land or only 5% more, the annual demand for cattle and grains would rise by 1×10^9 tons and 2×10^8 tons, respectively. We need to address the ecological and regulatory difficulties brought on by the worldwide agriculture business if we are to satisfy this goal in a sustainable manner. Given that the majority of farmland is already cultivated, boosting farm yields is necessary to attain these objectives. The backing of digital transformational technologies in agriculture will help find a solution to the challenge of feeding the globe.

Agricultural efficiency, productivity, and profit levels that were previously unattainable via conventional farming techniques are now accessible because of digitalization. The trip began with mechanization (1900–1930), continued with the green revolution that led to the creation of crop types that are novel and resistant as well as the use of agro-chemicals, and ended with genetic modification.

Digitalization has the greatest influence on the agri-food industry of all sectors, altering every link in the value chain. With the aid of agricultural technology improvements, the repair of the soil's nutrition as well as the control of pests and disease have been on track.

Additionally, the shorter agricultural cycle due to genetic innovation and improvement allowed us to increase crop yields. The farmers must contend with several weather variations despite the rising need for food production. These include things like changes in rainfall patterns, intense weather, a lack of water, and rising temperatures, all of which prevent the best yields. Innovation and productivity improvements on farms are urgently needed to fulfill this need. By using digital technology, which provide farmers with information and boost agricultural output, all of these difficulties may be solved.

These days, customers are just as concerned about the items they buy as farmers are, and they demand high-quality goods as well as complete transparency from the fields to their doorsteps. With the use of digital technology, it is possible to attain this degree of traceability, which will ultimately increase farm value and please customers. More than in the past, modern agriculture requires innovation. To guarantee transparency and sustainability in the goods that customers are eating, the agri-food business is now dealing with several issues. Among these difficulties include the higher expenditure of commodities, the labor scarcity, and alterations in customer purchasing habits. Corporations in the agricultural industry are creating the need for the solutions needed to address these problems. As a consequence, agricultural technology has seen tremendous development, with investments totaling $7.5 billion over the past five years and $2.3 billion over the previous years during a 10-year period.

Some of the important innovations that have had an impact on revolutionizing the agricultural industry include AI, current greenhouse technology [7], precision agriculture [8, 9], and blockchain [10–14]. A future where farming is a high-technology career and has a beneficial influence on resource management is possible, thanks to the digital revolution. By doing so, food waste could be eliminated and the rising need for food production might be met. Digital technologies like AI, blockchain, and the IoT [15–19] are transforming the way agriculture is done and are having an effect on every link in the value chain.

8.2 Survey

8.2.1 Agriculture is being transformed by digitization and AgTech

The world is now closer than it was 10 years ago because of the global adoption of digital technology. The main driving force behind the transformation is that mobile devices are becoming cheaper, quicker, more efficient, and smaller every day. Farmers often follow the trend of other industries in attempting to fully use the introduction of these digital technologies. These next-generation technological developments benefit farmers by giving them more control over everyday operations, which eventually enables them to better manage the amount of resources they utilize on a regular manner. Consequently, the idea of digital technology is presently having a wide range of applications in many agricultural operations, including livestock management and farm equipment.

The idea of "smart farming" is anticipated to quadruple in popularity over the next five years as a result of the advent of digitalization. The world's growing need for food has made it more important than ever for novel approaches to be widely adopted in order to benefit the ailing, aging agricultural sector. Some of these systems, such as the one that pairs GPS-enabled [20] sowing equipment with sophisticated digital monitoring tools for harvesting crops, must change the farmer's viewpoint on yield. Similar to agriculture, horticulture has benefited from digital technology that offers an unheard degree of management accuracy, including effective disease control, efficient water usage, and optimization of total production.

Another instance of the utilization of digital technologies that illustrate growth rate and feed conversion ratios, coupled with livestock uniformity and quantity, is the use of e-animal tracing with remote sensing capacity together with automated weighing and drafting. Even while all of these technologies have advanced, only a select number have succeeded in breaking through to the public.

8.2.2 Indian AgTech

Global demand and food security have been impacted by the significant increase in demand for sustainably produced food caused by the growing global population. India's agriculture economy relies mostly on abundant rainfall and labor from people to expand. While most industrialized nations in the globe created the contemporary technology-enabled equipment and machinery, farmers in India are still forced to toil their land using antiquated agricultural methods.

Unfortunately, they lack the skills and the most up-to-date technology to have their needs handled quickly and effectively. However, as India's connection infrastructure develops, the idea of digital farming is starting to gain traction among Indian farmers in order to help them more effectively with a variety of tasks relating to their crop safety and productivity. This idea still has not really penetrated the market, despite its participation. The primary cause of this is the absence of sufficient adoption and training of modern, cost-effective agricultural methods among farmers. This is one of the key causes for the difficulty the agricultural sector has in making a greater contribution to the country's total GDP.

The Indian government has begun to support new-age digital tech-based companies in this field via a variety of initiatives for the same reason. Following the announcement of Mr. Narendra Modi's "Digital India" initiative in 2015, there is an increase in the quantity of agreements between major software companies and the government to create different digital technologies for agriculture. In 2019, the NITI Aayog joined with a significant software company to establish a framework for yield projections utilizing AI so that agriculturists may learn about real-time warnings. This is an example of a cooperative partnership of this kind. Cisco reported that virtually by 2021, 4.6 billion people will be online, which means that agricultural entrepreneurs

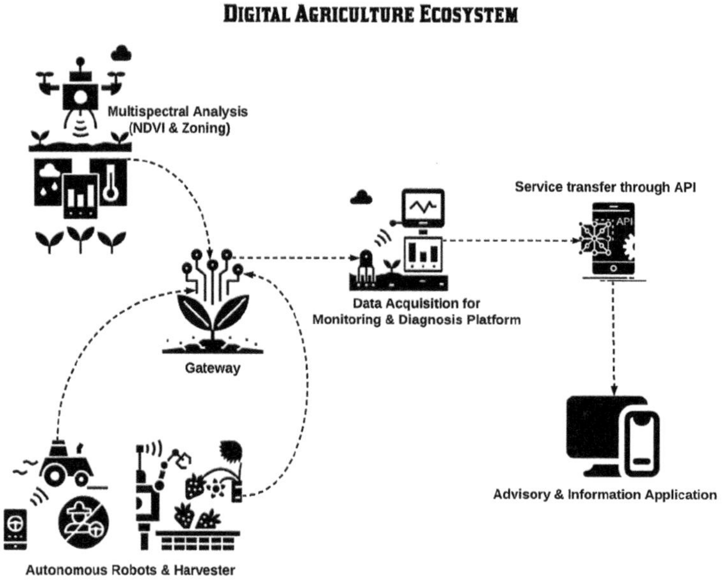

Figure 8.2 Ecosystem of digital farming.

throughout the globe will be able to unleash countless potential via ongoing innovation and change the face of agriculture like never before.

India's key goal is to maximize agricultural development via efficient frameworks, regulations, and programs, which may equip Indian agriculturists to allow a farming ecosystem (Figure 8.2) powered by technology. The Indian agricultural ecosystem (Figure 8.3) also faces issues with diminishing farm productivity, unsustainable resource use, a rise in the demand for safe and high-quality food, fragmented landholdings, and stagnant farm revenues.

The digitalization of agricultural-dependent businesses entails utilizing vast volumes of data from farming tools, gadgets, machineries, and soil, as well as applying contemporary technology to automate agriculture, which is essential for supplying the world's rising food demand. In addition to this, monitoring the health of crops and soil, remote sensing, geographic information systems, and livestock and farm management are some additional beneficial uses of digital technology in farms.

All of these apps provide the average farmer access to crucial information like crop recommendations and help acquiring insurance and loans, weather-related alerts, etc. Similar to this, the use of sophisticated analytics empowers farmers to make wise decisions about their resource consumptions and other crucial factors, eventually helping them to avoid making needless investments.

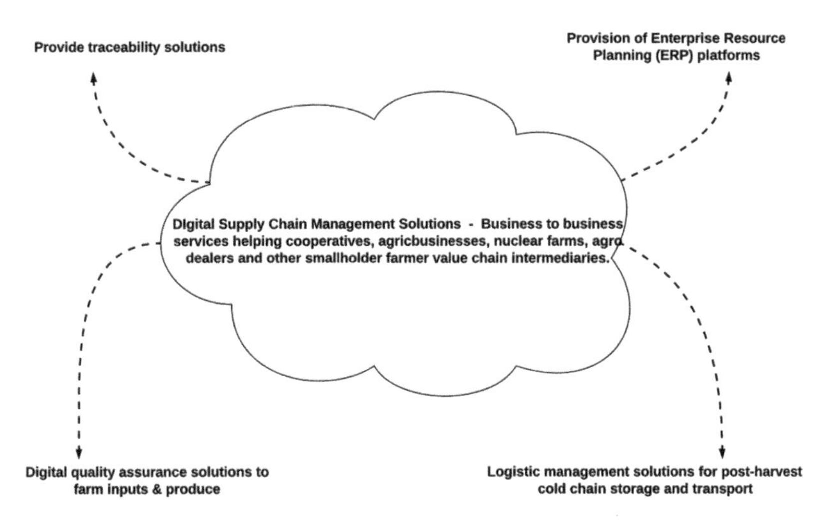

Figure 8.3 Agriculture supply chain management.

The agriculture industry in India is substantially flourishing solely as a result of the use of digital technology across various phases of farming, with over 500 startups in the agtech area. India can bring about a transformation in its agri industry at all stages – pre, post, processing, and logistical – with the use of low-cost, effective digital solutions from these businesses.

8.2.3 Landscape of agri-tech startups

Agricultural digitization initiatives were crucial for the acceptance and advancement of knowledge-based agriculture education. Through a variety of technologies and data analytics, researchers from all around the globe are promoting the concepts of smart and precision farming. Therefore, it seems to be very important to equip farmers with the most cutting-edge digital technologies. These technologies have the potential to spur more R&D, foster a culture of understanding, and hasten the adoption of cutting-edge technology in agricultural operations. In comparison to how they used to manage with traditional instruments, farmers can now deal with disasters and their aftermath far better, thanks to the most recent digital technology. Despite recent advancements in agriculture brought about by government engagement, it remains difficult to feed the country's steadily growing population. There has been a dramatic increase in the formation of various sorts of agtech-based companies in the nation since the introduction of the Digital India initiative in 2015.

The three primary types of them fall under the following categories:

The agricultural market

These types of companies are helpful in delivering services like local government vegetable market, weather forecast, price alerts, and agricultural news in their respective languages. Additionally, these firms assist in bridging the gap between farmers and the final consumer, hence eliminating the need for brokers or middlemen.

Digital farm management

These startups use cutting-edge technology like geospatial analytics, ML [21–25], and data analytics to provide a comprehensive perspective of their whole business, from manufacturing to inventory management.

Climate-smart precision agriculture

This startup category is in charge of offering improved data-driven agricultural choices utilizing on-farm sensing technology which combines with data science and AI to provide more accurate estimations for resource development and optimization. Generally speaking, these resources involve cost reductions for electricity, pesticides, water, and fertilizers to reduce product losses, automation, and receiving real-time notifications to alter their agricultural techniques appropriately. There has been a significant uptake of mobile technology and the Internet from the start of the Digital India initiative, which has unleashed a brand-new wave of development in this industry. As a consequence, the Indian government of India has already started a number of programs.

The India–Israel Innovation Bridge is one of the beneficial ones in this respect. It was created as a software system to allow bilateral interaction between Japan and Indian entrepreneurs, businesses, and tech hubs with the purpose of stimulating development in a variety of industries, including agriculture. The "Agriculture Major Competition" created by the Indian government's ministry of government to find fresh ideas and concepts in 12 different sectors was another helpful endeavor. It was created primarily to provide developing agri-based entrepreneurs in the digital domain with the infrastructure they need to become more integrated into the Indian agricultural industry. The challenge has many major issues, some of which include yield prediction, price forecasting, loss avoidance, the construction of an online marketplace, soil testing techniques, etc. By using robots, drones, and AGVs to increase production, the agricultural community will benefit from knowledge of digital farming solutions.

8.2.4 Startups using digital agtech

The significant agtech firms from across the globe that are employing digital technology to revolutionize the currently inert agricultural industry will be highlighted in this section.

1. Fasal is an AI-powered Internet of Things system with an Indian foundation that keeps track of various agricultural growth circumstances in order to help farmers make better choices. It makes use of AI and data science to provide a thorough overview and understanding of the methods farmers may take to maximize their production and gain control over the coming crop.
2. In order to deliver smart agriculture remedies for the optimization of agricultural techniques to boost yields, crop standards, and earnings in a sustainable manner, the computer behemoth IBM teamed with Yara, a worldwide pioneer in digital agricultural remedies from Norway. For providing farmers with fresh perspectives, they concentrated on several farm improvement verticals utilizing ML, AI, and data connected to farming. To increase traceability and supply chain proficiency and to find solutions to food ingenuity, food wastage, and sustainability, they developed a blockchain-based network of food chain systems. They start by supplying hyperlocal weather estimates along with real-time applicable suggestions, catered to the basic requirements of separate farms or crops.
3. One of the issues that prevent farmers from accessing financial services like financing and crop insurance is their inability to keep a proper transaction record of how much they get via intermediaries. Some time ago, a company called BanQu stepped in to help fix this issue. It is a SaaS [26] service platform built on the blockchain that generates a decentralized digital log of every transaction, assisting agriculturists in establishing credit and holding processors responsible. Every agriculturist got a digital payment via the blockchain network of BanQu rather than cash.

8.2.5 COVID Post-scenario

The COVID-19's devastating effects have been felt all around the globe. It had a significant negative impact on businesses as well as the whole country's economy. It was evident from the nation's total GDP's percentage decline in the first quarter of 2023. A hardest-hit nation, Italy, had a GDP decline of 16.5%, while India saw a GDP decline of 25.8% from the first quarter of 2019 to the similar quarter in 2020. The nation should initiate depending on a technologically oriented farming framework, which will revive the stagnant farming sector if it is to get back on track. The digitization of agriculture may prove to be the agro-based industry's best asset during this global epidemic. If a replacement for human-centric jobs is required, it may assist in managing transportation, supply chains, and finances much better.

In the farming supply chain, they serve cooperatives, agriculturists, whole-sellers, importers, exporters, contract agricultural businesses, retailers, and also consumers. Simply said, the idea or use of digital technology makes managing agricultural operations simpler in a number of ways. It handles standards, storage, sales, and everything else in between starting at the plantation. Because of this, agri-tech firms using enterprise resource planning platforms may end up receiving significantly more attention than the others in this market. Agriculturists are now more literate, cautious, equipped, and knowledgeable about the functionalities and advantages of incorporating digital technology in their agricultural operations; digitalization is transforming the agricultural scene.

The need for modernizing the traditional applications has grown over time. In order to create a hybrid framework that offers smarter data insights through agricultural analytics dashboards as well as the power of sensors, unmanned aerial drones, and robotics down to the micro-level, the majority of digital technologies are combining. In addition, analytics on the nutritional requirements of the farms, as well as water and crop timings, are supplied to the agriculturist for profit-making farming mechanisms.

8.2.6 COVID-19 Accelerates farming industry digitization

Despite having a devastating global effect, COVID-19 has given modern farmers the ability to use social and digital media channels to market their goods immediately to consumers. It has accelerated corporate digitization throughout the agri-sector at a breakneck pace. Some of the market-driven reasons are the rising necessity for farming food commodities, a change in consumer tastes toward higher grades of food safety and quality, and a shortage of workers during COVID-19. There is a noticeable growth in the quantity of data providing the demand estimation for every product, as millions of consumers increasingly depend on online grocery ordering. Therefore, COVID [27] has quickly accelerated the age of data in food yield and utilization, eventually necessitating the merging of more digitally dependent implementations in real time.

8.2.7 Various facets of digital agriculture technologies

In accordance with a research by the food and agriculture organization, the world's food consumption is anticipated to increase by 50% until 2050. Agriculture technology development is crucial to the future of food production. Providing farmers with affordable AgTech solutions can enable them to maintain their competitiveness in the next years. A significant portion of our worldwide food system will be digitized as smartphone ownership increases globally. This will facilitate farmers to acquire the assistance they need to preserve resources in a manner that is far more sustainable.

The market for global digital agriculture is expected to swell notably in the succeeding years because of the factors including fast reducing farm land, an increasing population having a rising demand for higher yield of fruits and grains, etc. Additionally, the rise in the number of green farms worldwide is moreover propelling the market for digital farming. Various agriculturists used digital agriculture to expand the productivity of their arable lands as a result of the higher order for higher-quality products. Thus, by 2025, it is predicted that the global agricultural market would be valued $12.8 billion USD. Increased command for mechanization and control systems, sensing devices, antennas, access points, and robots are other significant aspects that contribute to the market share's overall growth.

8.2.8 Agriculture telematics

The use of monitoring devices to find the location of the equipment for managing reasons is a major component of telematics services. The need for digital farming techniques has increased as a result of the rise in telematics-based applications in agriculture. A primary factor operating the expansion of the digital farming industry is the increasing challenge for such management services. Telematics is a system that collects valuable information from farm machinery or equipment operation and transmits the user data to the tracking dashboard in real time for additional applicable decisions for agriculture after appropriate analysis utilizing big data tools. In order to bring about efficiency in agricultural processes like sowing, it often includes robots, autonomous super-tractors, or a fleet of combine harvesters with cameras and sensors linked to the digital platform.

8.2.9 Vertical farming inside

The practice of growing agricultural products piled one on top of the other in a closed, regulated environment might be characterized as indoor vertical farming. According to [28], the arrangement of the shelves indicates that they are vertically erected in the same space as the field where the plants once grew. This approach has the potential to boost yields, reduce supply chain trip time, and get around the problem of scarce land. It is most helpful in metropolitan settings and cities where space is constantly at a premium. The crops mostly are hydroponic, where crops are grown in a nutrient-rich dish of water, and few of them are special in that they do not require soil to flourish. The majority of the time, artificial glow light is employed in installations for these vertical farming techniques. In addition to the benefits already described, this kind of farming also has the considerable benefit of using less water. Typically, it uses around 55% less energy than traditional farms. In addition to these factors, the use of robots for tasks like planting, transportation, harvesting, packaging, etc., lowers the cost of labor.

8.2.10 Automated farming

Generally speaking, farm automation is a technology that expands farming productivity and automates the cycle of upraising crops or animals. Numerous businesses are working on robotics innovation to create drones, robotic harvesters, and autonomous tractors, seeding, and automated watering robots in response to the rising need for automation. Farm automation technology's main objective is to handle simple, routine operations on farms in a manner that saves both time and money.

8.2.11 Animal agriculture technology

Aside from the significant technologies that were previously highlighted, the traditional livestock business is one of the underserved and ignored sections of the agricultural domain. According to [29], managing livestock has been recognized to sustain the operations of dairy farms, poultry farms, and other associated agribusinesses. The manager of such services is expected to oversee the relevant staff, ensure adequate animal feeding and care, and keep track of the financial records. But as technology advanced, this business also transformed agriculture to support a variety of crucial procedures and goals. These technical advancements were found to be the latest advances in genetics, digital technology, and nutritional innovations to have appropriate track of and manage cattle to increase and enhance productivity.

The installation of lone wearable sensors on cattle, which may assist to monitor their daily activities as well as health-related concerns, is one example of bringing something into the mainstream. The enormous volume of data gathered by these sensors is often transformed into insightful, usable knowledge that enables the producers to examine quickly and make swift, efficient management choices.

8.2.12 Precision farming

The agricultural industry is undergoing unprecedented change as a result of a variety of innovative technologies in farms. Companies that specialize in precision agriculture are developing new technologies to assist farmers maximize yields by maintaining control over the crop farming's changeable characteristics. The criteria mainly consist of things like soil and climatic situations, insect stress, moisture levels, etc. In addition, it aids farmers by offering precise methods for cultivating and planting crops in order to reduce costs and boost productivity.

8.3 The Main Digital Technologies used in Agriculture

A survey of several technologies in precision agriculture is given in this section.

8.3.1 Field inspection

Crop health monitoring

The normalized difference vegetation index is a well-accepted technique utilized to assess the health of crops using satellite data and drones. To do the necessary computations, it takes into account the various light wavelengths that lie in the visible and non-visible ranges. Simply put, this technology enables the farmer to identify crop variability as well as get an evaluation on the overall health of their crops.

Monitoring and prediction of yield

There are several methods to get data on the farm's productivity thanks to advancements in precision agricultural technology. Some of these include drones and satellite imaging, as well as sensors installed in tractors and harvesters used by farmers. All of these assist the farmer in obtaining crucial information on grain yield, moisture content, fertilization, and enable them to decide more wisely on their harvesting strategy. Figure 8.4 illustrates the key details in this respect.

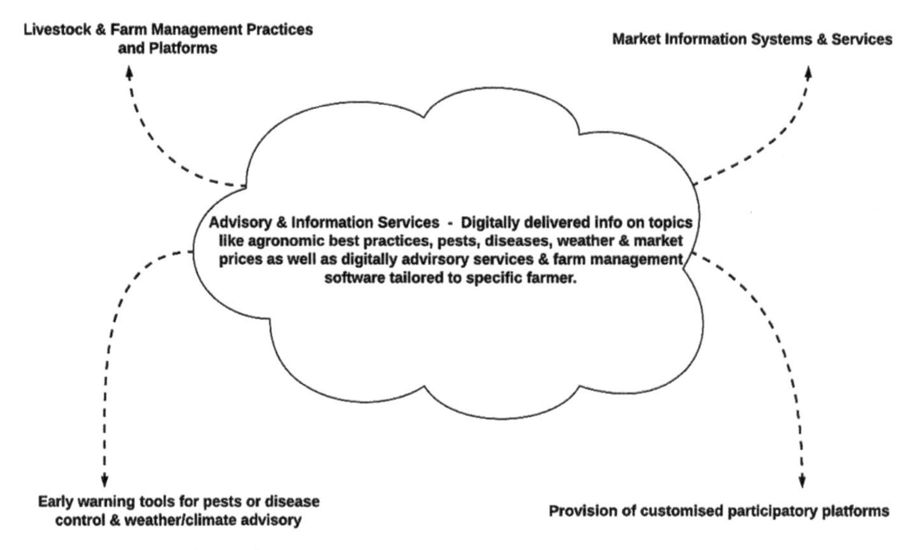

Figure 8.4 Agriculture advisory and information services.

Harvest scouting

The technologies mentioned before and this precision agricultural technology are pretty comparable. The single feature that sets it apart is the use of mobile devices like phones and tablets for gathering and tracking crucial information about crops like insect populations, the spread of invasive species, etc.

Pest, weed, and disease detection

It is observed that drones are playing a significant part in supplying information relating to the identification of weeds, pests, and disease from several applications of precision agriculture. The hyperspectral [30] imaging camera which is installed in it for scanning has made all of this feasible. It offers comprehensive information about the farm that a regular camera cannot supply.

Climate and soil conditions

Some plant- and ground-dependent sensors are used to gather vital data regarding soil and water in order to determine the crop's quality. The salinity, pH, soil texture, nutritional status, and organic matter levels are the main variables that these sensors primarily assess. The farmers often get weather reports through their cell phones from the weather stations.

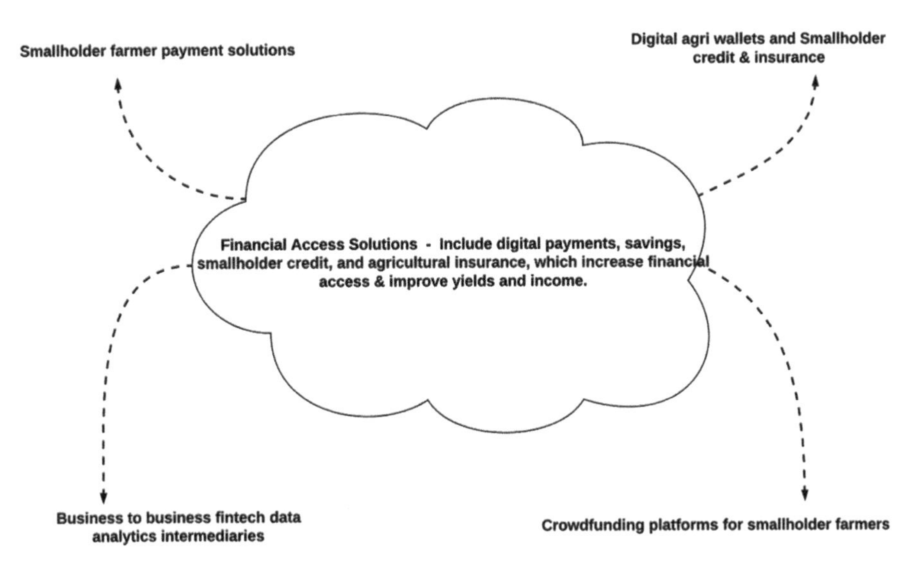

Figure 8.5 Agricultural digital fiscal assistances.

8.3.2 Management of data

Systems for managing farms

This type of software is a system that often has numerous hardware devices connected with it to assist farmers in managing their agricultural output effectively. As illustrated in Figure 8.5, the data gathered from various systems are combined into a single platform where it is processed and analyzed to aid agriculturists in controlling and improving their all-inclusive agricultural processes.

In addition, it simplifies crop production and harvesting schedules and makes it easier to monitor and analyze all connected processes. To put it briefly, the program allows the farmer access to financial data and environmental circumstances as depicted in Figure 8.6, allowing them to better manage record-keeping, business requirements, and farm output monitoring features.

This kind of software is demonstrating that it is a valuable and effective service to assist farmers worldwide, in a variety of environments and situations, by enabling them to operate more productively while economizing money, time, and resources. The comprehensive collection of features that

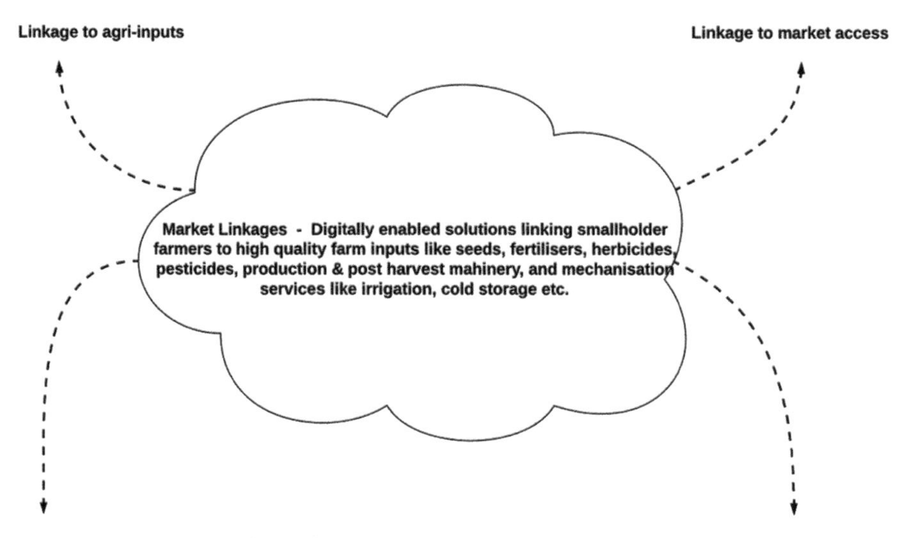

Figure 8.6 Market connections for farming.

are offered with it moreover enables specialist businesses in the agri-sector to better meet the demands of the agriculturist.

Variable rates applications (VRA)

A technique known as variable rate application focuses mostly on the automated implementation of materials such as seeds, herbicides, and pesticides depending on information gathered from sensors, maps, and several tracking sources. It entails the combination of many precision agriculture-related technologies.

Some of them include things like satellite images [31], hyperspectral imaging, harvester and tractor equipment, etc. In a nutshell, it aids in maximizing the use of seeds, fertilizers, and various pertinent chemicals.

8.3.3 Farm equipment automation

Farming robots

Robots have been used for many years in a variety of sectors to automate certain operations. Robots have proved helpful in a variety of applications since agriculture began to be automated. A significant utilization is the automation of weed control, for which 2 US businesses have created robots that utilize their cameras to recognize weeds in real time.

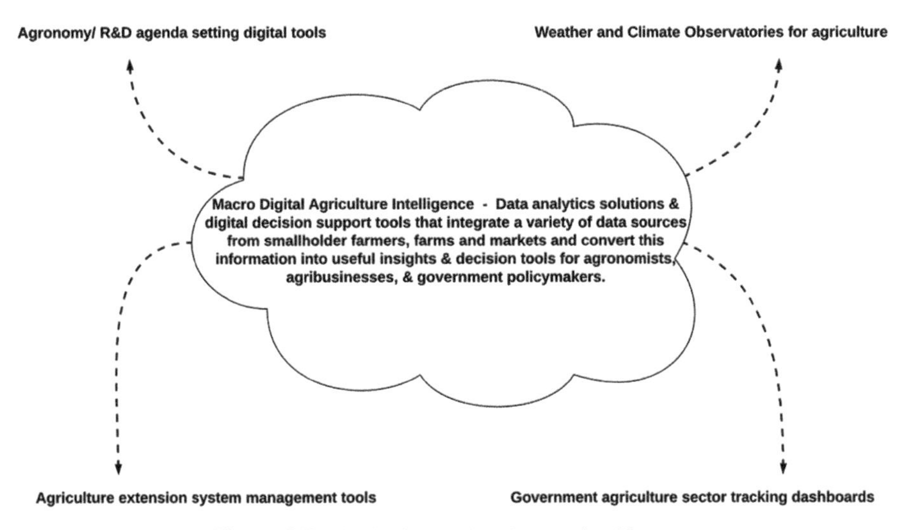

Figure 8.7 Agriculture-related macrointelligence.

Counseling methods using GPS

The GPS, sometimes referred to as the global positioning system, is a technology that is mostly utilized in the farming industry to direct automated equipment and vehicles including tractors, drones, and autonomous harvesters, among others.

Telematics

The technology mentioned in the preceding section is essential for creating machine-to-machine communication between the sensors and hardware used in the automation technique in agriculture. This may be seen, for instance, in the way that weeds are controlled on farms.

In this way, anytime a camera mounted on a piece of equipment finds a patch of weed, it notifies linked equipment that uses the information to locate the weed and pull it from wherever it is required.

Prudent planting

The precision planting idea seen in Figure 8.7 presents an automated approach that improves the depth control, seed spacing, and other aspects of seed planting. This idea often relies on other precision agriculture-based technology to give the best conditions for planting.

Blockchain

Simply said, a blockchain is a system that distributes digital information without allowing it to be replicated. There will only be one owner of each block of a certain piece of data. In order to address challenges like inefficient supply chains, food fraud, and traceability in the present food industry, blockchain [32] has significant potential. Its distinctive distributed form aids in the development of a variety of superior goods by guaranteeing the transparent verification of procedures and products. For all the activities for which this technology tend to be beneficial, food traceability stands out. The whole food business is at a dangerous point because of the prevalence of fake and perishable food. If a food-borne illness affects the general public's health, identifying the root of the problem becomes crucial.

However, relying heavily on paperwork to keep track of things takes a lot of time. Blockchain can help in this situation. Its design makes sure that each stakeholder in the supply chain produces and privately communicates the data points needed to create an accountable and traceable system. With no danger of tampering, this large quantity of points having labels indicating ownership may be readily recorded. As a consequence, it is possible to track the whole route of the food item in real time.

The blockchain technology enhances the current agro market by assisting in the speedy food tracking mechanism, creating a network ledger, and balancing market price. The cost of purchasing and selling agricultural products is determined by the opinions of the various market participants. Additionally, it allows confirmed transactions to be safely shared with each participant in the farm produce supply chain, resulting in a system with a high degree of transparency.

Machine intelligence

The quantity of data being gathered by remote sensors, satellites, UAVs [33], and other devices for various reasons has increased at an unheard-of rate with the emergence of newer digital technologies in the agricultural sector. These many goals mostly comprise activities such as monitoring of soil health, plant health, humidity, temperature, etc. The extensive data that is gathered will only be useful when the agriculturist comprehends the numerous aspects of running a farm better. In order to make sense of it, different artificial intelligence algorithms are used to process this data. This allows farmers to get a thorough grasp of numerous processes as a whole. By 2035, it is anticipated that the globe would require producing 50% more food than it does today. AI-enabled solutions may assist farmers in this circumstance by offering more production and more sustainability with the resource at hand.

8.3.4 Farming and agriculture using AI

Unmanned tractors

The agricultural industry will see a large increase in the quantity of driverless tractors, harvesters, and various pieces of equipment due to significant expenses in the growth of autonomous vehicles in recent times. Different AI algorithms are used to train these autonomous or driverless tractors to carry out a variety of duties in general. Some of these include finding where the fields have been plowed, among other things.

Robotic agriculture

Similar to how humans can effortlessly accomplish a variety of chores, AI [34] firms are creating farming-related robots. The majority of AI businesses today are creating robots that can automate a variety of duties for various agricultural methods. These jobs usually include inspecting the quality of the crops and looking for undesired plants, among other things.

Control of pest infestations

The worst enemies of farmers are often seen as pests like grasshoppers and locusts. Before the crops are harvested and kept for human use, they cause devastation around the world. To deal with it, fewer artificial intelligence firms have created a framework that supports agriculturists by alerting them via smartphone when these pests are likely to approach certain acreage or agricultural areas. These businesses are employing the most recent and previous satellite photographs of the similar region as the data for the artificial intelligence techniques in order to deliver valuable information.

Soil and crop health tracking

Fast deforestation and declining soil quality have been issues for a while. To deal with this, a German tech startup currently created an image recognition application based on deep learning, which may assist in recognizing nutrient deficits and possible soil faults, as well as the existence of pests and plant diseases, if any. Similar to this, Trace Genomic, a machine-learning-based business, offers soil analysis assistances to agriculturists, so they may respond appropriately.

Benefits of using AI in agriculture

By a category of instructions linked to water management, timely and optimal harvesting, insect assaults, nutrient management, and crop rotation, AI performs a variety of vital responsibilities in the agricultural industry.

Additionally, they aid in the study of agricultural sustainability, the forecasting of weather, and the inspection of lands for the existence of pests or plant diseases along with insufficient plant nutrition. Moreover, AI helps agriculturists in automating their agricultural apparatus or equipment to enhance crop yields and crop quality. With characteristics like precipitation, temperature, wind speed, and sun radiation, it helps farmers to comprehend crucial data insights for their crops. Modern farmers benefit from the use of AI in the agricultural industry in a variety of ways. It aids individuals in comprehending information such as wind speed, temperature, precipitation, and sun radiation. All of these data pieces' historical significance is often taken into account for examination by AI algorithms to provide superior results. The benefits of AI in agriculture are as follows:

- AI as a whole offers an effective approach to cultivate, grow, harvest, and market the crops.
- It aids in identifying crops with problems.
- AI boosts agriculture-based companies.
- It boosts crop management initiatives.
- It effectively addresses issues like weed and pest infestation, weather changes, etc.

8.4 Challenges

To satisfy the expectations of the stakeholders, cooperatives have long thought about incorporating digital technology into their functioning agricultural model. The technical breakthroughs and commercial and regulatory obstacles need to be effectively handled in order to seize this opportunity. The primary difficulty always resides in the correct practical execution in the majority, notwithstanding the increase of technology innovation. The requirement for proper working techniques to address various human difficulties inside the teams and governance is another major issue. Laws and rules governing data ownership are another issue to consider. It does not, however, seem to be a roadblock to further advancement. Numerous agtech businesses are working hard to integrate automation and intelligence to enhance agricultural production choices, but they all often confront a number of major hurdles. Each and every company in this industry faces the four main problems stated below.

8.4.1 Connectivity

Access to the Internet throughout the nation is one of the major obstacles preventing the widespread use of digital technology. This is because creating a heterogeneous Digitech-dependent agriculture framework will only be a pipe dream without the Internet.

Therefore, it is important to provide lower-cost internet remedies that may assist agriculturists in understanding how digitalization is affecting their output. Under the direction of Prof. Rajib Neogy, one such effort in this area was the "Gram Marg Solution for Rural Broadband" project at the IIT Bombay. The group came up with a clever and "indigenous" technique that makes use of idle TV airwaves to backhaul data from neighborhood Wi-Fi clusters to supply internet access. The project's primary goal was the ease with which the necessary infrastructure could be put up. The solution has so far been pilot-implemented in 25 communities.

8.4.2 Interoperability

The poor interoperability of diverse agtech-based systems and applications is the second biggest problem after connection. This difficulty primarily demands for the use of hybrid systems rather than the proprietary ones that were utilized for a long time, since they may operate in sync with one another to give a multiplicity of applications. These hybrid systems may help agriculturists get many advantages out of a little expenditure while lowering the danger of insufficient support assistances.

8.4.3 Portability

The problem of data portability often arises while working with different businesses. Data should be moved from one platform to another without degradation in order to boost agricultural efficiency. The whole purpose of this approach is to eliminate any obstacle brought about by data access limitations in several important areas like banking and energy.

8.4.4 Public–private sector collaboration

The involvement of the public and private sectors in commercialization and research and development is a final, crucial issue that should be taken into account in connection to the growth of digital technologies in farming. The incapacity of the public R&D organizations to identify suitable moments to transfer the outcomes of their job to private enterprises having the required consent so that they may create commercial technology which farmers may use is one of the major obstacles in this market. Therefore, there is a critical need for change in this area.

References

[1] V. Sunder M., A. Prashar, G. L. Tortorella and V. R. Sreedharan, "Role of Organizational Learning on Industry 4.0 Awareness and Adoption for Business Performance Improvement," in IEEE Transactions on Engineering Management, doi: 10.1109/TEM.2023.3235660.

[2] G. Singh and J. Singh, "A Fog Computing based Agriculture-IoT Framework for Detection of Alert Conditions and Effective Crop Protection," 2023 5th International Conference on Smart Systems and Inventive Technology (ICSSIT), Tirunelveli, India, 2023, pp. 537–543, doi: 10.1109/ICSSIT55814.2023.10060995.

[3] A. Kumar and B. Bhowmik, "Rice Cultivation and Its Disease Classification in Precision Agriculture," 2023 International Conference on Artificial Intelligence and Smart Communication (AISC), Greater Noida, India, 2023, pp. 200–205, doi: 10.1109/AISC56616.2023.10085072.

[4] S. Bhadula and S. Sharma, "Personalized Self Adaptive Internet-of-Things Enabled Sustainable Healthcare Architecture for Digital Transformation," 2023 Advanced Computing and Communication Technologies for High Performance Applications (ACCTHPA), Ernakulam, India, 2023, pp. 1–6, doi: 10.1109/ACCTHPA57160.2023.10083335.

[5] C. Huang, S. Zhou, J. Li and R. G. Radwin, "Allocating Robots/Cobots to Production Systems for Productivity and Ergonomics Optimization," in IEEE Transactions on Automation Science and Engineering, doi: 10.1109/TASE.2023.3270207.

[6] J. K. Kiruthika, T. Yawanikha, A. Kannika, C. Megaranjani, R. S. Rajalakshme and D. Kalaiyarasu, "User Interface of Blockchain-Based Agri-Food Traceability Applications," 2023 4th International Conference on Signal Processing and Communication (ICSPC), Coimbatore, India, 2023, pp. 258–261, doi: 10.1109/ICSPC57692.2023.10125777.

[7] T. T. H. Ngoc, S. Pramanik, P. T. Khanh, "The Relationship between Gender and Climate Change in Vietnam", The Seybold Report, 2023, DOI 10.17605/OSF.IO/KJBPT

[8] P. T. Khanh, T. H. Ngọc, and S. Pramanik, Future of Smart Agriculture Techniques and Applications, in Advanced Technologies and

AI-Equipped IoT Applications in High Tech Agriculture, Eds, A. Khang, IGI Global, 2023.

[9] T. H. Ngọc, P. T. Khanh and S. Pramanik, Smart Agriculture using a Soil Monitoring System, in Advanced Technologies and AI-Equipped IoT Applications in High Tech Agriculture, Eds, A. Khang, IGI Global, 2023.

[10] Reepu, S. Kumar, M. G. Chaudhary, K. G. Gupta, S. Pramanik and A. Gupta, Information Security and Privacy in IoT, in Handbook of Research in Advancements in AI and IoT Convergence Technologies, Eds, J. Zhao, V. V. Kumar, R. Natarajan and T. R. Mahesh, IGI Global, 2023.

[11] V. Veeraiah, V. Talukdar, Manikandan K., S. B. Talukdar, V. D. Solavande, S. Pramanik, A. Gupta, Machine Learning Frameworks in Carpooling, in Handbook of Research on AI and Machine Learning Applications in Customer Support and Analytics, Eds, Md S. Hossain, R. C. Ho and G. Trajkovski, IGI Global, 2023.

[12] V. Jain, M. Rastogi, J. V. N. Ramesh, A. Chauhan, P. Agarwal, S. Pramanik, A. Gupta, FinTech and Artificial Intelligence in Relationship Banking and Computer Technology, in AI, IoT, and Blockchain Breakthroughs in E-Governance, Eds, K. Saini, A. Mummoorthy, R. Chandrika and N.S. Gowri Ganesh, IGI Global, 2023.

[13] S. Pramanik, "An Effective Secured Privacy-Protecting Data Aggregation Method in IoT", in Achieving Full Realization and Mitigating the Challenges of the Internet of Things, Eds, M. O. Odhiambo and W. Mwashita, IGI Global, 2022, DOI: 10.4018/978-1-7998-9312-7.ch008

[14] S. Pramanik and S. Bandyopadhyay, Identifying Disease and Diagnosis in Females using Machine Learning, in Encyclopedia of Data Science and Machine Learning, Eds. John Wang, IGI Global, 2023, DOI: 10.4018/978-1-7998-9220-5.ch187.

[15] S. Pramanik and S. Bandyopadhyay, Analysis of Big Data, in Encyclopedia of Data Science and Machine Learning, Eds. John Wang, IGI Global, 2023, DOI: 10.4018/978-1-7998-9220-5.ch006

[16] R. Bansal, B. Jenipher, V. Nisha, Jain, Makhan R., Dilip, Kumbhkar, S. Pramanik, S. Roy and A. Gupta, "Big Data Architecture for Network Security", in Cyber Security and Network Security, Eds, Wiley, 2022, https://doi.org/10.1002/9781119812555.ch11

[17] D. Pradhan, P. K.Sahu, N. S. Goje, H. Myo, M. M.Ghonge, M., Tun, R, Rajeswari and S. Pramanik, "Security, Privacy, Risk, and Safety Toward 5G Green Network (5G-GN)", in Cyber Security and Network Security,Eds, Wiley, 2022, DOI: 10.1002/9781119812555.ch9

[18] R. Anand, J. Singh, D. Pandey, B. K.Pandey, V. K. Nassa and S. Pramanik, "Modern Technique for Interactive Communication in LEACH-Based Ad Hoc Wireless Sensor Network", in Software Defined Networking for Ad Hoc Networks, M. M. Ghonge, S. Pramanik and A. D. Potgantwar, Eds, Springer, 2022, https://doi.org/10.1007/978-3-030-91149-2_3
[19] S. Choudhary, V. Narayan, M. Faiz and S. Pramanik, "Fuzzy Approach-Based Stable Energy-Efficient AODV Routing Protocol in Mobile Ad hoc Networks", in Software Defined Networking for Ad Hoc Networks, M. M. Ghonge, S. Pramanik and A. D. Potgantwar, Eds, Springer, 2022, https://doi.org/10.1007/978-3-030-91149-2_6
[20] K. Dushyant, G. Muskan, A. Gupta and S. Pramanik, "Utilizing Machine Learning and Deep Learning in Cyber security: An Innovative Approach", in Cyber security and Digital Forensics, M. M. Ghonge, S. Pramanik, R. Mangrulkar, D. N. Le, Eds, Wiley, 2022, https://doi.org/10.1002/9781119795667.ch12
[21] S. Pramanik, An Adaptive Image Steganography Approach depending on Integer Wavelet Transform and Genetic Algorithm, Multimedia Tools and Applications, 2023, https://doi.org/10.1007/s11042-023-14505-y.
[22] R. Jayasingh, J. Kumar R.J.S, D. B. Telagathoti, K. M. Sagayam and S. Pramanik, "Speckle noise removal by SORAMA segmentation in Digital Image Processing to facilitate precise robotic surgery", International Journal of Reliable and Quality E-Healthcare, vol. 11, issue 1, DOI: 10.4018/IJRQEH.295083, 2022
[23] S. Pramanik, K. M. Sagayam and O. P. Jena, Machine Learning Frameworks in Cancer Detection, ICCSRE 2021, Morocco, 2021.
[24] S. Pramanik, M. G. Galety, D. Samanta, N. P Joseph, Data Mining Approaches for Decision Support Systems, 3rd International Conference on Emerging Technologies in Data Mining and Information Security, 2022.
[25] D. Samanta, S. Dutta, M. G. Galety and S. Pramanik, A Novel Approach for Web Mining Taxonomy for High-Performance Computing, The 4th International Conference of Computer Science and Renewable Energies (ICCSRE'2021), 2021, DOI:10.1051/e3sconf/202129701073.
[26] R. Bansal, A. J. Obaid, A. Gupta, R. Singh, S. Pramanik "Impact of Big Data on Digital Transformation in 5G Era", 2nd International Conference on Physics and Applied Sciences (ICPAS 2021), doi:10.1088/1742-6596/1963/1/012170, 2021.
[27] D. Pandey G. A. Ogunmola, W. Enbeyle, M. Abdullahi, B. K. Pandey and S. Pramanik, "Covid-19: A Framework for Effective Delivering of

Online Classes during Lockdown", Human Arenas, 2021, https://doi.org/10.1007/s42087-020-00175-x

[28] R. Ghosh, S. Mohanty and S. Pramanik, "Low Energy Adaptive Clustering Hierarchy (LEACH) Protocol for Extending the Lifetime of the Wireless Sensor Network", International Journal of Computer Science & Engineering, vol. 7, issue-6, pp. 1118–1124, 2019.

[29] S. Pramanik, AJ Obaid, N M, and SK Bandyopadhyay "Applications of Big Data in Clinical Applications", Al-Kadhum 2nd International Conference on Modern Applications of Information and Communication Technology, *AIP Conference Proceedings* 2591, 030086 (2023), https://doi.org/10.1063/5.0119414

[30] S. Pramanik, S Joardar, OP Jena, AJ Obaid, "An Analysis of the Operations and Confrontations of Using Green IT in Sustainable Farming", Al-Kadhum 2nd International Conference on Modern Applications of Information and Communication Technology, *AIP Conference Proceedings* 2591, 040020 (2023). https://doi.org/10.1063/5.0119513

[31] S. Pramanik, Niranjanamurthy M. and S. N. Panda, Using Green Energy Prediction in Data Centers for Scheduling Service Jobs, ICRITCSA 2022, Bengaluru, 2022.

[32] V. Vidya Chellam, V. Veeraiah, A. Khanna, T. H. Sheikh, S. Pramanik and D. Dhabliya, A Machine Vision-based Approach for Tuberculosis Identification in Chest X-Rays Images of Patients, ICICC 2023, Springer

[33] S. Praveenkumar, V. Veeraiah, S. Pramanik, S. M. Basha, A. V. Lira Neto, V. H. C. De Albuquerque and A. Gupta, Prediction of Patients' Incurable Diseases Utilizing Deep Learning Approaches, ICICC 2023, Springer

[34] B. K. Pandey, D. Pandey, S. Wairya, G. Agarwal, P. Dadeech, S. R. Dogiwal and S. Pramanik, "Application of Integrated Steganography and Image Compressing Techniques for Confidential Information Transmission", in Cyber Security and Network Security, https://doi.org/10.1002/9781119812555.ch8, Eds, Wiley, 2022.

[35] B. K. Pandey, D. Pandey, V. K. Nassa, A. S. George, S. Pramanik and P. Dadheech, Applications for the Text Extraction Method of Complex Degraded Images, in The Impact of Thrust Technologies on Image Processing, Nova Publishers, 2023.

[36] B. K. Pandey, D. Pandey, V. K. Nassa, A. S. Hameed, A. S. George, P. Dadheech and S. Pramanik, A Review of Various Text Extraction Algorithms for Images, in The Impact of Thrust Technologies on Image Processing, Nova Publishers, 2023.

9

Cucumbers in PH Disease Monitoring using an IoT-based Mobile App

Dinesh Singh[1], Feon Jaison[2], Vineet Saxena[3], Shilpa Sharma[4], Saurabh Raj[4], Ankur Gupta[5], and Sabyasachi Pramanik[6]

[1]Department of Agriculture, Sanskriti University, India
[2]Department of Computer Science and IT, School of CS and IT, Jain (Deemed to be) University, India
[3]College of Computing Science and Information Technology, Teerthanker Mahaveer University, India
[4]Symbiosis Law School, Nagpur Campus, Symbiosis International (Deemed University), India
[5]Department of Computer Science and Engineering, Vaish College of Engineering, India
[6]Haldia Institute of Technology, India

Email: dinesh.ag@sanskriti.edu.in; Feon.jaison@jainuniversity.ac.in; tmmit_cool@yahoo.co.in; shilpasharma@slsnagpur.edu.in; saurabhraj@slsnagpur.edu.in; ankurdujana@gmail.com; sabyalnt@gmail.com

Abstract

Agriculture is essentially a given in most nations because of how much or how little it affects their economies, from creating jobs to contributing to national revenue. Modern agricultural practices now include the novel and well-liked practice of PH farming. High yields may be achieved by using a PH that can provide water and nutrients in precisely the right quantities and under regulated conditions. Crops in PHs need to be closely monitored since stagnant air and a lack of air movement encourage the growth of insects and material loss. As a result, this chapter suggests a prototype IoT-dependent disease-tracking system for a farm/PH implementation. To detect the beginning of the illness in cucumbers, the prototype is created and put to the test. This research's first emphasis is on employing a NodeMCU and Raspberry-Pi-based hardware model to identify the serious cucumber infections in PHs. Using a specially created smartphone application, the farmers will be informed of the choices to be taken and any significant changes in the detected parameters.

9.1 Introduction

Our nation's economy relies heavily on agriculture, which employs roughly 60% of the population and accounts for around 26% of India's GDP. Recent worldwide climate changes have had a significant impact on Indian agriculture. The temperature has risen by around 2–3°C as a result of these climatic fluctuations, which has an effect on agricultural methods. On the other side, pests are causing illnesses that impact crops. In the tropics and semi-tropics, it has been estimated that weeds, illnesses, and pests cause around 40% of the losses. Other regions of the globe also experience these same circumstances but to varied degrees of severity. These issues need effective technology in order to significantly increase agricultural output, farming sustainability, and profitability under a variety of environmental situations. The PH technology is one such new technology.

With the aid of PH technology, plants may experience an appropriate atmosphere that encourages acceptable, unhindered development. Protecting the crops from unfavorable weather conditions brought on by changes in wind, rainfall, coldness, severe temperature, more radiation, illnesses, and insects is essential. Therefore, crop production in a PH is an appropriate option that also enables precision farming and gets around space constraints and the negative effects of climate change. Depending on the dimension of the PH structure, crop production economics, and revenue produced, crops that will be grown in PHs are carefully selected. To be more precise, Tamil Nadu, India has seen greater popularity for PH farming of the higher costing vegetable crops capsicum, cabbages, potatoes, and ladies finger. Additionally, they are raising marigold flower crops and hybrid cucumber varieties on a huge scale.

The Cucurbitaceae family of plants, which includes the cucumber (*Cucumis sativus*) is extensively farmed in India. According to [1], diseases in PH-cultivated cucumber cultivars are characterized by a high illness rate and quick and frequent infection. When a disease first appears, crop cultivation will be reduced, and product quality will suffer. Even for professionals, it is almost impossible to identify or categorize different plant diseases by sight alone. Additionally, manual inspection calls for qualified individuals to regularly check the crops. When done at large-scale farms, this will be time-consuming and expensive. Therefore, the creation of a disease tracking framework is very necessary to track alterations in ecological variables and spot the start of plant illnesses.

Precision agriculture, which can manage differences in production in a farm and maximize monetary returns, is recently emerging. This is achieved by using sensing and communication technologies to make strategic choices for farm management via the utilization of automatically programmed data collection, documentation, and application of the obtained data. IoT-enabled wire-free sensor networks are important for precision farming because they provide the following benefits:

- The capacity to efficiently manage irrigation, fertilizer supply, and weather in order to provide the optimal crop conditions, increase production efficiency, lower costs, and provide real-time information.
- Possibility of checking a greater region with more thorough sample.
- The capacity to create an agricultural system that is highly automated and has enhanced resolution.

IoT-dependent remedies in PH farming thus have a broader use. The primary emphasis of this study is on identifying the ecological factors and signs that contribute to the main cucumber illnesses, including red spider mites, white flies, aphids, and potassium shortage.

On the basis of monitoring the environmental factors that will cause the sickness to manifest, the diseases beginning may be anticipated. To find the illness infection, it is also possible to observe the outward physical signs. Table 9.1 lists the optimum ranges of these environmental factors for a variety of disorders.

Various illnesses' earliest physical signs may be recognized by analyzing a plant's photos. Early detection will stop the spread of the disease and the enormous loss that follows. Figures 9.1 and 9.2 display sample infected photos.

Red spider mites: Early leaf fall and spotted foliage. On the top leaf surface, there is little light mottling. The foliage seems discolored, pallid, and sometimes golden.

- White flies: Adult and nymphal whiteflies, especially on younger leaves' lower surfaces, may cause distortion, wilting, and leaf loss.
- Aphids: Plants seem feeble and yellow in hue. Young fruit buds and floral buds will fall. Young shoots perish and the afflicted area looks black due to the growth of mold germs.

Table 9.1 A spectrum of environmental factors that are favorable for certain illnesses.

	Diseases				
Parameters	**Red spider mites**	**White flies**	**Aphids**	**Lack of boron**	**Lack of potassium**
Temperature	30.24°C	19–30°C	25–28°C	30°C	30°C
pH	4.8–6.7	4.7–5.6	4.1–5.4	< 6.5	< 6.3
Humidity	81.36%	97%	78%–90%	90%–99%	95%–99%
EC	4.3 dS/m	4.2-5.4 dS/m	4.0–5.5 dS/m	3.8 dS/m	3.5–4.0 dS/m
Light intensity	9.7% (D)	>12% (D)	15% (D)	7% (D)	12.7% (D)

Figure 9.1 Examples of pests seen on diseased cucumbers include: (a) white flies; (b) red spider mites; and (c) aphids.

- A lack of boron results in the deformation of younger leaves and the development of a wide yellow edge around the edges of the older leaves. Fruits that are mature have stunted growth and mottled longitudinal yellow streaks.
- Potassium deficit: Potassium deficiencies first manifest as a slowing of development, shoot die-back, and brittle branch formation. On leaves, a few potential symptoms include a fading bluish-green hue, notably in the leaf interveinal areas, and leaf edges and tips exhibit a dull overall chlorosis. The mature leaves have tip burning and marginal scorch, and the leaves are curled downward or upward.

Data collection, processing, analysis, and decision-making are all included in the plan of an automated framework to diagnose illnesses. Vital characteristics including humidity, temperature, light intensity, pH, soil moisture, and electrical conductivity are monitored using sensors. Additionally, image-dependent monitoring of the farming fields is now possible because of the development of digital technology and the accessibility of inexpensive cameras. Tracking these variables can aid in effective agriculture by giving a hint of when to apply fertilizers, pesticides, and water. Additionally, the ability to identify plant illnesses using image processing has benefits since it reduces the labor required to monitor extensive agricultural fields and enables early

Figure 9.2 Examples of disease-infected cucumbers include those with deficiencies in boron and potassium.

disease diagnosis. In this effort, the camera is connected to a Raspberry Pi, and the analog sensors are connected to a NodeMCU. To determine the proper symptoms and related disorders, the photos are processed and examined. An IoT gateway is used to transfer the sensed data and the derived results into the cloud, enabling real-time data display. IoT is nothing more than a network of physically linked things, like sensors or gadgets that will communicate with one another through the Internet and the cloud. IoT gateways allow physical devices like microprocessors, microcontrollers, sensors, and actuators to communicate over the Internet. Farmers will get the conclusions and corrective recommendations from ThingSpeak through MIT App Developer2.

The following sections make up this chapter. Section 9.1 contains the introduction. For the relevant background information, see Section 9.2. The suggested method for illness detection is described in Section 9.3. Section 9.4 discusses the system implementation and illustrates the simulation outcomes that go along with it. Section 9.5 explains the potential future work, and Section 9.6 wraps up this chapter.

9.2 Background

Crop production is done on a commercial basis under cover in more than 50 different nations. The United States of America has over 4000 acres in total covered with PHs used mostly in floriculture. Additionally, it contributes an extra 2.8 billion US dollars in annual revenue, and it is anticipated that the area covered by PHs would grow significantly. The PHs that are mostly utilized for growing vegetable plants like potato, chili, cabbages, brinjals, ladies finger, and bean kinds are situated on an area of 25,000 ha in Spain and 18,500 ha in Italy. In UK, PHs are built utilizing simple tunnel constructions without the use of sophisticated environmental control technology, and their covering is often made of UV-stabilized polyethylene film. The PH market is well-established in Canada for both the flower and off-season vegetable industries. Tobacco, cucumber, and capsicum are the primary vegetable crops grown in Canadian PHs.

The globe imports vegetables and fruits that are cultivated in polyhouses built in the Sweden. The Swedish have the most sophisticated PH industry, which covers 89,600 acres. Glass frames are used in the construction of Dutch PHs in order to withstand the year-round, oppressive cloud cover. Cucumbers and tomatoes rank among Saudi Arabia's most significant crops, accounting for roughly 94% of the nation's output. Israel, the world's greatest supplier of cut flowers, covers a vast array of cut flowers and vegetables on its 15,000 ha of land, whereas Turkey covers 10,000 ha. Evaporative cooling is the most widely used cooling technique for all of them. In around 1000 acres, PHs are mostly built in Egypt employing plastic-covered tunnel-like structures. A natural ventilation system is used to control the humidity and temperature levels. Cucumbers, tomatoes, melons, and other nursery plants are some of the principal crops planted in these PHs. Asia's two biggest markets for PHs are China and Japan. China has advanced in PH technology far more quickly than any other nation. In their PHs, in addition to vegetables, fruits like pineapple, watermelons, oranges, guava, litchis, and lemon are also grown. The majority of Chinese PHs is reportedly unheated, and the country employs straw matting to resist the heat. In Japan, there are more than 40,000 ha of PHs, of which 7500 ha are used for fruit orchard production. Japan used to cultivate a lot of vegetables and fruits in its PHs, and the majority of the country's vegetable needs were satisfied by their production. Even South Korea has more than 21,000 acres of land covered with PHs for the cultivation of fruits and flowers. Under severe weather, PHs are a good option for crop production. PHs is the finest option for crop production in areas with harsh winters like Canada and Russia and harsh, unpleasant summers like the UAE, Kuwait, and Israel. PHs are constructed in the Philippines, where there is an abundance of rainfall, to produce plants and commodities. Numerous nations with mild climate conditions also utilize PHs. Overall, it is clear that PH horticulture is performed all over the world and in all conceivable meteorological and climatic circumstances, and that decision support systems for such environments are developing.

A number of contributions have been made to the development of agricultural decision-support systems.

In the State of Washington, two separate WSNs in agricultural applications were created. The initial wireless sensor network design focuses on site-specified implementations like weather networks, whereas the second design is for on-field applications like temperature sensors to prevent frost. In order to service the intended applications, the designs, notably AgWeatherNet for the first and AgFrost-Net for the second, were successfully implemented. In order to assure scheduled plant watering, a WSN-based system was developed in Scotland for wireless data capture and aggregation of many meteorological factors together with soil moisture. The use of several wireless data gathering devices fueled by sunlight in soil moisture readings has improved irrigation efficiency.

To regulate the fertilizer dosage and application rate, Hegedus et al. [2] created an automatic fertilizer employing unit. Through the input module, it detects environmental factors and communicates them to the DSS to determine the fertilizer amount and spreading pattern. The choice is determined after assessing the real-time data detected.

The development and deployment of a multi-hop wireless sensor network framework with 65 nodes were placed over the course of six months in a vineyard by Kasimati et al. [3]. The system's data was used to aggregate the damage from heat and frost during wine production. Additionally, it has been utilized to combat insect and fungal issues. Zang et al. [4] developed a plan to combat the Phytophtora disease that affects potato crops. By constantly monitoring the temperature and humidity, their device lessens the impact of this sickness. A PH system with appropriate protocols was developed by Rani et al. [5] to cope with environmental variations and guarantee proper energy consumption. It was utilized to keep track of the CO_2 concentration, relative humidity, and interior brightness.

For the creation of seedlings for PHs, a decision-driven system was created. In order to obtain the proper dry weight of the seedlings while moving them, a growth monitoring model was implemented. This was accomplished by regularly taking temperature-based decisions and continually tracking the temperature, light conditions, and other environmental factors. In order to accomplish the anticipated sapling development within the allotted time frame, it assisted in maintaining and managing the situations within the PH.

Dearden et al. [6] created an AgriSense decentralized framework in India to track temperature, relative humidity, leaf wetness, and soil moisture. The main goal of this experiment was to identify the disease known as bud necrosis virus that affects groundnut crops. A WSN system was created by Deka and Pradhan [7] utilizing eKo nodes to track the soil moisture constituent in orange fields.

Ladakh also has several PHs built to increase the growing season for crops from three to eight months. Both vegetable and fruit seedlings are produced in PHs in the northern plain areas in an effort to both increase the quality of the seedlings and gain early market share. Additionally, PHs will be a key factor in making floriculture operations profitable.

Even if other regions of the globe have made many achievements, India has a lot of room for developing and deploying these technologies. Additionally, IoT-enabled [8] technologies are urgently needed. Therefore, the focus of this study is on using an IoT-based mobile app to identify illnesses based on environmental factors and symptoms present on leaves.

9.3 System for Detecting Disease

The suggested disease detection system scans the area using a variety of sensors and cameras to gather environmental data and pictures, which are then processed. Through two steps, the proper identification of illness onset, disease types, and pests that spread infections is accomplished. The gathering of ecological parameter estimates, efficient farm coverage by sensors, and transferring the gathered data to the cloud are all part of the initial phase. Based on the data that was taken out of the cloud, the second step involves data processing, analysis, and decision-making at the server. The suggested method is shown in Figure 9.3.

9.3.1 Data gathering

Environmental conditions

Using an innovative, low-cost, low-power IoT sensor network [9], the environmental parameters are gathered. The server stores the gathered data. The system architecture comprises a number of sensors deployed in the agricultural field that are associated with NodeMCU. They compile the physical data gathered from the environment. Along a safe IoT hub, the data is sent to the cloud. This makes it possible to see and comprehend data in many ways.

Images

The act of collecting an image from a hardware-dependent source for processing is called image acquisition. In this work, the digital camera and Raspberry Pi [10] are interfaced. Through Raspberry Pi, the data that has been gathered is sent to the cloud.

Processing of data

Through Raspberry Pi, the environmental characteristics are sent to the cloud. Nevertheless, the photographs are altered, the details are taken out, and they

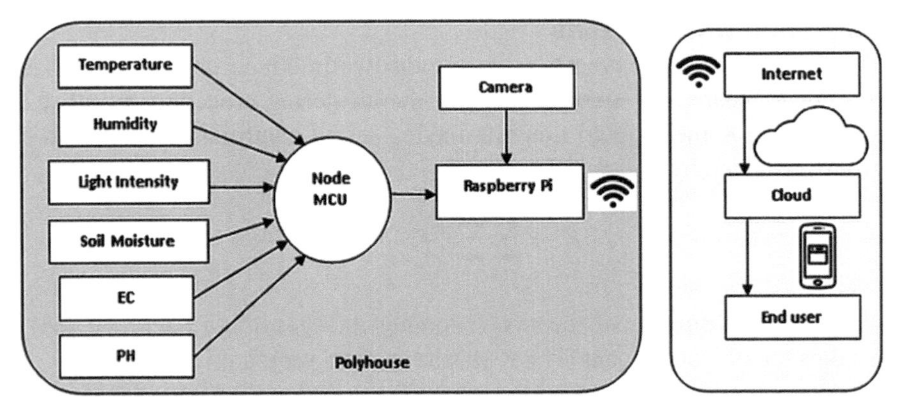

Figure 9.3 System block diagram for detecting disease.

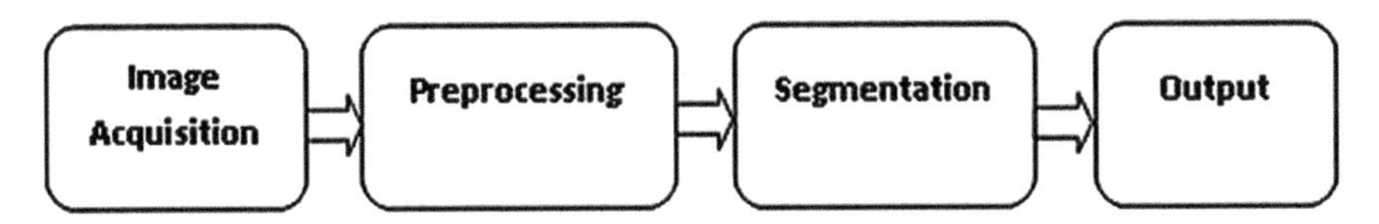

Figure 9.4 Steps in image processing.

are then uploaded to the cloud. Figure 9.4 depicts the image processing [11] approach.

Pre-processing

Image scaling is a part of the picture pre-processing [12]. Additionally, it incorporates color conversion or color space conversion. Red, green, and blue alone were insufficient to render color distinctions as precisely as the human visual system in color image processing. The HSI (hue, saturation, and intensity) color space is created by converting RGB [13] to this format. The proportion of pathogen infection in leaves is indicated by HIS [14].

Segmentation

Image segmentation is a technique for dividing a picture into its components, which will be comparable in certain aspects or properties. To extract the object features or related data from the digital photos, the pixels with the same attribute will be labeled identically. The outcome will be a collection of contours that, when integrated, will depict the whole picture. The color, intensity, or texture may be used to explain the similarities. *K*-means clustering [15] is used in this study to divide the photos depending on the characteristics.

Clustering with *K*-means

The *K*-means method creates clusters with well-defined physical boundaries. For *K*-means, the *n* components of the clustering procedure are e1, e2, and e3. The *K*-means goal function having initial centroids, divided into *K* groups, is given by

$$K = \sum_{k=1}^{l} \sum_{m=1}^{p} \| f_i - E_j \|,$$

where *K* is the number of clusters, *E* denotes the centroid of the group, and *f* denotes the object element. The *K*-means method protocol is described.

1. Set up the centroid.
2. Find the distance between the centroid and the object element.
3. Give the object's components to the group with the shortest distance.
4. Calculate the group's new mean.
5. Until a threshold level is achieved, repeat steps 2–4.

The *K*-means clustering function is shown below.

definition of centroid_histogram (clt):

In order to generate a histogram, count the number of various clusters. For example, numLabels = np.arrange(0, len(np.unique(clt.labels_) + 1) (hist, _) = np.histogram(clt.labels_, bins = numLabels).

Make the histogram normalized so that it adds to 1.

Hist = Hist.astype("float"), and Hist /= Hist.sum()

Return the histogram

Cloud storage

Utilizing Python [16] and Lua Script, the sensor data and features are gathered and thereby posted in ThingSpeak (a cloud platform). A free Internet of Things application is called ThingSpeak. It is an API, which aids in storing and retrieving data over the Internet utilizing the HTTP protocol [17]. You may see the uploaded values as graphical notations. As a result, the time series data collected from the field may be efficiently represented. Channels that have data fields, location farms, and a status field may be included in ThingSpeak activity. With the use of MATLAB [18] code, the detected data

Table 9.2 Variety of thresholds for various illnesses.

Diseases	Threshold range		
	Hue	S	I
Red spider mites	≥ 228	95–156	45–70
White flies	≥ 235	≥ 236	≥ 232
Aphids	207–225	218–234	90–100
Potassium deficiency	215–235	190–225	72–126

can be published on a channel, analyzed, and evaluated, and any differences may be addressed with tweets and other warnings. The channel offers tabs for data import and export as well as private and public viewing. The read/write API and the channel ID are necessary for publishing the data detected by the nodes on-field.

Decision-making and data analysis

The cloud is used to extract segmented data, which is then further classified using thresholding. The collected color values from the clusters are contrasted to a threshold. The threshold is painstakingly adjusted (after several trials across varied photos) to accurately detect the illnesses as indicated in Table 9.2.

1. The choice is taken in light of the analysis, and the end users (farmers) are informed of the countermeasures through a mobile app. MIT App Developer created the app. It is a tool for block-dependent programming that makes it simpler to create Android apps.

9.4 Installation of the System and Results

In a polyhouse next to Thiruporur Sipcot, Payyanur, the whole prototype is created and tested. Figure 9.5 depicts the general configuration of the experiment.

The sensor data are transferred to the cloud platform ThingSpeak. Figure 9.6 displays the O/P estimates for sensors (such as humidity, soil moisture, temperature, EC, pH, and light intensity) connected to a NodeMCU [19] utilizing the ArduinoIDE [20].

The output of ThingSpeak, which provides graphs of temperature, light, soil moisture, humidity, pH, and electrical conductivity measurements, may also be analyzed by users. The charts shown in Figures 9.7–9.10 show parametric data from different sensors and a decision about disease infection.

Figure 9.5 Overall experimental configuration.

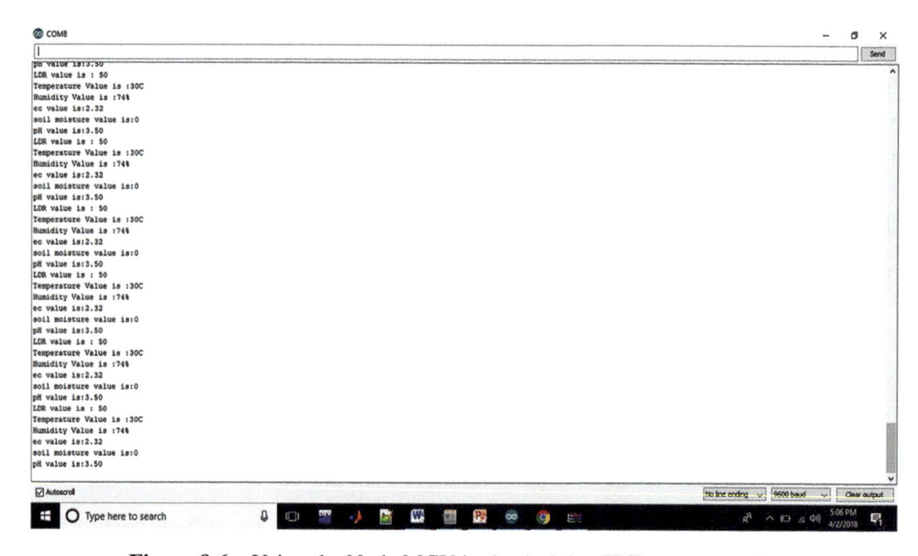

Figure 9.6 Using the Node MCU in the Arduino IDE, get sensor data.

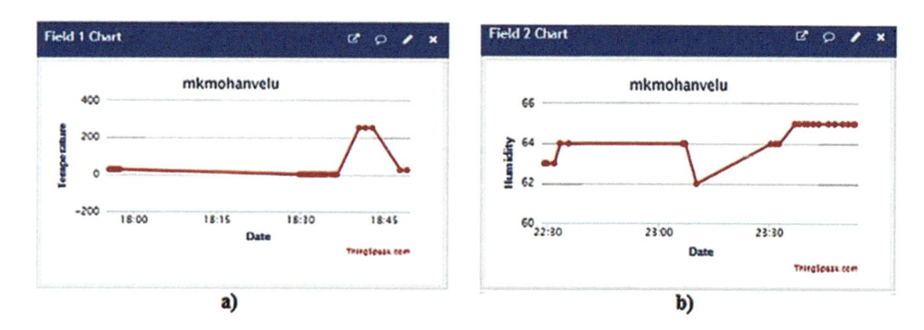

Figure 9.7 Temperature and humidity are the sensed parameters.

Figure 9.8 Sensed parameters: (a) pH; (b) soil moisture.

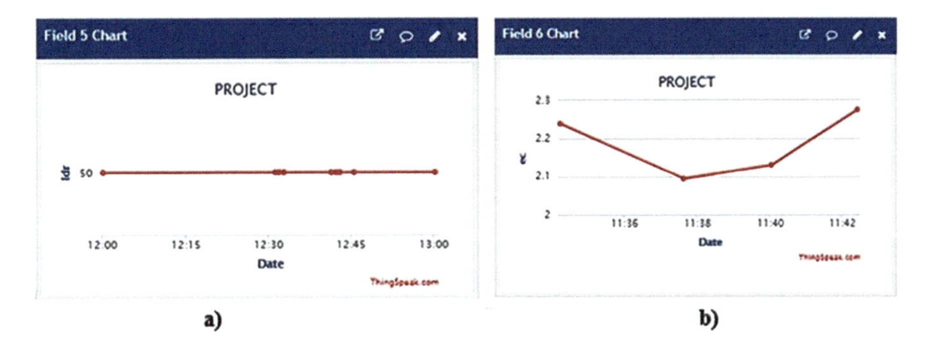

Figure 9.9 LDR and EC are the sensed parameters.

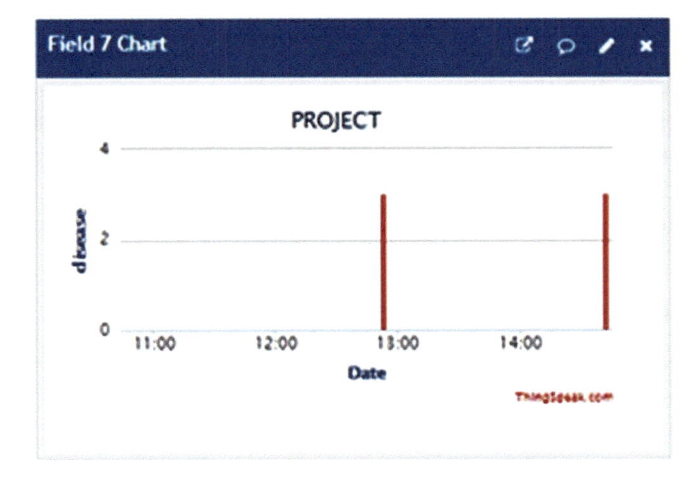

Figure 9.10 Choice about illness infection.

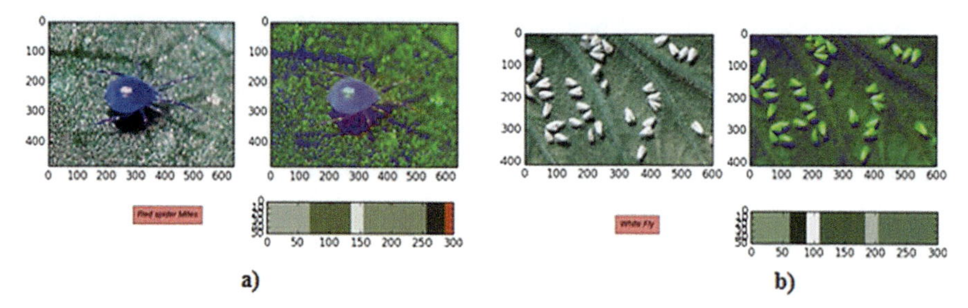

Figure 9.11 Classification of segmented images: (a) red spider mites; (b) white flies (white).

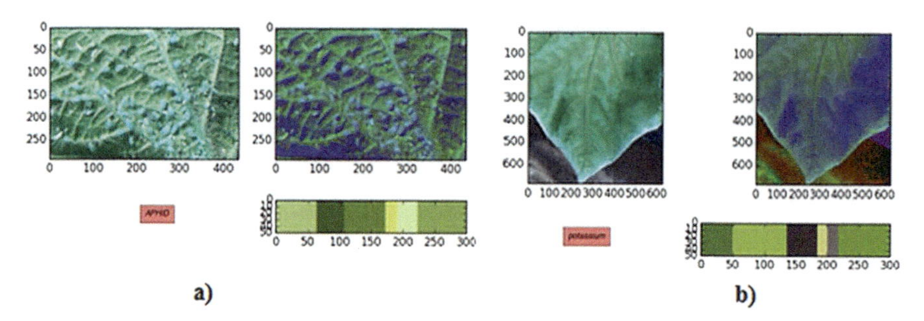

Figure 9.12 Classification findings for segmented images: (a) aphids (whitish yellow); (b) potassium deficiency (yellow).

Analysis of images:

Figures 9.11 and 9.12 depict the segmented picture that is used to distinguish between five cucumber illnesses, including red spider mites (red), white flies (white), aphids (whitish yellow), and yellow disease (potassium shortage).

Smartphone application:

The Android [21] application displays the detected parametric values and the specifics of the diagnosed condition. The information seen in the mobile app is retrieved from ThingSpeak.

Figure 9.13 displays the alternatives that are accessible.

9.5 Future Directions for Research

The created on-farm disease estimation framework would aid in preventing the incorrect assumptions provided by agriculturists in certain situations. Additionally, it may lower their reliance on local chemical agents and lower the costs associated with agricultural production. By eliminating the usage of pesticides via other preventative measures, the system may also pave the

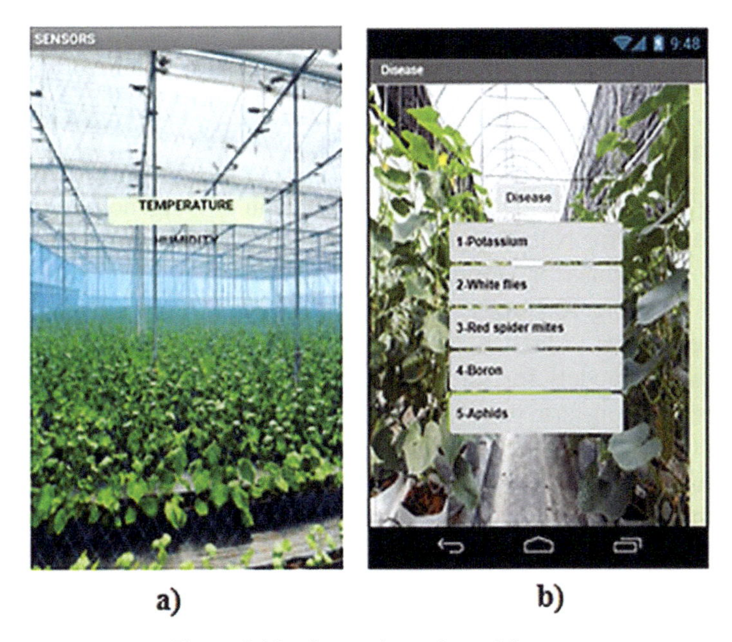

Figure 9.13 Screenshots of a mobile app.

road for organic farming. By enhancing the segmentation process, utilizing ML [22–26] and expanding the system's capacity to identify every potential illnesses of the cucumber plant, more system improvement may be achieved. The system may also be modified to fit any other important plants. This module may be improved further by being transformed into a mobile drone that can photograph the whole field from any angle. Future smart cities will be built around such technologies, improving the quality of our main foods.

9.6 Conclusion

The suggested approach makes it possible to detect illness in PH cucumbers early on. Measurement of sensor parameters, picture capturing, real-time realization, and data analysis are all made possible by IoT-based architecture. Using NodeMCU and ESP8266, environmental parameters including temperature, humidity, soil moisture, pH, EC, and light intensity are monitored and uploaded to the cloud. Cameras connected to a Raspberry Pi take pictures of the visual details, which are then segmented utilizing *K*-means clustering. Outcomes from segmentation are transferred to the cloud. The thresholding approach is used for further categorization. In order to effectively monitor user behavior and take appropriate corrective action in PH, the gathered data are shown via a specially developed application on end users' mobile devices. The system's prototype is created and put to the test in a real-world setting.

References

[1] Singh, S., Singh, D.R. (2023). Andaman's Indigenous and Exotic Vegetables for Nutrition and Entrepreneurship. In: Singh, B., Kalia, P. (eds) Vegetables for Nutrition and Entrepreneurship. Springer, Singapore. https://doi.org/10.1007/978-981-19-9016-8_12

[2] Hegedus, P.B., Maxwell, B.D. & Mieno, T. Assessing performance of empirical models for forecasting crop responses to variable fertilizer rates using on-farm precision experimentation. Precision Agric 24, 677–704 (2023). https://doi.org/10.1007/s11119-022-09968-2

[3] Kasimati, A., Psiroukis, V., Panoutsopoulos, H., Mouseti, S., Mylonas, N., Fountas, S. (2023). Management Information Systems and Emerging Technologies. In: Vougioukas, S.G., Zhang, Q. (eds) Advanced Automation for Tree Fruit Orchards and Vineyards. Agriculture Automation and Control. Springer, Cham. https://doi.org/10.1007/978-3-031-26941-7_9

[4] Zang, C., Kong, T., Liang, B. et al. Evaluation of imide substance from Streptomyces atratus PY-1 for the biocontrol of Phytophthora blight. Eur J Plant Pathol 165, 725–734 (2023). https://doi.org/10.1007/s10658-023-02648-5

[5] Rani, N., Kaur, G., Kaur, S. et al. Plant Growth-Promoting Attributes of Zinc Solubilizing Dietzia maris Isolated from Polyhouse Rhizospheric Soil of Punjab. Curr Microbiol 80, 48 (2023). https://doi.org/10.1007/s00284-022-03147-2

[6] Dearden, A.E., Wood, M.J., Frend, H.O. et al. Visual modelling can optimise the appearance and capture efficiency of sticky traps used to manage insect pests. J Pest Sci (2023). https://doi.org/10.1007/s10340-023-01604-w

[7] Deka, M.K., Pradhan, Y. (2023). Trends of Tea Productivity Based on Level of Soil Moisture in Tea Gardens. In: Saraswat, M., Chowdhury, C., Kumar Mandal, C., Gandomi, A.H. (eds) Proceedings of International Conference on Data Science and Applications. Lecture Notes in Networks and Systems, vol 551. Springer, Singapore. https://doi.org/10.1007/978-981-19-6631-6_15

[8] D. Samanta, S. Dutta, M. G. Galety and S. Pramanik, A Novel Approach for Web Mining Taxonomy for High-Performance Computing, The 4th International Conference of Computer Science and Renewable Energies (ICCSRE'2021), 2021, DOI:10.1051/e3sconf/202129701073.

[9] R. Bansal, A. J. Obaid, A. Gupta, R. Singh, S. Pramanik "Impact of Big Data on Digital Transformation in 5G Era", 2nd International Conference on Physics and Applied Sciences (ICPAS 2021), doi:10.1088/1742-6596/1963/1/012170, 2021.

[10] S. Pramanik, An Adaptive Image Steganography Approach depending on Integer Wavelet Transform and Genetic Algorithm, Multimedia Tools and Applications, 2023, https://doi.org/10.1007/s11042-023-14505-y.

[11] B. K. Pandey, D. Pandey, V. K. Nassa, A. S. George, S. Pramanik and P. Dadheech, Applications for the Text Extraction Method of Complex Degraded Images, in The Impact of Thrust Technologies on Image Processing, Nova Publishers, 2023.

[12] B. K. Pandey, D. Pandey, V. K. Nassa, A. S. Hameed, A. S. George, P. Dadheech and S. Pramanik, A Review of Various Text Extraction Algorithms for Images, in The Impact of Thrust Technologies on Image Processing, Nova Publishers, 2023.

[13] S. Pramanik, "An Effective Secured Privacy-Protecting Data Aggregation Method in IoT", in Achieving Full Realization and Mitigating the Challenges of the Internet of Things, Eds, M. O. Odhiambo and W. Mwashita, IGI Global, 2022, DOI: 10.4018/978-1-7998-9312-7.ch008

[14] R. Jayasingh, J. Kumar R.J.S, D. B. Telagathoti, K. M. Sagayam and S. Pramanik, "Speckle noise removal by SORAMA segmentation in Digital Image Processing to facilitate precise robotic surgery", International Journal of Reliable and Quality E-Healthcare, vol. 11, issue 1, DOI: 10.4018/IJRQEH.295083, 2022.

[15] S. Pramanik and S. Bandyopadhyay, Identifying Disease and Diagnosis in Females using Machine Learning, in Encyclopedia of Data Science and Machine Learning, Eds. John Wang, IGI Global, 2023, DOI: 10.4018/978-1-7998-9220-5.ch187.

[16] S. Pramanik and S. Bandyopadhyay, Analysis of Big Data, in Encyclopedia of Data Science and Machine Learning, Eds. John Wang, IGI Global, 2023, DOI: 10.4018/978-1-7998-9220-5.ch006

[17] S. Pramanik, AJ Obaid, N M, and SK Bandyopadhyay "Applications of Big Data in Clinical Applications", Al-Kadhum 2nd International Conference on Modern Applications of Information and Communication Technology, AIP Conference Proceedings 2591, 030086 (2023), https://doi.org/10.1063/5.0119414

[18] R. Anand, J. Singh, D. Pandey, B. K.Pandey, V. K.Nassa and S. Pramanik, "Modern Technique for Interactive Communication in LEACH-Based Ad Hoc Wireless Sensor Network", in Software Defined Networking for

Ad Hoc Networks, M. M. Ghonge, S. Pramanik and A. D. Potgantwar, Eds, Springer, 2022, https://doi.org/10.1007/978-3-030-91149-2_3
[19] P. T. Khanh, T. H. Ngọc, and S. Pramanik, Future of Smart Agriculture Techniques and Applications, in Advanced Technologies and AI-Equipped IoT Applications in High Tech Agriculture, Eds, A. Khang, IGI Global, 2023.
[20] T. H. Ngọc, P. T. Khanh and S. Pramanik, Smart Agriculture using a Soil Monitoring System, in Advanced Technologies and AI-Equipped IoT Applications in High Tech Agriculture, Eds, A. Khang, IGI Global, 2023.
[21] Reepu, S. Kumar, M. G. Chaudhary, K. G. Gupta, S. Pramanik and A. Gupta, Information Security and Privacy in IoT, in Handbook of Research in Advancements in AI and IoT Convergence Technologies, Eds, J. Zhao, V. V. Kumar, R. Natarajan and T. R. Mahesh, IGI Global, 2023.
[22] K. Dushyant, G. Muskan, A. Gupta and S. Pramanik, "Utilizing Machine Learning and Deep Learning in Cyber security: An Innovative Approach", in Cyber security and Digital Forensics, M. M. Ghonge, S. Pramanik, R. Mangrulkar, D. N. Le, Eds, Wiley, 2022, https://doi.org/10.1002/9781119795667.ch12
[23] M. Sinha, E. Chacko, P. Makhija and S. Pramanik, Energy Efficient Smart Cities with Green IoT, in Green Technological Innovation for Sustainable Smart Societies: Post Pandemic Era, C. Chakrabarty, Eds, Springer, 2021, DOI: 10.1007/978-3-030-73295-0_16
[24] V. Veeraiah, V. Talukdar, Manikandan K., S. B. Talukdar, V. D. Solavande, S. Pramanik, A. Gupta, Machine Learning Frameworks in Carpooling, in Handbook of Research on AI and Machine Learning Applications in Customer Support and Analytics, Eds, Md S. Hossain, R. C. Ho and G. Trajkovski, IGI Global, 2023.
[25] V. Jain, M. Rastogi, J. V. N. Ramesh, A. Chauhan, P. Agarwal, S. Pramanik, A. Gupta, FinTech and Artificial Intelligence in Relationship Banking and Computer Technology, in AI, IoT, and Blockchain Breakthroughs in E-Governance, Eds, K. Saini, A. Mummoorthy, R. Chandrika and N.S. Gowri Ganesh, IGI Global, 2023.
[26] A. Gupta, A. Verma and S. Pramanik, Security Aspects in Advanced Image Processing Techniques for COVID-19, in An Interdisciplinary Approach to Modern Network Security, S. Pramanik, A. Sharma, S. Bhatia and D. N. Le, Eds, CRC Press, 2022.

10

New Technologies for Sustainable Agriculture

Kuldeep Maurya[1], Pawan Kumar[2], Shambhu Bhardwaj[3], Rushil Chandra[4], Siddharth Ranka[4], Dharmesh Dhabliya[5], and Gautam Srivastava[6]

[1]Department of Agriculture, Sanskriti University, India
[2]Department of Computer Science and IT, School of CS and IT, Jain (Deemed to be) University, India
[3]College of Computing Science and Information Technology, Teerthanker Mahaveer University, India
[4]Symbiosis Law School, Nagpur Campus, Symbiosis International (Deemed University), India
[5]Department of Information Technology, Vishwakarma Institute of Information Technology, India
[6]Computer Science, Brandon University

Email: kuldeep.ag@sanskriti.edu.in; pawan.kumar@jainuniversity.ac.in; shambhu.bharadwaj@gmail.com; rushilchandra@slsnagpur.edu.in; siddharthranka@slsnagpur.edu.in; dharmesh.dhabliya@viit.ac.in; srivastavag@brandonu.ca

Abstract

Over the last several decades, there have been tremendous technical advancements in the agricultural industry. Modern technology is converting conventional agricultural concepts into digital agriculture. These technologies have allowed traditional agriculture to develop into an intelligent farming framework. Farmers may gather data and use the results of the analysis to fertilize and care for their crops in a smart farming framework. The smart agricultural system offers less expensive, more precise solutions to forecast and safeguard crop development. By incorporating this technology, the agriculture sector becomes more profitable, less wasteful, more efficient, and sustainable. This chapter examines the cutting-edge technology used in the agricultural industry and suggests a smart agriculture paradigm.

10.1 Introduction

The earliest form of human civilization is agriculture. Modern technology and digitalization have recently disturbed it. The likelihood of contracting a disease, experiencing unexpected weather, or experiencing a pandemic may grow with current technology rising and delivering food for a rising global population is a newer challenge in the current world. This problem requires expanding food supply and ensuring food security. Agriculture and food production must increase to keep up with the growing global population improving productivity and able to produce large crops in a short amount of time. Farmers and agricultural groups must innovate beyond the bounds of their present techniques in order to satisfy these demands.

A resource that has an effective food production system that takes sustainability into account is required. Utilizing water effectively, for instance, prevents soil erosion, assures minimal deterioration, and decreases energy input. The top difficulties facing agriculture today, according to Melanie McMullen, are

- Farmers are not properly trained or equipped to employ current technology.
- The majority of agriculture businesses is small and has few workers.
- Farmers' dwellings are placed distant from the fields.

Temperature, weather, and climatic conditions are less predictable due to global warming.

Less money is available to the producers to implement the technologies on their yearly earnings.

- Pests may completely destroy a crop.

The contemporary agricultural revolution might leave a big vacuum in the infrastructure due to a lack of digital knowledge and skills. These circumstances permit the introduction of several approaches for integrating digital technology into agriculture. Every farmer aspires to do all of their tasks for the least amount of money and time possible. However, these aims impose certain conditions that cannot be addressed by conventional farming practices.

10.1.1 Motivation

For instance, planting seeds is a strenuous manual operation in the agricultural system. Humans often utilize a dispersed approach for this activity. When seeds are planted outside of the ideal location, this procedure may be incorrect and wasteful. Consequently, it needs a successful seeding technique to provide the best possible plant development.

Modern agriculture uses seeding machines instead of the scatter technique to cover more ground more quickly than a person can. A lot of the uncertainty surrounding the cultivating cycle is removed by precision seeding equipment, which is developed using a variety of elements (such as combined geomapping, sensor data, soil grades, density, moisture, and fertilizer levels). Seeds have the chance to develop and grow, which might result in a higher overall yield. Existing precision seeders may go along with autonomous work vehicles and frameworks that are enabled by innovation as conventional farming progresses into the future. A single person might plant a whole field while monitoring the process on a digital control as many equipment travel around the area.

10.1.2 New approaches to agriculture

Ranchers and partners must invest heavily in information and more automated agricultural equipment due to the increase in food demand and the necessity for agribusiness sustainability. The question of "how to expand the number and quality of farming commodities" is more urgent than ever in the current day. Utilizing cutting-edge technology in farming, or smart farming, is the solution. Technologies that are predicted to have a substantial impact on the agricultural industry may be identified through emerging technologies. Examining cutting-edge agricultural technology and comparing them to selection criteria is the goal. The further research inquiries are:

- How have developing technologies affected agriculture?

What obstacles prevent these new technologies from being used in agriculture?

Solution:

New technologies have created a wide range of options for the agricultural industry. By reducing the need for additional time and effort, these technologies turn conventional agriculture into technology-based farming (also known as AgriTech), which is currently essential. Recent technological developments have made it possible to create "sensing solutions," which spontaneously gather data from the farms. Online farming assistances may lower costs for agriculturists, begin supply chains, and make an agriculturists' network for business. By utilizing technological advances, agriculturists have improved command over raising crops and caring for their livestock. These data may aid in decision-taking, sanction untimely estimation of cattle's health, and apply appropriate corrective husbandry uses. Innovative farming practices adhere to strict requirements for profitability, sustainability, and intelligence.

An intelligent (or smart) agricultural system is one that incorporates smart technology into agriculture. For enhancing the quality and quantity of farming products, a smart agriculture framework blends modern ICT [1] technologies with conventional agricultural methods. Currently, technology like sensors [2] and actuators [3], robots [4], GPS [5], big data [6] and analytics, drones [7], etc. are employed to digitize agricultural operations. By using these technologies, farmers may enhance agricultural production while spending less money and effort, while also increasing crop output.

Other organizations (such as governments, private sector businesses, and small agripreneurs) as well as issues like village livelihood, labor, and youth joblessness may be impacted by the digital transformation technique in farming. Additionally, adopting these technologies comes with a number of difficulties that might have an adverse impact on the social, economic, and environmental spheres.

10.1.3 Contribution

Here, we contribute by introducing a variety of cutting-edge technology to the system of contemporary agriculture. A technological model for a smart agricultural system is also put out. These include mobile technology (GPS), data science, data analytics, data management, drones, robots, the IoT, ML [8–12], sensors, etc. With less work required, this model's predicted cost-reduction and production-improvement benefits contemporary agriculture.

The chapter is structured as follows. Section 10.2 discusses how contemporary agriculture has been impacted by cutting-edge technology. Section 10.3 has a methodology and a review of the literature. The topic of new technologies for smart agriculture that support sustainable agriculture is covered in Section 10.4. Finally, Section 10.5 brings this chapter to a close having a few future improvements.

10.2 Current Technologies Combined with Smart Farming

Because of digitization, expansion, technology, and innovativeness, the agrifood business is facing challenges unlike any other time in history. The agrifood sector's transition to digitization will be a significant issue. Digital farming as a global paradigm will need significant transformations in conventional farming systems, rural markets, cultures, and natural resource management to realize its full potential.

The foundation of the current agricultural system continues to be innovation. The first major human technical development worldwide is agriculture. The first man discovered how to grow and collect food, and they ceased foraging for food by moving about. The development of agriculture paved the way for the development of communities, commerce, and languages. In order to prepare for droughts, cyclones, and floods that can endanger crops and way of life, farmers have grown food in accordance with early seasonal patterns and predictive information. Food yield and agriculture must become more productive and proficient of producing large harvests in a short span of time as a result of the growing population in the globe. Farmers and agricultural groups must innovate while advancing their present techniques in order to satisfy these demands.

10.2.1 New approaches to agriculture

Modern agricultural technology has always been entwined with "smart agriculture," which advances several businesses. To better comprehend the state of the land, the newest technologies – including microcontroller, cloud, web, cameras, sensors, and devices – are being used.

ICT use is still growing quickly in the agricultural sector. Roles and duties in ICT [13–16] are now crucial. For instance, RFID tags that automatically identify and collect data are used in supply chain management and retail, while biometrics and personal identity are required for accessing smart equipment and buildings. Additionally included in the definition of technology are AI [17–21], big data, cloud computing [22–25], data management, IoT, VR [26–28], and coding of audio, image, multimedia, and hypermedia data. All aspects of cultivation, from increasing yields to providing ranger services, may benefit from these developments. Precision farming and automated/computerized farming are two significant areas of horticulture that innovation has the potential to transform.

- Precision agriculture: Exactness horticulture, also known as exactness cultivating, may improve the control and accuracy of cultivation. Cows, plants, and crops may all get treated more precisely. The effectiveness of insecticides and manures on farms may be backed up by the ranchers.

- Drones used in agriculture: Robots may be used for a variety of tasks, including crop health evaluation, water system maintenance, crop overseeing, crop splashing, planting, soil testing, and more.
- Internet of Food, also known as Farm 2022: It looks at how the Internet of Things may improve gardening. IoT enables robotic dynamics in development and horticulture.
- Intelligent greenhouses: Smart nurseries that are driven by innovation may smartly screen, manage the environment, and eliminate human intervention.
- TGR or the Third Green Revolution: IoT, sensor-driven farming, and smart cultivation are laying the groundwork for the so-called TGR. The TGR combines the use of data analysis with precise farming. Current agricultural technologies also offer improved food traceability, which promotes food safety. Because they utilize water efficiently and provide agricultural goods efficiently, these technologies are good for the environment. They increase the precision and resource efficiency of agricultural output that is sustainable.

10.2.2 Electronic agriculture

Unpredictable levels of digital technology might emphasize the many difficulties. Additionally, it may aid in the development of digital agriculture and prevent challenges to the global agrifood industry (such as the digital divide) system. The digital gap is no longer linked to rural areas and poverty. The digitalization of agriculture has widened the economic and sectoral disparity.

Through the accessibility of linked devices and computational technologies, agriculture with digital technology presents prospects. Every link in the agrifood value chain is profoundly altered by digital agriculture. Digital agriculture has an impact on how farmers act, how input suppliers do their duties or processes, and how retail businesses promote, price, and sell their goods. All facets of agricultural systems may be affected by digital agriculture, including resource management for personalized, intelligent governance that is real-time, web-dependent, and data-driven. Higher productivity, safety, readiness, and weather adaptation will be the outcomes, providing food security, profitability, and sustainability.

The benefits of digital farming are significant, and the move to digitalization must address the difficulties.

10.2.3 Effective farming

Modern and cutting-edge technologies are incorporated into agriculture via smart agriculture to improve crop management, enable accurate soil analysis, and provide fresh approaches to crop production and harvesting. Precision and green agriculture are two other concepts associated with smart agriculture. The monitoring of every aspect of crop production, harvest, weed control, and soil analysis is made possible through smart agriculture. These specifics aid in smart agriculture's ability to provide more efficient and cost-effective outcomes. With the utilization of an alert management framework, the quality of the plants may be significantly improved.

Numerous cutting-edge technologies, like big data, data analytics, GPS, the IoT, and linked gadgets, are used in smart agriculture. An automated farming system and smart agricultural system assist in gathering data from the field. IoT sensors, cameras, drones, and actuators can handle the field data. The information is thereby examined to help the farmer make informed choices about how to develop high-quality plants and products after being supplied to him over the Internet.

Remote sensors used in precision agriculture use drones and UAVs to photograph crops. Sustainable agriculture makes greenhouse farming possible.

The application of sustainable development to contemporary agriculture is known as "green agriculture."

10.2.4 Wise farming

A software management system called "smart farming" is designed to provide the agriculture sector the infrastructure it needs to take use of contemporary technologies. Software-managed and sensor-monitored smart farming are both possible. Farming activities may be tracked, monitored, automated, and analyzed using smart farming. The technologies that are regarded to be elements of smart farming include sensors, location, robots, and software. It has expanded because of the necessity to utilize natural resources effectively, the increased utilization of Information and Communication Technologies, and the rising need for climatic situations. Connectivity (cellular, LoRa, etc.), location (GPS, satellite, etc.), robotics (autonomous tractors, processing, etc.), sensors (soil, water, light, humidity, temperature management,

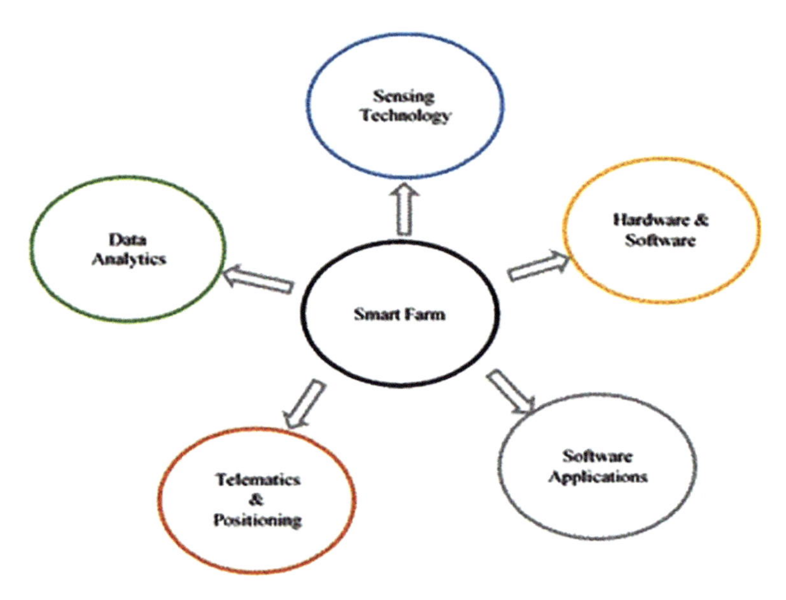

Figure 10.1 Modern tools for smart agriculture.

etc.), software, and data analytics are some of the technologies that are now accessible.

Management information systems, precision agriculture, farming automation, and robotics are some of the technologies that are associated with intelligent agriculture. By the aid of smart agriculture, agriculturists may track the nutritional demands of every animal and make the necessary adjustments. It improves herd health and aids in disease prevention. Modern technology is used in smart farming to boost agricultural output. Figure 10.1 shows technologies utilized in smart farming. Intelligent agriculture connects intelligent devices and sensors to provide agricultural operations data-driven and data-enabled.

Smart agriculture also makes machine-to-machine generated data more accessible. A decision-support system may be supplied with this deduced data. Farmers can now more precisely understand what is occurring thanks to this data. Pesticides, natural catastrophes, pests, and other risks are concerns in traditional farming. They provide problems for both the live animals used in agriculture and the food that comes from farms. For managing crops, machinery, and data, smart farming provides farmers with greater data and resources. The distinctions between conventional and modern farming are shown in Table 10.1.

Table 10.1 Conventional and smart farming: differences.

Conventional agriculture	Smart agriculture
Implementation of fertilizers and pesticides in the farm	Farm and financial data obtainable in similar location
Geo-tagging and region tracking impossible	It may estimate the areas in agriculture by the aid of satellites
Physical maintenance, distinct financial data, and causing faults	Timely tracking and implementation in the concerned locations
Weather prediction impossible	Weather analysis and estimation may be achieved
Similar set of use for cultivation of a crop all over the place	A diverse set of practices in cultivation of crops and optimized water needs

10.2.5 Lifecycle of smart agriculture

The sensors and gadgets used in smart farming may gather and analyze data repeatedly in order to improve the agricultural technique. They enable agriculturists to respond to problems and alterations in the environment. A cycle of observation, diagnosis, decision-making, and action governs smart farming.

- Observation: Rural sensors and technology may capture the observed data from the crops, metered animals, soil, and weather.
- Diagnostics: The recorded data is managed to a dazzling growing phase using pre-defined decision rules and models that may be used to assess the state of an item and identify its deficiencies.
- Decisions: When problems are found, the stage's clients and artificial intelligence techniques may select how explicitly to address the region.
- Action: After the last customer evaluation and stage, the ranch cycle starts again from the very beginning.

Smart farming helps ranchers well-track the requirements of each animal and modify their diet to prevent illness and improve population health.

10.2.6 Benefits

Agriculture is more efficient when it uses automation and decision-support systems. By using superior production techniques, automated agricultural services improve product quality and volume. They help reduce costs and waste in agricultural operations. They enhance farming's connectivity and intelligence. Smart agriculture lowers overall costs, raises product quality, boosts output, promotes sustainability, and enhances the customer experience. The following is a list of advantages of technology for smart and precise agriculture:

- Reliable forecasting: Reliable forecasting enables agriculturists for maintaining and ensuring a targeted level of production throughout the year.
- Efficiency: Automation in agriculture may lower overall cost, human labor, and error rates.
- An increase in output: The health of crops can be improved, sickness can be avoided, and requirements can be monitored.
- Data in real time: Real-time information helps speed up the decision-making process.
- Remote observation: Farmers may be able to make choices in real time from any location, thanks to remote monitoring of several fields.
- Sustainable farming: Costs and waste levels may be efficiently handled using production control.

A clever farming system may enable careful management of the interest rate and delivery of goods to the market to reduce waste. This system can manage sensor data, operate the growing system, and transmit far-reaching information for quick decisions.

10.2.7 Sustainable farming

The availability of growing knowledge is essential for sound horticulture and clever farming. Smart farming supports profitable agriculture via the combination of satellite information and land perception data, which ranchers may take into account while farming.

Growing technology is considered to be a subset of information research. Ranchers have access to a variety of information research techniques. Agribusiness maintainability may be achieved by properly using data in the dynamic framework. Farmers may run across a range of issues, from soil development to ecological alterations. Such types need an audit that will be useful for the cultivation method. Smart farming emphasizes the need of dynamic large-scale information innovation and meets predetermined creation goals. As an example, with significant information innovation:

- Farmers may determine the productivity of their farms by looking at satellite images and researching ranch information.
- Sensible agricultural practices have ensured the availability of valuable satellite data and earth perception data.
- The use of GPS technology in work trucks may be demonstrated in smart agriculture. Ranchers may use GPS to convey information about the location of the farm vehicle, develop land regularly, and save gasoline.
- Ranchers may use sensors to help them decide how, where, and when to distribute certain assets in order to increase yield in both biological and monetary terms.
- Pesticide usage is reduced in sustainable agriculture so as to protect the health of farmers and consumers.

Sustainable agriculture is related to the following essential ideas:

- It generates productive agricultural systems that are effective, self-sufficient, and affordable.
- It improves the effectiveness of food production and market distribution.
- It controls soil, water, and air quality to improve output.
- It makes the best possible utilization of natural resources for manufacturing.
- It safeguards territorial integrity and biodiversity.

10.2.8 Associated terms

Agribusiness improvement and computerized innovation have led to the emergence of new concepts, such as advanced cultivation, cunning cultivation, exactness cultivation, and smart cultivation. ICT and information breakthroughs are used in clever farming to simplify farming systems. However, some of the concepts related to shrewd farming are defined below. Ranchers may use smart devices to get continual information regarding the condition of the soil and plants, the region, environment, climate, asset usage, work force, finance, etc.

Both precision farming and smart farming are integrated into the practice of digital farming.

Big data analytics are used using web-based data systems.

iFarming, or intelligent farming: According to M.D. Ashifuddin Mondal (2018), intelligent farming provides constant insight and grasp over a smart field. It has elevated it to the forefront of innovations that enhance agricultural techniques.

Precision agriculture is another name for precision farming. Agricultural production processes are monitored and optimized using digital tools, according to Michell Christopher's 2018 analysis of contemporary farming management. To optimize fertilizer usage, save expenses, and minimize environmental effect, precision farming evaluates field soil changes and fertilizer strategy.

10.2.9 Technologies for agriculture

As stated by Meghan Brown in 2018, new advancements in planting, watering, treating, and collecting might benefit keen farming. It must handle concerns including population growth, environmental change, land erosion, and labor problems brought on by a rise in innovation. The majority of agricultural innovations may be divided into three categories: self-sufficient robots, robots or UAVs, and sensors or IoT. The developments are expected to become the smart ranch's mainstays.

- Automatic irrigation and watering: Ranchers may regulate when and how much water is given to their crops using the water system approach. Complex IoT-enabled sensors may enable the farming framework to function autonomously by relying on data from sensors in the farms to carry out water system changing.
- Tractors without drivers: Ranchers employ work trucks for a variety of agricultural tasks. With time and advancements (such as cameras and machine vision frameworks, GPS, IoT, and LiDAR), self-governing farm vehicles get more capable and autonomous. These developments fundamentally reduce the need for individuals to successfully operate these devices.
- Drones: Robots that are equipped with cameras may be used to quickly produce ethereal images. Ranchers may improve every area of their land and harvest the board by using these photographs.
- Harvesting: In turn, the sophisticated technology with sensors, IoT devices, and automated machinery may start the reaping when circumstances are optimal and release the rancher for other undertakings. For instance, the apple-picking robot from Abundant and the tomato-picking

robot from Panasonic with the help of an IoT framework can continuously monitor fields, monitor plants with their sensors, and collect ready harvests in the right quantities.

- Analysis and monitoring: Robots may do distant verification, field inspection, and harvest data collection. By monitoring on sound plants, ranchers may gather and examine the data to avoid every excursion as opposed to sitting down and exerting themselves.
- Planting: Model robots are being tested to replace challenging labor in farming and planting, such as DroneSeed and BioCarbon. They have the ability to direct-fire seedpod packages holding manure and nutrients into the ground using compressed air. Robots are also giving us the chance to automate difficult tasks. They can carry out crop splashing tasks more effectively, with greater prominence of accuracy, and with lesser waste.
- Decreasing labor costs and enhancing production: The goal of integrating self-governing mechanical technology into farming is to reduce physical labor and increase product production and quality. They direct the ranchers to carry out various tasks, such as repairing, troubleshooting robots, decomposing data, and planning ranch operations.
- Automated work: AgBots, often known as agrarian robots, automate human labor in lieu of it. It may be seen on ranches doing errands like planting, watering, harvesting, and organizing. With less labor, ranchers can produce more and better food.
- IoT and sensors: It is possible to implant sensors at any time throughout the growing cycle. Overhead sensors collect data on weather, soil conditions, water systems, air quality, contaminants, and illumination levels. These details will increase production of greater yields, reduce costs, and improve the availability and quality of food.
- Maintenance of the crop and weeding: Basic plant support functions like weeding and irritation management are perfect for autonomous robots. These robots can autonomously monitor a farm using advances in video, LiDAR, satellite, and GPS technology. These robots may also be used to identify annoyances and apply insecticides. It recognizes weeds using artificial intelligence (AI) before getting rid of them.

Despite the aforementioned, business intelligence plans expect that IoT devices adopted in agriculture will need access to additional data. The spine of the tech-savvy household of the future will be made up of this enormous knowledge and other data created by farming technologies. Ranchers may choose to see and examine every aspect of their business so that they can make informed decisions.

10.3 Review of the Literature and Methodology

ICT technology use is still increasing in the agricultural industry. ICT now plays a more important function and has greater obligations in this industry. Smart devices may access smart farms using ICTs for farming business and client applications using automated ID and RFID tags.

Digital breakthroughs and disruptive technologies are being driven by the Fourth Industrial Revolution, which also affects the food and agricultural industries. In 2020, it was challenging to get data regarding smallholder agriculturists, their fundamental requirements, and issues including access to inputs, pricing, markets, and financing. The next wave of mobile technology is anticipated to originate in rural areas where people regularly engage in agriculture-related activities. Remote sensing services and mobile technology (3G/4G/5G) have created new potential for smallholder farmers to be included in cutting-edge digital agrifood systems. The most recent "revolution" to affect the agricultural industry has been dubbed technological advancements. Recent studies have underlined the significance of leveraging and the yield improvements that contemporary technology can provide. Smart technologies play a crucial part in attaining higher eco-efficiency and productivity.

Precision planting, precision fertilizing, and precision spraying all contribute to precision agriculture's ability to boost production. Real-time land monitoring and proactive response to changes are made possible by these technologies. Smart farming is built on IoT and offers automated farming techniques, high precision crop management, and data collecting. The temperature and humidity of the soil may be monitored using an intelligent agricultural field monitoring system.

Farmers may open and stop the ventilation system, sprinklers, manage irrigation, and apply manures with the use of smart farm apps. As a result of higher labor productivity, farmers are more productive, produce more crops, save time, and reduce costs. The intelligent farm can aid in preventing events like cold climate damage.

A novel automated aeroponics system is developed and put into use utilizing IoT devices. A linked farm using IoT was created to offer smart agricultural systems for the end user. Growing plants without utilizing soil is possible with the well-organized and successful method of aeroponic farming.

Cloud computing may assist identify adoption behavior barriers and improve the current agricultural. In Ireland, young farmers are more likely than older farmers to use cloud computing.

According to [29], robotic process automation (RPA) will enable agriculturists and merchants to automate routine tasks by 2020, freeing them up to concentrate their energy and efforts on more important tasks that call for particular skill and attention. The goal of the e-commerce platform is to become a major player in the agriculture industry. Farmers and merchants of agri goods (Kisan Market, farMart, and Agroman) will find this platform to be of tremendous value.

In the long run, ML and DL approaches may be coupled with computer vision technologies. Massive datasets from the agriculture management system may be used to apply deep learning. This management approach encourages the use of more sophisticated agricultural automation tools and methods.

Drones and satellite data may be collected by agricultural data processing systems, which are then stored in the management system. Farmers will be able to boost their output with the use of this data by using better production and harvesting strategies.

10.4 Emerging Technologies: Methods

Utilizing cutting-edge technology, smart agriculture, digital agriculture, and precision agriculture are essential to sustainable agriculture. By using fewer energy and inputs (such as fertilizers, phytosanitary treatments, and water), they make it easier for the farmers to collect as many products as they can. Crop rotation and diversification, no-till farming, combined pest control, agroforestry, natural animal husbandry, utilization of renewable energy sources, etc., are examples of sustainable farming techniques.

10.4.1 Factors examined

Farmers may improve the performance and income of their intelligent farms with the use of smart gadgets and technology. As a result, some elements must be taken into account while offering a smart farming solution. Hardware, brain, upkeep, mobility, and infrastructure make up this group.

- Hardware: The agricultural industry must decide which sensors and smart devices to use when creating a technology solution. The effectiveness of trustworthy data from the lands may be necessary for the solution. The success of the product also depends on the quality of the sensors.

- Infrastructure: It requires strong internal support and infrastructure to guarantee the smart agricultural application.
- Maintenance: The sensors utilized in the land may readily be destroyed for a variety of causes. Hardware is thus long-lasting, simple to maintain, and replaceable. For IoT devices in agriculture, hardware upkeep is often a problem.
- Mobility: Systems for smart farming may be modified for usage in fields and farms. A farmer may view the data remotely or on-site using smart devices like smartphones and tablets or personal computers. These gadgets could have sufficient wireless range to talk to other instruments and transfer information to a central server.
- The mind: Every smart agricultural solution may be built on data analytics and scientists. In order to get useful insights from the acquired data, smart farming systems must thus incorporate sophisticated data analytics tools, predictive algorithms, and ML and DL.

Additionally, internal farming systems need to be safer. The likelihood that someone may break into the agricultural system, steal the data, or even take control of the autonomous tractors increases if it is not secured.

10.4.2 Wise farming

The usage of smart technology and digitization in farming is growing. The performance of sensors and other equipment equipped with intelligent technologies is important to the success of intelligent farming systems. The smart farming solution using cutting-edge technology is shown in Figure 10.2. Smart devices or IoT sensors are used to gather data from agricultural areas. These measurements include sunshine, temperature, soil acidity, moisture, CO_2 levels, water quality, and other variables. The acquired data is subsequently examined by scientists and data analysts, who then relay the results to farmers for increased agricultural output.

As a result, processing of the data obtained is necessary. The farm framework absorbs the data, analyzes it, and compares it to past data. Farmers and agriculturists may also investigate the data to determine the weather pattern, soil fertility, crop quality, necessary quantity of water, etc. The data is then

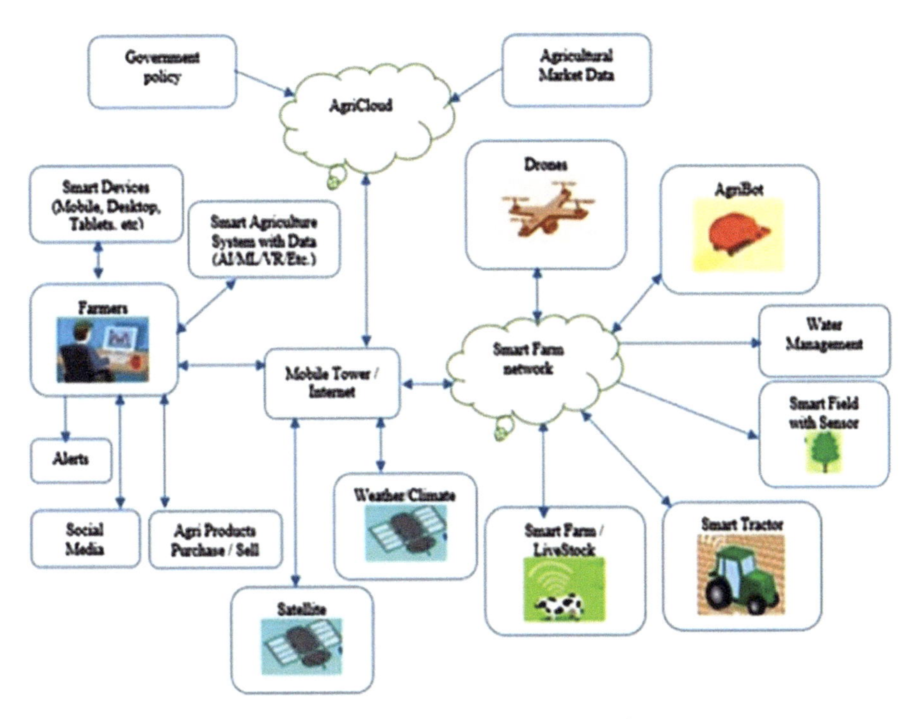

Figure 10.2 Digital agriculture solution.

made usable by technologies, giving farmers insightful information on how to better care for their crops. The smart systems allow watering schedules to be set based on the size of the operation, quantities, duration, light intensity, CO_2 levels, pest control, etc. Every element of a regulated environment, including HVAC, fertilizer injection systems, pH, heating and cooling, etc., may be automated by the intelligent farming controller system.

Self-governing reaping, autonomous farm trucks, hereditary changes, domestic animal observation, and astute water systems are examples of smart farming solutions.

- Self-sufficient harvesting: Mechanization of reaping ensures less burden on the company and more revenue from crops for social gatherings. It protects the worker from any accidents or snakebites. For instance, Saga Robotics simulates a number of tasks, such as gathering food that has grown on the ground.
- The autonomous tractor: As a result of computerized driving and self-sufficient work cars, ranchers have more time to focus on other

horticulture features. Ranchers may benefit from self-driving work vehicles like Agribot, which has a few perks.

- Genetic modification: The most advanced ways of pest control in agriculture may be hereditary. These processes, such as CRISPR/Cas9, genomic selection, or C4 photosynthesis, may change the yield quality. By changing the plant's genetic makeup, plasticomics techniques may enhance yields.
- Livestock observation: Animals like cows, lambs, pigs, goats, and poultry are essential to human nutrition. Animal husbandry is the daily conscious breeding, nurturing, and monitoring of domesticated animals. Desamis, an AI-dependent sophisticated dairy stage screen, recognizes and analyzes cow behavior so that the rancher may make wise decisions.
- Intelligent irrigation: An excellent water system provides the best water delivery to crops. It ensures that a large amount of the water used for agricultural is wasted. For example, One Water, an IoT-dependent smart water system framework, can identify temperature, humidity, and soil moisture to activate a trickle water system on the ranch, saving valuable resources.

Depending on their unique concepts and methods of collaboration, the ranchers may identify the most significant clever farming arrangements.

10.4.3 Ingenious technologies

The need for smart agriculture is anticipated to increase as commercial farming becomes more popular. Smart technologies enable farming to follow greenhouse farming. Smart technologies are used to better understand their land's state. Examples include a microcontroller, cloud platform, web-dependent platform, intelligent devices, sensors, and cameras. Automated farming and field data collecting are both benefits of smart agriculture. Devices, sensors, cameras, microcontrollers, and actuators are used to acquire field data. The information is sent via the Internet, where it is then analyzed to help cultivate a high-quality crop.

Precision farming makes use of clever innovations to provide the tools and data needed for an in-depth analysis of farming practices. In this way, farmers may make better informed decisions regarding the homestead and what information sources are typically suitable for their soil and harvest, increasing agricultural efficiency. With the help of cooperatives, drones, sensors, IoT, LoRa, blockchain, cloud computing, dashboards, and other intelligent technologies, comprehensive and smart agricultural solutions are offered. These innovations each have a specific role to play and influence on the smart farming and the agrifood supply chain. Depending on the complexity and maturity level of the chain, integrating technology into the agrifood value chain may be necessary. As a result, the structure, complexity, and distribution of digital technologies in the field of smart farming and agrifood are categorized in this chapter.

- Cloud computing, big data analytics.
- A module for coordination and integration that incorporates systems for blockchain, data management, ERP, microfinancing, and insurance.

Drones, robots, autonomous systems, artificial intelligence, machine learning, and deep learning are all examples of intelligent systems.

- Social networking platforms, mobile technologies, and mobile devices.
- IoT, UAV, and satellite sensor technologies, as well as precision agriculture.

The following list of technologies is classed as:

AI, or artificial intelligence: AI researches several environmental situations to determine the crop's water needs. The risk of illness and insect infestation at their farms may be decreased using AI technology. Using fertilizers and pesticides responsibly is also ensured by AI systems. As a consequence, AI technology lowers the total cost of farming and boosts crop output.

Platform for big data: A variety of information regarding farmers' labor on the farms is gathered. These data are quite large in size. Agriculture researchers may fill out the aforementioned data. The effect of smart farming may be accelerated and increased by offering a big data platform. Coordination and analysis of data on a large scale are possible. The aforementioned information on rainfall, fertilizer usage, crop kinds, and productions is used by farmers to monitor their fields.

Bitcoin technology: By facilitating better access to funding and data exchange, blockchain technology improves the value chains in the agriculture industry. It includes the verification of ranchers, their exchanges, increasing food detectability, connecting ranchers and merchants, and doing away with the need of experts. IoT and blockchain both have the potential to significantly improve farming worldwide. Using blockchain and AI, the Hello Tractor platform, for instance, enables farmers to obtain timely and pertinent data to boost harvests. This helps with time savings, income growth, and portfolio management.

Cloud computing: Through a single device, cloud technology manages all agricultural activities as well as additional tasks. Horticultural businesses may collect, deconstruct, and handle all the homestead activities to fork using information analysis and satellite symbolism. To provide pieces of knowledge with sharp accuracy and speed, cloud framework saves enormous amounts of data relating to climatic cycles, crop designs, soil quality, gathering, and satellite symbols. The customers may quickly access all the ranch-related information that has been stored in the cloud. The board's database makes all types of information available for the homeowner, enabling dynamic. Before providing the perfect estimate of cultivating, water, and pesticide for a homestead, meteorological information, market information, ranch information, GIS, and water accessibility information are thoroughly broken down. When irregularities in crop growth are noticed, the ready framework may issue an alert, providing ranchers with important information. Data storing is also significant in the precise analysis. Access to all stored information is available by phone, computer, and tablet.

Cooperatives: Cooperatives must be included in the precise agribusiness selection process. Cooperatives may provide the financial power required to invest in shrewd cultivation and make the benefits of shrewd developments available to everyone.

Dashboard: The smart agriculture or smart farm solution's dashboard is an example of smart farming data visualization. Application-programming interfaces (APIs) are used to gather real-time data from sensors and display it on the dashboard. The sophisticated tooltips and legend, low-latency updates, and chart zooming are all highlighted on the dashboard for smart farming.

Analysis of data: Farmers now place greater importance on forecast. The agriculture sector and farmers can more correctly estimate the results, thanks to data analysis. Predictions, however, may be demanding and time-consuming tasks. The predictions are becoming increasingly important due to rising output expectations and extreme weather that has a negative impact on crops in certain locations.

Drones: Drones capture yield images of crops and ensure that people make well-informed decisions. Drones examine standardized vegetation distinguishing files to determine if a region has living, green vegetation. Robotic reconnaissance may provide early warning of medical issues and harvest pressure by transmitting crop images that provide precise and reliable observations. Drones equipped with sensors may identify weeds, estimate yields, gauge water availability (or lack thereof), impede incursions, and detect nutritional deficiencies. Sensor-equipped drones took off in a legal vacuum and flew into clearly defined airspace.

Differentiating technology for evidence: The dairy industry makes extensive use of the distinctive proof innovation sensor. Each nursing animal is regularly encouraged to visit one region to be drained by the radiofrequency ID (RFID) framework, which also takes into account individualistic management. Dairy bovine RFID currently relies on large, high-recurrence transponders, or neck collars. Scaled-down low-recurrence devices, such as ear labels, injectable devices, or rumen boluses, are less costly and easily accessible.

IoT: IoT innovation consists of sensors, robots, and robots connected to the Internet that carries out activities to increase consistency and effectiveness. Ranchers are starting to take agricultural robots (Agribots) and mechanization in agriculture more seriously. The Internet of Things (IoT) employs sensors to collect data about social events, which is then transferred to research apparatuses for analysis. As farm sectors become more digital, the IoT with AI and ML helps them to automate procedures, increase productivity, and save costs.

LoRa system: Using LoRa technology, agribusinesses may expand their operations at a low cost. Among the various services offered by LoRa technology are fill level detection for water tanks and animal waste bins. It collects information from sensors to keep track of the salinity, moisture, and temperature of the soil. It follows cattle to keep an eye on their well-being and find out where they are. It keeps an eye on the whereabouts of agricultural machinery and equipment.

Technologies for remote sensing: These innovations are used to monitor agricultural conditions remotely. These developments rely on electromagnetic radiation interaction with soil and plant matter photographed using sensors attached to various platforms (such as satellites, aircraft, and autonomous airplane frameworks). Accuracy farming involves identifying and taking pictures using reflectance data from the visible and near-infrared wavelengths from either exposed soil or yield shades. To achieve the highest spatial target, ranchers and yield scouts may use autonomous aircraft systems (UAS).

A discussion on flora: Crop evaluation may be improved by using satellite and UAS sensors. These include thermal imaging, multispectral imaging, hyperspectral imaging, radar, and light detection and ranging (LIDAR).

Automated robotic process (RPA): There are extremely few qualified workers in agriculture, which necessitates a high workload and several repetitive jobs. Automation is the answer to freeing humans from these monotonous duties. RPA enables businesses and farmers to automate the majority of repetitive chores so that they may focus their precious time on more important tasks that need concentration and expertise. These duties may result in more production and, therefore, greater profit.

Sensors: Smart farming uses sensors to obtain accurate data. The farm control centers may get this data. These centers provide information on pest management and fertilizer needs. The efficiency of the plants' growth may be determined using these sensors. A common agricultural sensor gives farmers access to particular information through Wi-Fi or a mobile phone and provides a variety of field data. Crop production estimations are made possible by sensors utilizing data on the actual health of the plants and weather predictions.

Advanced logistics: Farmers may use IoT to guarantee that the environmental conditions during travel were kept within predetermined standards, revolutionizing the agricultural business. The challenges (such as temperature, gas and O_2 levels, and humidity) are addressed by innovative smart logistics solutions, which also improve the quantity, quality, sustainability, and cost-effectiveness of farming output. It continuously checks the material's temperature and immediately reports any variations from predetermined criteria.

Wearable technology: These tools are being utilized in dairy herds on a regular basis. The majority of commercially available reproductive management methods (such as oestrous mating and calving) rely on accelerometers to record some kind of activity. They may choose to wear leg pedometers, nose halters, neck collars, or ear tags. Some gadgets provide data about feeding habits, such as eating, drinking, and ruminating.

Utilizing the aforementioned technologies in farming enables farmers to produce food more effectively and sustainably raise the quality of their total output. Given that agriculture is the most significant sector of the global economy, it is more crucial than ever for the sector to adopt developing technologies.

10.4.4 Solution for sustainable agriculture

The automated sprinkler system, drone monitoring, livestock tracking, remote equipment monitoring, sensor-based field and resource mapping, smart greenhouses, smart pest management, etc., are all included in the proposed sustainable agriculture. The following lists these characteristics:

Sprinkler automation system: There is less of a chance of crops being harmed by overwatering because of the automated or regulated water dispersal. It could use IoT sensors or gadgets.

Climate forecasting and monitoring: Climate monitoring and determining may alert the rancher of impending changes and ensure that preventative steps are taken. The crops may be saved from extinction thanks to the IoT sensors' ability to predict and analyze the climate.

Drone surveillance: Robots can look at the harvests' vegetation list and pace of growth. Drones with IoT capabilities may collect data and determine crop health using heat markings.

Geofencing and livestock tracking: For the maintenance of any ranch, domesticated animals are kept for use as goods and products. Ranchers benefit from a geofencing that is perpetual.

Crop and livestock predictive analytics: Future-looking research may predict knowledge to enhance cultivation practices.

Remote monitoring of soil and crops: Through constant movement and visual representations made possible by mobile devices, the moisture, soil richness, and harvest development rate may be remotely watched. It influences the rancher to make wise decisions for the farm.

Remote monitoring of equipment: Sensors from the IoT are integrated into hardware like pickup trucks, farm equipment, and collecting devices. It should be feasible to remotely ingest, manage, curate, and investigate IoT data.

Resource mapping and sensor-based fields: Sensors may be used by ranchers to control and keep an eye on the whole property. It includes HR, equipment, and institutional resource insights.

Intelligent greenhouses: In savvy nurseries, plants grow and thrive with an increase in quality and production. The size and capacity of nurseries to create foods produced on the ground has been industrialized.

Smart warehousing and logistics: Large structures are often built on homesteads. Reap times result in a yield that is a coordinator's worst nightmare. In distribution facilities, capacity and preparation should be simple to do with careful agricultural arrangements in place.

Smart pest control: Pesticides assist in preventing invasions. However, a certain unacceptable level might result in completely destroyed yields. The board's keen vermin performs a detailed inspection that foretells swarm examples and raises alerts about the crops' health.

Statistics on produce and livestock feeding: Taking care of steers as examples helps us to frequently foresee potential diseases. The quantity and quality of the animals' use determines the quality of the milk and protein produced.

10.4.5 Framework for sustainable agriculture

A technology model's framework enables users and customers to recognize the following:

1. The components that make up contemporary agriculture's digital transition.
2. Making progress in the methodology's structure as the agricultural industry embraces digital transformation.

This technique enables structure of the factors that support the implementation of digital transformation, including technology, incentives, legislation, and agricultural models. Based on fundamental needs, enablers, and the influence of technology, the structure is streamlined.

1. Conditions essential for enabling digital agriculture include connection (such as cell subscriptions, network coverage, and Internet access), educational systems, job opportunities, and legislative frameworks.
2. Utilizing digital technology, developing digital skills, and innovating are all enablers for adoption of digital technologies.
3. Utilizing the benefits of technology to enhance economic, social, cultural, and environmental consequences with various sorts of resources is part of the application of digital technologies.

10.4.6 Scientific model

The revolutionary approach for sustainable horticulture is shown in Figure 10.3. Ranchers must implement and encourage a variety of tools to

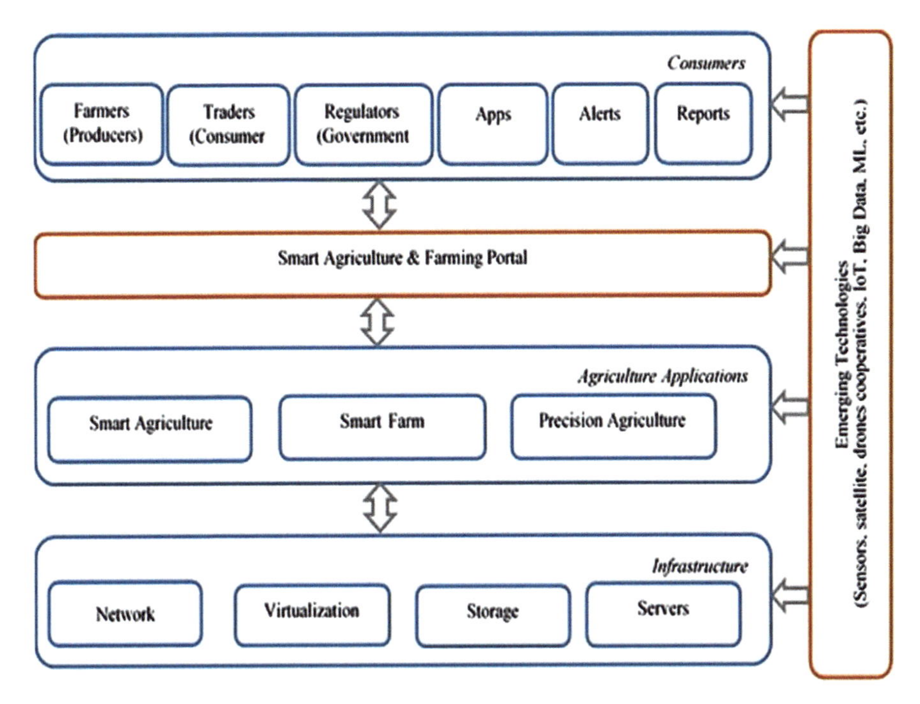

Figure 10.3 Model for sustainable agriculture using technology.

manage all aspects of homestead activities. The suggested model contains a universal device that can handle all of these activities and more with only one piece of equipment. Ranchers and agribusiness firms may collect, deconstruct, and deal with all the homestead activities to fork using information investigation and satellite symbolism.

The finest use of clever farming is information gathering. It collects and retains a larger number of data on crop designs, soil quality, gathering, and satellite systems in order to provide knowledge quickly and precisely. The cloud has been used to store all ranch-related information, making it instantly accessible.

Ranch-information-based management encourages a higher degree of activity. Before providing the most accurate estimation of the cultivation, water, and pesticide requirements for a ranch, all the information from over a long period of time is thoroughly evaluated.

All of the information is always available and is accessible through clever devices. Information storing plays a significant role in the precise inquiry as well. The more information that is available about ranches, the more accurate the identification of climate wonders, problems, crop yields, and advantages will be. The application of modern technologies and the suggested model to farming promises increased efficiency, expansion, usage of fewer resources, cleaner processes, agility, and improved product quality. Because of this, the aforementioned causes may ultimately result in increased revenue. As an example, cows may unintentionally get pregnant or mistakenly become separated from their herd. These problems can be solved using IoT devices. The farmer has to be cautious of this as cows require specific care for being healthy and give birth to healthy calves.

IoT hardware and sensors are combined in a farm management system. It has built-in accounting and reporting capabilities and may be implemented on-premises as a dashboard. The promise of technology may be held by companies, farmers, and landowners for greater revenue and productivity. With the aid of a farm management system, existing strengths can be maximized quickly and at a low cost. There is a growing need for more food as the population grows. The development of affordability and sustainability in contemporary agriculture may be strengthened by smart technology.

New avenues are opened by agricultural technology to address typical problems faced by farmers. Monitoring climatic conditions, managing crops more effectively, monitoring livestock, and using farm management systems all present some challenges. Sensors or IoT devices positioned across fields gather useful data from the atmosphere and transfer it to the Agri Cloud for later use. When the technology or equipment malfunctions, this work might be difficult. Crop management sensors and IoT devices can learn a lot from real-time data. These statistics help to understand crop health. This information might change depending on the region. The health of cattle must be closely monitored. IoT gadgets can keep an eye on things like exercise, diet, and body temperature. In the event that the apparatus malfunctions, it is difficult to view.

10.4.7 Applications and use cases in real time

The above advantages have particular applications in the actual world. The applications include crop tracking, drones, automated greenhouses, animal tracking, and climate tracking.

Crop surveillance: Crop tracking may gather data on temperature, humidity, crop health, precipitation, and other factors. Farmers may notice any anomalies in the data obtained in advance and take appropriate action. Farmers may use IoT technology to assist them choose the optimal time of year to grow and harvest their crops. For instance, Semios keeps track of crops to provide farmers remote management over disease, pest, and climate monitoring.

Drones: Robots can identify drier areas in a farm, and actions could then be taken to branch such territories having superior strategies. Drones may provide agriculturists with time and area-specific information about parasitic contaminations, crop well-being, development bottlenecks, etc. For exceptionally effective harvest watching and research, an eBeeSQ agricultural robot may go over many plots of land in a single trip.

Automation in nurseries: Climate stations may naturally alter the soil's properties to match the existing obstacles and provide the ideal nursery environment. For instance, GreenIQ, a clever sprinkler controller, interacts remotely with the lighting structures and water system.

Animals are watching: Ranchers may acquire information on the location, health, and welfare of their farmed animals with the use of sensors and IoT devices. They are encouraged to differentiate between the states of their animals by this information-checking. For example, finding weaker animals to separate from the group prevents the disease from spreading to the whole herd of cows. For any health warnings, infection pointers, estrus identification, and feed advancement, ML transmits critical information such as movement, temperature, rumination, and behavior.

Display the weather conditions: Climatic stations may collect climatic data and provide a rancher with useful information. Programs for climate observation analyze this data and deliver it to the rancher so that they can prevent crop damage. For instance, allMETEO monitors weather patterns and provides early warnings about ice, high temperatures, and severe weather on the farm fields.

10.5 Conclusion and Results

The modules of the proposed model are covered in great depth in this section. It goes into further detail on the technology model and its elements. However, it draws attention to a potential security issue with the model. The difficulties, restrictions, and possibilities were also depicted. This chapter reveals several technical issues stifling innovation. These technical elements provide technological self-efficacy, cooperativity, and interaction, as well as resource-enriched interaction.

10.5.1 Advantages and drawbacks

Sustainable agriculture has several benefits, including improved public health, ecological preservation, and economic and social fairness. Environmental protection, corrosion prevention, resource conservation, better air and water quality, increased biodiversity, and a reduction in carbon emissions are all benefits of sustainable agriculture. Pesticides and fertilizers that are hazardous are not used in sustainable agriculture. Farmers are able to produce less objectionable and healthier food as a consequence for customers and nearby communities.

Food production, resource availability, and food security are a few of the difficulties. Sustainable farming produces a requirement to address food security by yielding much food in lesser time and with less estimated resources due to the expanding population, the high levels of hunger, and malnourishment. Without a doubt, the fast deterioration and depletion of natural resources is a problem for green agriculture.

10.5.2 Issues with smart agriculture

The whole agricultural industry may be saved by a smart agriculture system that makes use of contemporary technology. However, there are certain challenges in integrating technology with conventional farming practices, such as connection, design and durability, as well as time and resource constraints.

- Connectivity: The connectivity has to be dependable in all weather and open-space situations.
- Design: Any IoT framework used in agriculture should have the capability to handle open-air space conditions. Robots, sensors, Internet of Things (IoT) devices, and weather stations should all operate with a certain amount of enthusiasm and follow a straightforward practical strategy.
- Resources are scarce. Rural organizations that plan and build IoT for farming must take into account rapidly changing environmental conditions, evolving climatic limits, working with limited land accessibility, and terrible elements like pollinator transmission.
- Cost: Due to the interoperability of sensors and equipment, the cost of a smart field is a considerable obstacle. However, this may be improved by simplifying the device, enhancing data flow across devices, and turning the information gathered into something that is useful and available.

Agricultural digital technology highlights potential areas for development and acceleration. These zones specifically relate to the economic, social, and environmental spheres.

1. Economic: Technologies may boost output, lower logistical expenses, cut waste, expand market opportunities, provide sustainability, value chains, etc.
2. Cultural and social: Through the gadgets, technologies may influence people on a social and cultural level.
3. Environmental: Agricultural processes, value chains, and delivered goods may all be monitored and optimized using smart, precision, or digital agriculture. The optimal use of natural resources is made possible by the application of digital technology, which also enables prevention and adaptation.

10.6 Conclusion and Potential Improvements

The systematic utilization of crop rotations, agricultural innovations, the green process, and the routine usage of pesticides and fertilizers manufactured by humans are only a few of the revolutions that have occurred in traditional agriculture. The growing use of ICT in farming has the potential to bring about another revolution in agriculture. In order to feed a rising population, ICT alters agriculture. Farmers may raise the value of their produce by using relevant data, adding nutritional value, and being sustainable. Because of its efficiency, durability, and scalability, employing current technology in agriculture has a bright future. In this chapter, new technologies for smart agriculture were reviewed. The results of this chapter allow for a number of inferences on the cutting-edge technical applications in smart agriculture. There are still a few obstacles to go through. High-quality farming data that trust all parties in contemporary agriculture are guaranteed by agricultural business models and technology design. In order to prevent data in the realm of agriculture from being misused, a security system is needed. This will be the chapter's ongoing work.

References

[1] S. Saide and M. L. Sheng, "ICT Team Dual-Innovations in the Microlevel of Circular Supply Chain Management: Explicit-Tacit Knowledge, Exchange Ideology, and Leadership Support," in *IEEE Transactions on Engineering Management*, doi: 10.1109/TEM.2022.3166763.

[2] R. Fu, X. Ren, Y. Li, Y. Wu, H. Sun and M. A. Al-Absi, "Machine Learning-Based UAV Assisted Agricultural Information Security

Architecture and Intrusion Detection," in IEEE Internet of Things Journal, doi: 10.1109/JIOT.2023.3236322.

[3] S. Mishra, S. Nayak and R. Yadav, "An Energy Efficient LoRa-based Multi-Sensor IoT Network for Smart Sensor Agriculture System," 2023 IEEE Topical Conference on Wireless Sensors and Sensor Networks, Las Vegas, NV, USA, 2023, pp. 28–31, doi: 10.1109/WiSNeT56959.2023.10046242.

[4] A. K. Nuhel, M. M. Sazid, D. Paul, E. Hasan, P. H. Roy and F. P. Sinojiya, "A PV-Powered Microcontroller-Based Agricultural Robot Utilizing GSM Technology for Crop Harvesting and Plant Watering," 2023 IEEE International Students' Conference on Electrical, Electronics and Computer Science (SCEECS), Bhopal, India, 2023, pp. 1–5, doi: 10.1109/SCEECS57921.2023.10062995.

[5] G. Singh and J. Singh, "A Fog Computing based Agriculture-IoT Framework for Detection of Alert Conditions and Effective Crop Protection," 2023 5th International Conference on Smart Systems and Inventive Technology (ICSSIT), Tirunelveli, India, 2023, pp. 537–543, doi: 10.1109/ICSSIT55814.2023.10060995.

[6] S. Pramanik, "An Effective Secured Privacy-Protecting Data Aggregation Method in IoT", in Achieving Full Realization and Mitigating the Challenges of the Internet of Things, Eds, M. O. Odhiambo and W. Mwashita, IGI Global, 2022, DOI: 10.4018/978-1-7998-9312-7.ch008

[7] S. Pramanik and S. Bandyopadhyay, Identifying Disease and Diagnosis in Females using Machine Learning, in Encyclopedia of Data Science and Machine Learning, Eds. John Wang, IGI Global, 2023, DOI: 10.4018/978-1-7998-9220-5.ch187.

[8] S. Pramanik and S. Bandyopadhyay, Analysis of Big Data, in Encyclopedia of Data Science and Machine Learning, Eds. John Wang, IGI Global, 2023, DOI: 10.4018/978-1-7998-9220-5.ch006

[9] K. Dushyant, G. Muskan, A. Gupta and S. Pramanik, "Utilizing Machine Learning and Deep Learning in Cyber security: An Innovative Approach", in Cyber security and Digital Forensics, M. M. Ghonge, S. Pramanik, R. Mangrulkar, D. N. Le, Eds, Wiley, 2022, https://doi.org/10.1002/9781119795667.ch12

[10] A. Gupta, A. Verma and S. Pramanik, Security Aspects in Advanced Image Processing Techniques for COVID-19, in An Interdisciplinary Approach to Modern Network Security, S. Pramanik, A. Sharma, S. Bhatia and D. N. Le, Eds, CRC Press, 2022.

[11] D. K.aushik, M. Garg, Annu, A. Gupta and S. Pramanik, Application of Machine Learning and Deep Learning in Cyber security: An Innovative

Approach, in Cybersecurity and Digital Forensics: Challenges and Future Trends, M. Ghonge, S. Pramanik, R. Mangrulkar and D. N. Le, Eds, Wiley, 2022, DOI: 10.1002/9781119795667.ch12

[12] M. Sinha, E. Chacko, P. Makhija and S. Pramanik, Energy Efficient Smart Cities with Green IoT, in Green Technological Innovation for Sustainable Smart Societies: Post Pandemic Era, C. Chakrabarty, Eds, Springer, 2021, DOI: 10.1007/978-3-030-73295-0_16

[13] A. Mandal, S. Dutta, S. Pramanik, "Machine Intelligence of Pi from Geometrical Figures with Variable Parameters using SCILab", in Methodologies and Applications of Computational Statistics for Machine Learning, D. Samanta, R. R. Althar, S. Pramanik and S. Dutta, Eds, IGI Global, 2021, pp. 38–63, DOI: 10.4018/978-1-7998-7701-1.ch003

[14] Y. Meslie, W. Enbeyle, B. K. Pandey, S. Pramanik, D. Pandey, P. Dadeech, A. Belay, A. Saini, "Machine Intelligence-based Trend Analysis of COVID-19 for Total Daily Confirmed Cases in Asia and Africa", in Methodologies and Applications of Computational Statistics for Machine Learning, D. Samanta, R. R. Althar, S. Pramanik and S. Dutta, Eds, IGI Global, 2021, pp. 164–185, DOI: 10.4018/978-1-7998-7701-1.ch009

[15] S. Pramanik, "Carpooling Solutions using Machine Learning Tools", in Handbook of Research on Evolving Designs and Innovation in ICT and Intelligent Systems for Real-World Applications, K. K. Sarma, N. Saikia and M. Sharma, IGI Global, 2022, DOI: 10.4018/978-1-7998-9795-8.ch002.

[16] G. Taviti Naidu, KVB Ganesh, V. Vidya Chellam, S Praveenkumar, D. Dhabliya, S. Pramanik, A. Gupta, Technological Innovation Driven by Big Data, in "Advanced Bioinspiration Methods for Healthcare Standards, Policies, and Reform", Hadj Ahmed Bouarara IGI Global, 2023, DOI: 10.4018/978-1-6684-5656-9

[17] R. R. Chandan, S. Soni, A. Raj, V. Veeraiah, D. Dhabliya, S. Pramanik, Ankur Gupta , Genetic Algorithm and Machine Learning, in "Advanced Bioinspiration Methods for Healthcare Standards, Policies, and Reform", HadjAhmedBouarara,IGIGlobal,2023,DOI:10.4018/978-1-6684-5656-9

[18] D. Mondal, A. Ratnaparkhi, A. Deshpande, V. Deshpande, A. P. Kshirsagar and S. Pramanik, Applications, Modern Trends and Challenges of Multiscale Modelling in Smart Cities, in Data-Driven Mathematical Modeling in Smart Cities, IGI Global, 2023, DOI: 10.4018/978-1-6684-6408-3.ch001

[19] B. K. Pandey, D. Pandey, V. K. Nassa, A. S. Hameed, A. S. George, P. Dadheech and S. Pramanik, A Review of Various Text Extraction

Algorithms for Images, in The Impact of Thrust Technologies on Image Processing, Nova Publishers, 2023.

[20] B. K. Pandey, D. Pandey, V. K. Nassa, A. S. George, S. Pramanik and P. Dadheech, Applications for the Text Extraction Method of Complex Degraded Images, in The Impact of Thrust Technologies on Image Processing, Nova Publishers, 2023.

[21] V. Veeraiah, V. Talukdar, Manikandan K., S. B. Talukdar, V. D. Solavande, S. Pramanik, A. Gupta, Machine Learning Frameworks in Carpooling, in Handbook of Research on AI and Machine Learning Applications in Customer Support and Analytics, Eds, Md S. Hossain, R. C. Ho and G. Trajkovski, IGI Global, 2023

[22] S. Pramanik, An Adaptive Image Steganography Approach depending on Integer Wavelet Transform and Genetic Algorithm, Multimedia Tools and Applications, 2023, https://doi.org/10.1007/s11042-023-14505-y.

[23] R. Jayasingh, J. Kumar R.J.S, D. B. Telagathoti, K. M. Sagayam and S. Pramanik, "Speckle noise removal by SORAMA segmentation in Digital Image Processing to facilitate precise robotic surgery", International Journal of Reliable and Quality E-Healthcare, vol. 11, issue 1, DOI: 10.4018/IJRQEH.295083, 2022

[24] S. Pramanik, R. P. Singh and R.Ghosh, "Application of Bi-orthogonal Wavelet Transform and Genetic Algorithm in Image Steganography", Multimedia Tools and Applications, https://doi.org/10.1007/s11042-020-08676-2020, 2020.

[25] S. Pramanik, K. M. Sagayam and O. P. Jena, Machine Learning Frameworks in Cancer Detection, ICCSRE 2021, Morocco, 2021.

[26] S. Pramanik, M. G. Galety, D. Samanta, N. P Joseph, Data Mining Approaches for Decision Support Systems, 3rd International Conference on Emerging Technologies in Data Mining and Information Security, 2022

[27] D. Samanta, S. Dutta, M. G. Galety and S. Pramanik, A Novel Approach for Web Mining Taxonomy for High-Performance Computing, The 4th International Conference of Computer Science and Renewable Energies (ICCSRE'2021), 2021, DOI:10.1051/e3sconf/202129701073.

[28] R. Bansal, A. J. Obaid, A. Gupta, R. Singh, S. Pramanik "Impact of Big Data on Digital Transformation in 5G Era", 2nd International Conference on Physics and Applied Sciences (ICPAS 2021), doi:10.1088/1742-6596/1963/1/012170, 2021.

11

Agriculture Automation

Pankaj Kumar Tiwari[1], Mir Aadil[2], Praveen Kumar Singh[3], Nuvita Kalra[4], Karun Sanjaya[4], Ankur Gupta[5], and Kassian T.T. Amesho[6]

[1]Department of Agriculture, Sanskriti University, India
[2]Department of Computer Science and IT, School of CS and IT, Jain (Deemed to be) University, India
[3]College of Agriculture Science, Teerthanker Mahaveer University, India
[4]Symbiosis Law School, Nagpur Campus, Symbiosis International (Deemed University), India
[5]Department of Computer Science and Engineering, Vaish College of Engineering, India
[6]National Sun Yat-sen College of Engineering, Taiwan

Email: pankaj.ag@sanskriti.edu.in; mir.aadil@jainuniversity.ac.in; dr.pksnd@gmail.com; nuvitakalra@slsnagpur.edu.in; karunsanjaya@slsnagpur.edu.in; ankurdujana@gmail.com; kassian.amesho@gmail.com

Abstract

To significantly boost agricultural yields using agricultural automation, many newer proficiency and technological usage are needed. Technology may be utilized in intelligent agriculture to track, monitor, and analyze different agricultural processes. The IoT system is made up of specialized hardware and software, cameras, drones, edge devices, cloud servers, sensors, actuators, and communication technologies. The hardware and software technological components which may be employed in agricultural automation will be covered in this chapter. There are four parts in the chapter. An overview of smart agriculture and precision farming is given. The literature review of recent studies is included in the second part. For the implementation of smart agriculture, the third part discusses IoT methods, sensors, and cloud and edge computing remedies. Case studies of IoT use in smart agriculture are presented in the next part. The chapter will go into detail on the IoT framework remedy for whole agricultural automation.

11.1 Introduction

Agricultural automation automates crop production and animal management to increase farm productivity and efficiency. It does this by using technology, the Internet of Things (IoT) [1], and robots [2]. As the population grows, so does the quantity of food consumed, and people's concerns about the quality of their food have grown through time. Additionally, the effects of environmental changes have compelled scientists to consider developing novel solutions to the problems facing the agricultural industry. All of these changes have caused precision agriculture to emerge as a new paradigm.

Recently, precision agriculture has received a lot of attention. Basically, information technology is used in farming via precision agriculture. Making better, more educated decisions in farming has a great deal of promise with the use of information technology. It results in the development of intelligent farms that will improve output while also being environmentally responsible and sustainable. The management of crops, animals, and other agricultural resources are also included in the scope of precision agriculture products of the arts. Precision farming has the potential to increase agriculture's overall effectiveness, output, and profitability.

In order to practice sustainable agriculture, the soil's quality is crucial. Organic matter and other nutrients must be present in the soil. Composting has been shown to be more effective than artificial fertilizers in maintaining the necessary minerals and nutrients in the soil. The IoT has the ability to enhance composting, raising the general soil quality in the process. This chapter shows a variety of smart agriculture and composting system ideas. The research that has been done recently on smart farming is highlighted in the section that follows.

11.2 Literature Review

Recent research on smart farming is included in the literature review. A smart IoT monitoring system was presented by Wen et al. [3]. The program facilitates a tracking analysis. As noted in the research, information on various crops may be accumulated via a WSN. The crops of potato and tomato are utilized in this investigation. The ML approaches show a warning message utilizing the data which was acquired.

When growing potato crops, machine learning algorithms may be used to identify illness early on and take the necessary precautions to stop the disease's spread. The method proposed in this study uses regression analysis to identify the presence of the late blight disease in crops like potatoes and provides the farmer with the necessary guidance through a graphical user interface. The benefit of the suggested approach is that it would use fewer pesticides than would otherwise be required.

An IoT-enabled compost monitoring system was proposed by Onu et al. [4]. The technology shows real-time tracking of major soil characteristics. An IoT-dependent approach was used here to successfully proceed with various composting operations. The system's CPU is a Raspberry Pi, and its LoRa-dependent communication module. Temperature and relative humidity was tracked utilizing a DHT11 sensor. The proposed approach will aid the agriculturists in selecting the prime time to turn the compost.

In [5], suggested a survey on the utilization of the IoT in farming. This study surveyed the IoT hardware platforms for agriculture, consisting of those created on the Arduino, Raspberry Pi, and Intel architectures. The report moreover surveyed numerous farming-based software platforms. The software platforms are cloud-dependent like the IoT platforms from Amazon Web Services (AWS), Salesforce, SAP, Google Cloud Platform, Oracle, IBM Watson, and Microsoft Azure. Irrigation and environmental tracking are one of the IoT implementations highlighted in the report. Moreover, highlighted are a number of mobile applications for duties connected to agriculture. The normalization of IoT devices, sensors, and communication protocols, security and privacy, as well as data storage and analysis, are cited as the issues with the usage of IoT in agriculture.

Nguyen et al. [6] proposed an approach for utilizing IoT data collection for precision agriculture in decision-making. There are several sources from which the data might be assembled. For example, it will consist of information on the climate, crops, land, and farming gear. The choice of the perfect sensors is needed for data collecting. Common sensors are DHT11, LM35, and SHT3x-DIS temperature and humidity sensors. A decision model is then needed for decision-making. A decision tree-dependent implementation is given in this work.

Another approach for IoT-based precision agriculture was put up by Kumar and Rikendra [7], who claimed it to be an energy-efficient solution. It is a system based on a wireless sensor network (WSN). There are tens of thousands of low-cost sensor nodes in a WSN system. These sensor nodes, however, often run on batteries. As a result, when the nodes' battery is depleted, the nodes might ultimately die. In this study, an energy-collecting strategy is proposed. Using a solar panel, energy is gathered. The system also uses clustering methods for energy-efficient routing. In this work, the use of data aggregation algorithms like LEACH is covered. The network simulator NS2 was used to carry out the experiment. It was established that the suggested model was more effective than the alternatives.

An intelligent irrigation approach was presented by Ko et al. [8]. They utilized IoT to compute the ambient temperature, humidity, soil moisture, CO_2 level, and sunshine intensity for the suggested method. There are several locations when the computing sensor nodes are utilized. In this work, an ideal irrigation usage which is attained by exact measurements is provided. Neural networks may be utilized to estimate the quantity of water which the soil will require, and this information may be utilized to manage the water valves. Moreover, the unvaried water need for the whole field may be computed utilizing this information to determine the regions of farms having water scarcities. This study recommended using artificial neural networks (ANNs) and fuzzy logic to make predictions.

An overview of research projects pertaining to smart farming was given by Tahir et al. [9]. According to this report, the majority of research efforts are focused on using farming robots, UAVs to gather aerial photography to monitor the growth of disease, and WSNs to gather data. The absence of connection and access to energy provide the biggest obstacles to smart farming in remote locations. The acquisition of data necessitates the usage of cloud storage, which necessitates constant connection. Lack of knowledge and technological know-how among agriculturists, and the expensive expenditure of installing sensors, robotics, and UAVs, are other significant obstacles to the implementation of smart farming.

In order to enable precision agriculture, Trivedi et al. [10] suggested an IoT-dependent prediction platform dubbed PLANTAE. By foreseeing plant illnesses, this platform aims to regulate the environment on plantations. Photos of leaves are gathered, and machine learning techniques are then used to classify the diseases in the photos. The suggested platform moreover gathers data and uploads it to a web server, much like existing systems. The categorization process is then carried out with a ML model. The suggested framework makes use of a pH sensor, temperature and humidity sensors, as well as a soil moisture sensor. When soil moisture falls below 30%, the sensor data is utilized to activate a water pump, and when the temperature rises over 37°C, a fan is activated. Images are taken using a mobile device's camera and then sent to a web server. Deep learning is used to forecast plant diseases.

The IoT-based smart farming solution for rural regions was presented by Deepa et al. [11]. The suggested system significantly lowers network latency by using fog computing. On WSN gateways, calculation for the fog and clouds is done.

IoT was suggested as a precision agriculture option by Memon et al. [12]. This research specified the utilization of wearable sensors for tracking cattle health along with employing sensors to track the climate. An IoT-dependent smart agricultural framework for controlling irrigation and managing agriculture was presented by [38]. The device utilizes less energy. A web application and an Android application are built for users in this research. Larger devices employ RF communication using the LoRa protocol.

A weather station and tracking system for precision farming was suggested by Sharma et al. [13]. The purpose of this research is to determine if a local weather station is necessary to monitor the local environment can be exactly known. The DHT22 sensor that detects temperature, humidity, dew point, absolute pressure, light intensity, and the quantity of precipitation, is employed in the suggested framework. MATLAB has provided the data in a visual approach.

An IoT-dependent framework for agricultural field tracking was recommended by Javaid et al. [14]. The suggested method intends to spontaneously address the field's watering needs, making the automated system more successful than the present hand-operated approach. The device retains monitoring of soil temperature, humidity, and light intensity. Both the soil moisture sensor and the temperature sensor LM35 were utilized. On a cloud server, data is assembled.

A review of IoT-dependent interdisciplinary approaches for smart agriculture was provided by Akpabio et al. [15]. Big data, WSNs, cyber–physical systems, and cloud computation were mentioned in the report.

For smart agriculture, Saha et al. [16] suggested a tracking system. Here, sensors and a CC3200 single chip are used to track the temperature and humidity of the farm. The chip also has a camera interface attached to it for picture capture.

An IoT-dependent precision farming framework for groundnut manufacturing was proposed by the authors in [17]. Using IoT-dependent strategies, this research fosters to boost agricultural productivity. Agriculturists may presently recognize the characteristics of their crop and rapidly initiate the appropriate steps thanks to IoT-dependent approaches. WSNs were employed in this investigation. The network's sensor nodes collect information on farming parameters. The Android app may be utilized to alert the agriculturists. But the data which was assembled and surveyed in this research is on soil moisture.

Utilizing IoT, Riaz et al. [18] suggested crop tracking in farming. Furthermore to temperature and humidity sensors, other research publications have been studied in this research that mention the usage of leaf wetness sensors, soil moisture sensors, soil pH sensors, and atmospheric pressure sensors.

For waste management, Abdalzaher et al. [19] presented an IoT-dependent solution. This research highlights the significance of wet and dry wastes for the proper management of both kinds of trash. Prior waste management methods are then examined. To get the data, the previous systems used a lot of sensors. Some of them suggested using sensors to monitor trash levels in order to prevent garbage from overflowing. To monitor garbage levels, weight sensors and height sensors were used. The usage of ultrasonic sensors, moisture sensors, and flame sensors is suggested in this research. As a result, the suggested system gives users the ability to determine if the trash can is full. If the trash is moist then it can be utilized as a moisture sensor or if there is a fire in the trash it can be utilized as a flame sensor. The website notifies users of any of these occurrences and sends an SMS to their mobile device as well.

An IoT service-based farm monitoring system was presented by Reepu et al. [20]. Utilization of a WSN is made by the framework. The suggested system is for sophisticated, large-scale agricultural operations. It shows the integration of the Internet of Things, aerial photography, and Service-Oriented Architecture (SOA). IoT sensors, aerial sensors, and irrigation controls are all used by the system. For gathering and storing the data, it contains network administration middleware. Additionally, the system contains a MySQL database storage. A rule engine container and a rule engine core are present in the framework. The Rete pattern matching algorithm is used by the system. The rule container stores the parameter rules for irrigation use. The weather rules are also included in these guidelines. The condition portion and the action section are both included in every rule. The rule keyword is used to define rules, which are then followed by their names. A multimedia browser may be used to see the system's output.

An IoT-dependent framework for finding soil moisture level was proposed by Veeraiah et al. [21]. Here, a WSN yielding as a wireless moisture sensor network is employed. Temperature, humidity, and soil moisture are all calculated by the sensors employed in this network. Comparisons between the system's efficiency with and without a sensor network have been made. It has been noted that there is a typical 1500 mL save per day of water per tree. The proposed approach is helpful for figuring out the correct quantity of fertilizer and water to use.

An IoT-dependent framework for intelligent composting monitoring and control was also suggested by Khanh et al. [22]. According to this research, there are several types of organic wastes, such as plant leaves, food scraps, and meal leftovers that may be used to create organic manure. The soil's quality may be improved by turning these biodegradable elements into stable compost. The temperature rises significantly when the biodegradable substance is being broken down. The temperature drops after that. The breakdown process may thus be detected using a temperature sensor. Since there should be 60% moisture at this point in the breakdown process, a moisture sensor is also necessary. As a result, there are a number of variables that need to be watched when composting. These variables include the following: temperature, humidity, oxygen content, pH, carbon-to-nitrogen ratio, etc. A Precision farming technique depending on IoT and WebGIS was suggested by Pramanik [23]. There are many components of the precision agriculture management system that was employed in this research. The communication systems, sensor networks, and picture transmission systems make up the information infrastructure module. Image and sensor data are stored in databases that are maintained by the database module. The WebGIS module gathers, analyzes, and presents the geographical data of agricultural land. The neighborhood system module gathers and examines the sensor information. Data monitoring by the user is made possible through the mobile client module.

The many approaches to use IoT platforms to build smart farming solutions are covered in the next section.

11.3 Authorities of Smart Farming

11.3.1 Platforms for IoT in smart agriculture

11.3.1.1 IoT platforms for monitoring the weather

The micro-climate study is carried out to identify agricultural production risks and vulnerabilities to harsh weather. It is necessary to protect crops against meteorological events including heat and cold waves, hailstorms, cyclones, floods, and draughts. Utilizing IoT in agriculture aids in micro-climate control to lessen agricultural yield loss. The methods described below may be used to modify the microclimate.

1. Measurement of humidity and temperature
2. The presence of enough sunshine
3. Rainfall measurement
4. Detection of Soil Moisture

The required sensors and a data processing system are components of an IoT platform for micro-climate study.

11.3.1.2 Crop management IoT platforms

Crop management techniques include soil analysis, determining watering needs, checking crops for disease as they develop, applying fertilizers, and spraying pesticides. The IoT platform offers the sensors needed to measure numerous parameters, analyze the data, and take the necessary action.

Farming automation includes a big contribution from robotics. Robots are often used to gather fruits and vegetables. Machines for autonomous seeding and planting may stand in a larger area in a shorter amount of time. These devices employ sensor data to assess the density, moisture, and nutrition of the soil.

11.3.1.3 IoT platforms for managing live stock

Various sensors, including temperature sensors and other biosensors, may be used to monitor the health of farm animals. These are wearable sensors that enable remote animal health monitoring.

The robotic pills may be used to keep tabs on agricultural animals' health. The health of cows has an impact on how much milk they produce. Cows can accurately deposit robotic tablets in their stomachs by utilizing AI [24–28]. The tablets transmit location information together with health-related data. The location information assists in separating the ill cows from the rest of the herd so they may get treatment separately.

11.3.1.4 IoT platforms for managing logistics

The harvesting process is more effective because to the automation of tractors and agricultural equipment. AI may be used to pre-program self-driving cars or to make them autonomous. Sensors may be added to autonomous tractors to monitor the harvesting operation. These self-driving cars can be seen remotely. Along with their main duty, they also enable us to remotely gather the data for further analysis.

The use of autonomous self-driving tractors allows for the effective harvesting of huge tracts of land without the need for outside labor.

11.3.2 Sensors for smart agriculture

The use of several sensors makes precision agriculture possible. Farmers may learn a lot from the data the sensors offer regarding the crops and the environment. Agriculturists may enhance agricultural productivity when the use of sensors while utilizing fewer resources like water, fertilizer, and seeds. In smart agriculture, farming sensors are necessary. With the use of sensors, agriculturists may alter their crop yields to the nearby environment. Moreover, weather stations, UAV [29–32] or drones, or robots may all possess sensors connected to them. Mobile applications may be utilized to operate the sensors. Most of the optical sensors are connected in robots or drones since they can survey the whole area and take the required photos.

The following is a list of the sensors used in precision agriculture:

1. A position sensor is an example, like the NJR NJG1157PCD-TE1. With the use of satellites, location sensors help in the field's position. These sensors furnish longitude, latitude, and altitude information.
2. In precision agriculture, soil characteristics are crucial. The optical sensors can assess soil characteristics. These sensors can keep an eye on organic matter, plant disease, clay, and soil moisture. Drones that can be used to monitor the whole field may be equipped with these sensors.
3. The chemical composition of the soil may also be used to assess the soil's quality. It provides information on the soil's pH and nutrients. The chemical characteristics of soil are discovered using electrochemical sensors. These sensors look for certain ions in the soil.
4. By installing mechanical sensors in tractors, it is possible to assess the pulling demand in soil. A tensiometer, like the Honeywell FSG15N1A, is highly helpful in irrigation, for instance.
5. Utilizing a dielectric soil moisture sensor, dielectric constant evaluates soil moisture.
6. Airflow sensors are capable of measuring soil air permeability. These sensors aid in identifying crucial soil qualities.

Sensor usage may also result in higher crop yields. Applications for these sensors include guiding systems, variable rate fertilizers, weed mapping, variable spraying, border management, and yield monitoring.

The sensors can keep an eye on the health and conduct of cattle. For example, the sensors keep track on the health, behavior, and amount of activity of cows with inclusion of sensors, a variation of smartphone tools are obtainable to aid with smart agriculture. Among these instruments are a phone camera for photographing plants and illness for spotting a possible disease, a GPS for finding weeds, sun radiation, the requirement for fertilizing, and the placement of paste. It is possible to preserve farming machines utilizing microphones. Gyroscope and accelerometer are extra attributes on smartphones.

Various smartphone implementations are obtainable to help with smart agriculture. The following is a collection of few examples of these clever applications.

1. Applications for disease detection. These implementations employ plant and leaf photos to analyze disease occurrence utilizing ML [33–37] techniques.

2. Compute for fertilizer needs. These implementations calculate the requirement for fertilizers on a given region utilizing input data from various soil sensors.

3. Apps for soil analysis. These applications find soil quality by acquiring data from sensors which calculate the soil and the climate.

4. Apps which predict water needs. The applications receive picture data from selected fields and data from soil moisture sensors, and they find the amount of water required in a specific region.

Thus, the sensors provide farmers a variety of advantages. These make it possible to use less fuel and energy. As a result, there is a reduction in carbon dioxide emissions. The emission of nitrous oxide is decreased by using fewer fertilizers. It is possible to manage pests exactly. As a result, there is a reduction in chemical spraying. Additionally, the amount of water used may be reduced.

Thus, we can sum up that for precision farming, we can find fruit diameter and leaf moisture. The diameters of the stem and the trunk may also be measured. We can gauge the moisture of the soil for irrigation purposes. We can gauge temperature, humidity, and sun radiation for greenhouses. In irrigation systems, weather stations are also necessary. Therefore, we must record the temperature and humidity. Thus, we are able to use an anemometer, a wind vane, and a pluviometer.

In smart agriculture, sensors offer several advantages. They do, however, have certain shortcomings. The primary disadvantage of employing sensors is the constant need for Internet access. Electricity and Internet access may not always be available in remote locations. This condition makes smart farming less practical. Additionally, smart farming depends on the presence of necessary infrastructure, such as smart grids and cell towers, which may not be practical in rural parts of underdeveloped nations. Another disadvantage is that agriculturists are often not literate fully to use IoT methods.

Smart farming uses sensors, which creates a tremendous amount of data each day. Therefore, cloud storage is also necessary for smart farming. Again, cloud storage requires constant Internet connectivity. Edge computing may be used to get around the need for cloud storage. Edge computing enables the local storage and analysis of data at the network's edge that has been gathered by sensors. The centralized cloud storage should only be used to preserve processed data which will be used for upcoming analysis and usage. As contrast to centralized computing with cloud servers, edge computing enables the utilization of distributed computing.

The sensor technology utilized in smart agriculture brings about feasibility in AI and ML to get predictions. The usefulness of present crop categories and the possibility of future types may both be explored utilizing the data collected throughout time. Agriculturists may effectively finish their irrigation responsibilities thanks to sensor technology that aids them save time and resources.

11.3.3 Solutions for cloud and edge computing

The usage of sensor nodes has increased dramatically as a result of the IoT's introduction in a variety of precision agriculture applications. Irrigation systems, composting, environment tracking, soil scrutiny, plant disease tracking, animal health tracking, irrigation equipment, and logistics management are some of the areas where the Internet of Things might be useful. Each IoT applications for smart agriculture needs several sensor nodes linked to one another to create a WSN. These networks, which are designed for certain uses, routinely gather data. Each application transfers the data gathered by sensor networks to a cloud server for additional processing. It shows the requirement of both cloud server storage and constant Internet access. As a result, daily vast volumes of data that are often unnecessary are gathered by sensor networks and stored on cloud servers.

Assume, for example, that a robot is photographing plant leaves in order to look for illness. It is possible that a significant portion of the photos recorded show typical leaf texture. Only a few photos need more analysis. There is no need to upload all of the photographs to the cloud server in this circumstance. Before uploading to the cloud server, the photographs required to be filtered.

Similar to this, action and data storage are only necessary in environment monitoring applications when there is a rapid increase in temperature or rainfall. The summary data may be kept in cases of typical weather. Similar to this, every sensor data collection requires local network edge analysis and filtering before being sent to a central cloud server. It will do away with the necessity for constant Internet access and substantial storage. Additionally, as the amount of data sent decreases, bandwidth usage decreases as well, accelerating and significantly lowering the cost of the data transmission.

Edge computing will thus become a key tendency in smart agriculture. The following are some benefits of using edge computing versus centralized cloud computing:

1. If cloud computation is the sole option, all data must initially be sent to the cloud server before any more processing can be done. The processing results are sent back to the field equipment so that it may take the necessary action. Shifting raw data securely to a cloud server and avoiding data theft are challenges. The system will be susceptible to prospective cyberattacks because of the raw data flow. Sending the cloud server merely the summary data will stop this from happening. If the temperature doesn't fluctuate much during the day, for instance, merely the average temperature for that time period may be transmitted to the cloud server. The likelihood of cyberattacks is much decreased as a result of the lower traffic to and from the cloud server. Additionally, even if the data is somehow exposed, the attackers will not have all the information. Thus, edge computing technologies increase system security.

2. Imagine that an irrigation duty is being carried out by many sensor nodes in a WSN. Every sensor node in the network gathers data and delivers it to the cloud server on a regular basis. It will hold up analysis and answer sending. The system will be rendered useless if the agricultural equipment are monitored and managed remotely since prompt action cannot be performed when a problem like an animal's health or a fire breakout is detected. Since edge computing speeds up data flow, a cloud server's speedy response may be attained.

3. Using edge computing will also result in lower costs since it would eliminate the need for pricey cloud storage. Additionally, processing times are shorter, which lowers the cost of cloud servers.

A few application scenarios for smart farming are presented in the next section.

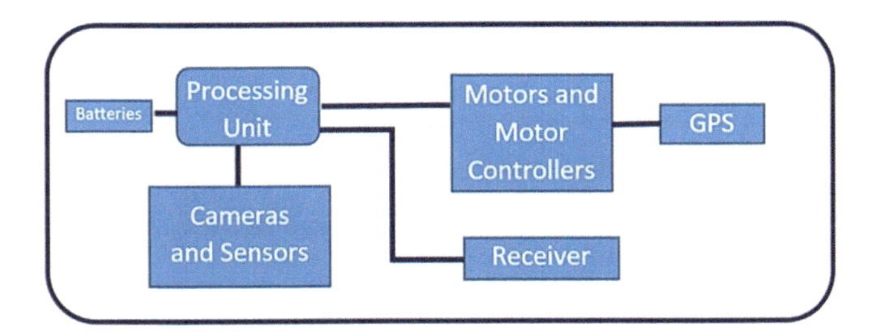

Figure 11.1 An unmanned ground vehicle block diagram.

11.3.4 Smart farming use cases in the IoT

11.3.4.1 Robotic agricultural vehicles (RAVs) or unmanned ground vehicles (UGVs)

Farms may use autonomous unmanned ground vehicles as sensor bearers. In order to identify the spread of a disease, the UGV may be fitted with cameras that can take pictures of the soil and plants and leaves.

Processing units, batteries, motors, GPS, motor controllers, remote control receivers, laser scanners, ultrasonic scanners, and cameras are the main components of a UGV (as shown in Figure 11.1).

The localization sensor used to locate objects is managed by the processing unit. The whole region is captured by cameras. The obstructions in the route are found using a laser scanner. Collisions are prevented using ultrasonic sensors. Three-dimensional posture estimation is produced by the accelerometer, gyrometer, and magnetometer. Direct connection between a human operator and the remote control receiver is made possible.

11.3.4.2 Drones or unmanned aerial vehicles (UAVs)

UAVs have a lot more uses in smart agriculture than unmanned ground vehicles (UGVs). UAVs equipped with cameras may be used to monitor the crop and the condition of the field by flying over it and taking pictures. Additionally, insecticides and fertilizers may be sprayed using UAVs. Depending on the situation, different spray dosages might be used.

Additionally, UAV is utilized to geolocate crops and vehicles and aids in the comprehension of soil variability. A microcontroller like the ATMega328 may serve as the processing component of a UAV (as shown in Figure 11.2).

11.3.4.3 Smart greenhouse

Vegetable, fruit, and other crop yields have been raised via the use of greenhouse farming. A smart greenhouse (as shown in Figure 11.3) uses sensor technologies to keep an eye on the development of the plants. A smart

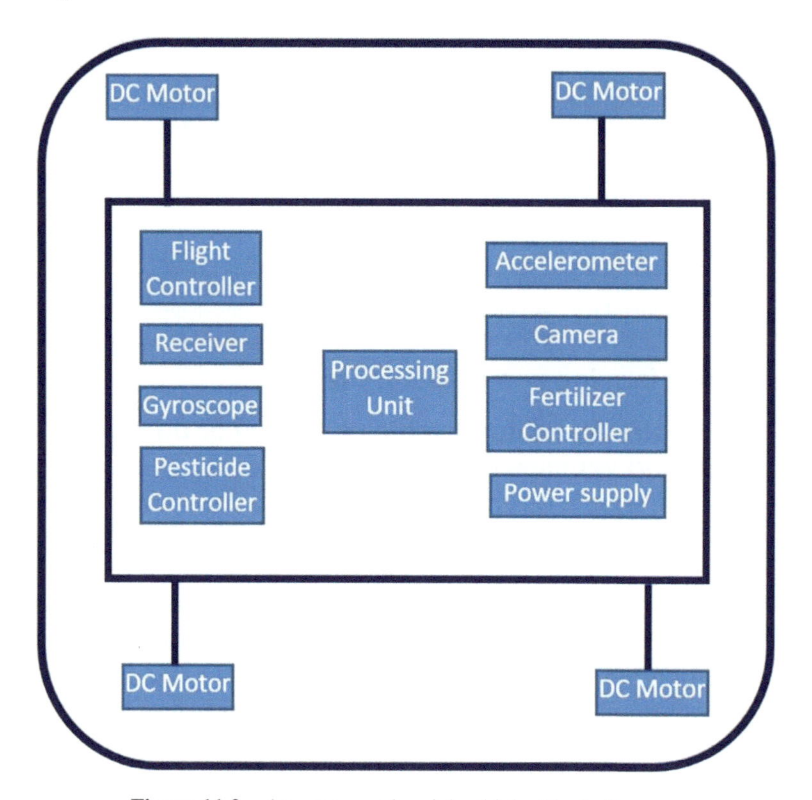

Figure 11.2 An unmanned aerial vehicles block diagram.

greenhouse monitors and regulates a number of climatic variables common sensors in smart homes sensors for temperature and humidity, light, pressure, and greenhouse sent to the cloud server for additional processing is the sensor data. The reaction produced by data processing may include activating an actuator to open a window, switch on a heater or fan, or perform any other activity. IoT utilization may contribute to a 100% increase in output.

11.4 Benefits of Smart Agriculture

11.4.1 Consumer advantages

With the use of technology, farmers can now more thoroughly monitor their crops. By using them sparingly, they may lessen the waste of various resources including water, pesticides, and fertilizers. Consequently, the total cost of farming may be decreased. Farmers may minimize crop loss by constantly monitoring crops for illnesses. Consequently, a higher agricultural

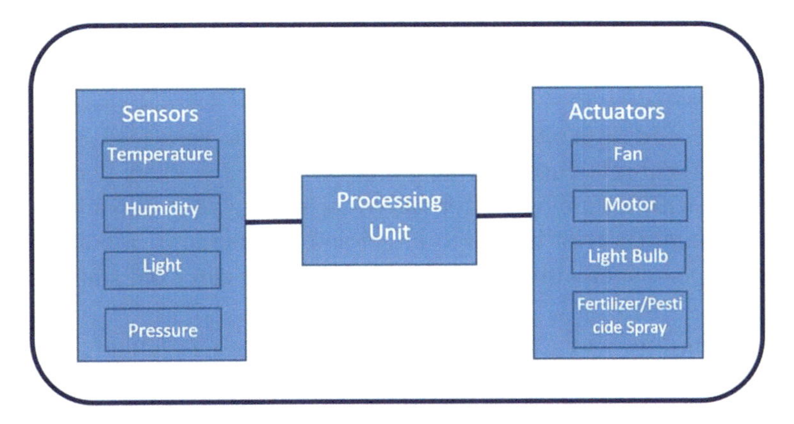

Figure 11.3 An illustration of a smart greenhouse.

production may be produced for less money. As a result of increased agricultural productivity brought about by automation, customers will pay less for farm products. Additionally, they will be delivered quicker and fresher.

11.4.2 Effective farm workforce

Instead of spending their time on the tiresome activities of watering, fertilizing, and pesticide spraying, farmers may spend more time studying agricultural data and developing new methods for collecting crops. They have much simpler labor because to the sophisticated machinery and trucks utilized in farming. Additionally, fewer personnel are needed in farming at a given time. Automation may thus help to relieve the labor shortage.

11.4.3 Environment-friendly agriculture

The usage of pesticides and fertilizers will be decreased, which will result in a decrease in greenhouse gas emissions. Environmental cleanliness will result from less chemical use. We can easily create compost using the smart farming method, which will benefit the crop.

11.5 Difficulties in Smart Agriculture

The practice of smart farming comes with a number of difficulties.

It is expensive, to start. Installing many sensors around the field is necessary for smart farming. These sensors need regular maintenance and observation. It's time to get new batteries. This all raises the price of farming.

Robots, drones, self-driving cars, and other autonomous technologies are also necessary for smart farming. As a result, it demands a large sum of money at first.

The second difficulty with smart farming is the need for laborers with a variety of skill sets. Smart farming requires skilled workers that are adept at working with technology, information technology, and agriculture. They should be familiar with both web-based and mobile apps. Finding a large number of personnel with the necessary abilities to carry out smart farming is difficult.

The third difficulty is that applications for smart agriculture need a lot of data. They gather a lot of data, which has to be stored and examined. Cloud-dependent remedies are required. But, it will result in a delay in the data being sent to the cloud server for processing and receiving the result, rendering the implementation useless.

Electricity and a constant Internet connection are also necessary for smart farming. It is a significant issue that has to be addressed in many areas.

11.6 Conclusion

The researchers require a remedy to the issue of having food accessible to all as the world's population continues to rise each year. The existing agricultural practices are insufficient to meet the rising food demand. By improving agricultural productivity, smart agriculture with the utilization of technology may provide a solution. The utilization of sensor technologies, robotics and autonomous systems, cloud and edge computing remedies, and online and mobile implementations, makes smart farming viable.

References

[1] A. K. Sangaiah, A. Javadpour, F. Ja'fari, H. Zavieh and S. M. Khaniabadi, "SALA-IoT: Self-Reduced Internet of Things with Learning Automaton Sleep Scheduling Algorithm," in IEEE Sensors Journal, doi: 10.1109/JSEN.2023.3242759.

[2] I. Kapelyukh, V. Vosylius and E. Johns, "DALL-E-Bot: Introducing Web-Scale Diffusion Models to Robotics," in IEEE Robotics and Automation Letters, vol. 8, no. 7, pp. 3956–3963, July 2023, doi: 10.1109/LRA.2023.3272516.

[3] W. Wen et al., "Health Monitoring and Diagnosis for Geo-Distributed Edge Ecosystem in Smart City," in IEEE Internet of Things Journal, doi: 10.1109/JIOT.2023.3247640.

[4] P. Onu, C. Mbohwa and A. Pradhan, "Artificial intelligence-based IoT-enabled biogas production," 2023 International Conference on Control, Automation and Diagnosis (ICCAD), Rome, Italy, 2023, pp. 1–6, doi: 10.1109/ICCAD57653.2023.10152349.

[5] G. Singh and J. Singh, "A Fog Computing based Agriculture-IoT Framework for Detection of Alert Conditions and Effective Crop Protection," 2023 5th International Conference on Smart Systems and Inventive Technology (ICSSIT), Tirunelveli, India, 2023, pp. 537–543, doi: 10.1109/ICSSIT55814.2023.10060995.

[6] K. -V. Nguyen, C. -H. Nguyen, T. V. Do and C. Rotter, "Efficient Multi-UAV Assisted Data Gathering Schemes for Maximizing the Operation Time of Wireless Sensor Networks in Precision Farming," in IEEE Transactions on Industrial Informatics, doi: 10.1109/TII.2023.3248616.

[7] K. Kumar and Rikendra, "Recent advancements of Internet of Things in Precision Agriculture: A Review," 2023 International Conference on Disruptive Technologies (ICDT), Greater Noida, India, 2023, pp. 482–485, doi: 10.1109/ICDT57929.2023.10150981.

[8] K. Il Ko, M. Hun Lee and H. Yoe, "CWSI-based Smart Irrigation System Design," 2023 International Conference on Artificial Intelligence in Information and Communication (ICAIIC), Bali, Indonesia, 2023, pp. 037–040, doi: 10.1109/ICAIIC57133.2023.10066983.

[9] S. Tahir, Y. Hafeez, F. Qamar and G. N. Alwakid, "Intelli-farm: IoT based Smart farming using Machine learning approaches," 2023 International Conference on Business Analytics for Technology and Security (ICBATS), Dubai, United Arab Emirates, 2023, pp. 1–6, doi: 10.1109/ICBATS57792.2023.10111232.

[10] S. Trivedi, T. Anh Tran, N. Faruqui and M. M. Hassan, "An Exploratory Analysis of Effect of Adversarial Machine Learning Attack on IoT-enabled Industrial Control Systems," 2023 International Conference on Smart Computing and Application (ICSCA), Hail, Saudi Arabia, 2023, pp. 1–8, doi: 10.1109/ICSCA57840.2023.10087713.

[11] R. Deepa, M. Sankar, R. R, C. Sankari, Venkatasubramanian and R. Kalaivani, "IoT based Energy Efficient using Wireless Sensor Network Application to Smart Agriculture," 2023 International Conference on Intelligent Data Communication Technologies and Internet of Things (IDCIoT), Bengaluru, India, 2023, pp. 90–95, doi: 10.1109/IDCIoT56793.2023.10053446.

[12] K. Memon, F. A. Umrani, A. Baqai and Z. S. Syed, "A Review Based On Comparative Analysis of Techniques Used in Precision Agriculture," 2023 4th International Conference on Computing, Mathematics and

Engineering Technologies (iCoMET), Sukkur, Pakistan, 2023, pp. 1–7, doi: 10.1109/iCoMET57998.2023.10099182.

[13] D. R. Sharma, V. Mishra and S. Srivastava, “Enhancing Crop Yields through IoT-Enabled Precision Agriculture,” 2023 International Conference on Disruptive Technologies (ICDT), Greater Noida, India, 2023, pp. 279–283, doi: 10.1109/ICDT57929.2023.10151422.

[14] S. Javaid et al., “Communication and Control in Collaborative UAVs: Recent Advances and Future Trends,” in IEEE Transactions on Intelligent Transportation Systems, vol. 24, no. 6, pp. 5719–5739, June 2023, doi: 10.1109/TITS.2023.3248841.

[15] E. S. Akpabio, K. F. Akeju, K. O. Omotoso, F. Ohunakin and R. O. Ogundokun, “Climate Smart Agriculture and Nigeria’s SDG 2 Prospects: A Review,” 2023 International Conference on Science, Engineering and Business for Sustainable Development Goals (SEB-SDG), Omu-Aran, Nigeria, 2023, pp. 1–6, doi: 10.1109/SEB-SDG57117.2023.10124557.

[16] P. Saha, V. Kumar, S. Kathuria, A. Gehlot, V. Pachouri and A. S. Duggal, “Precision Agriculture Using Internet of Things and Wireless Sensor Networks,” 2023 International Conference on Disruptive Technologies (ICDT), Greater Noida, India, 2023, pp. 519–522, doi: 10.1109/ICDT57929.2023.10150678.

[17] A. K. D B, D. N, B. Bairwa, A. K. C S, G. Raju and Madhu, “IoT-based Water Harvesting, Moisture Monitoring, and Crop Monitoring System for Precision Agriculture,” 2023 International Conference on Distributed Computing and Electrical Circuits and Electronics (ICDCECE), Ballar, India, 2023, pp. 1–6, doi: 10.1109/ICDCECE57866.2023.10150893.

[18] M. H. Riaz, H. Imran, H. Alam, M. A. Alam and N. Z. Butt, “Crop-Specific Optimization of Bifacial PV Arrays for Agrivoltaic Food-Energy Production: The Light-Productivity-Factor Approach,” in IEEE Journal of Photovoltaics, vol. 12, no. 2, pp. 572–580, March 2022, doi: 10.1109/JPHOTOV.2021.3136158.

[19] M. S. Abdalzaher, M. M. Salim, H. A. Elsayed and M. M. Fouda, “Machine Learning Benchmarking for Secured IoT Smart Systems,” 2022 IEEE International Conference on Internet of Things and Intelligence Systems (IoTaIS), BALI, Indonesia, 2022, pp. 50–56, doi: 10.1109/IoTaIS56727.2022.9975952.

[20] Reepu, S. Kumar, M. G. Chaudhary, K. G. Gupta, S. Pramanik and A. Gupta, Information Security and Privacy in IoT, in Handbook of Research in Advancements in AI and IoT Convergence Technologies,

Eds, J. Zhao, V. V. Kumar, R. Natarajan and T. R. Mahesh, IGI Global, 2023.

[21] V. Veeraiah, V. Talukdar, Manikandan K., S. B. Talukdar, V. D. Solavande, S. Pramanik, A. Gupta, Machine Learning Frameworks in Carpooling, in Handbook of Research on AI and Machine Learning Applications in Customer Support and Analytics, Eds, Md S. Hossain, R. C. Ho and G. Trajkovski, IGI Global, 2023.

[22] P. T. Khanh, T. H. Ngọc, and S. Pramanik, Future of Smart Agriculture Techniques and Applications, in Advanced Technologies and AI-Equipped IoT Applications in High Tech Agriculture, Eds, A. Khang, IGI Global, 2023.

[23] S. Pramanik, "An Effective Secured Privacy-Protecting Data Aggregation Method in IoT", in Achieving Full Realization and Mitigating the Challenges of the Internet of Things, Eds, M. O. Odhiambo and W. Mwashita, IGI Global, 2022, DOI: 10.4018/978-1-7998-9312-7.ch008

[24] S. Pramanik and S. Bandyopadhyay, Identifying Disease and Diagnosis in Females using Machine Learning, in Encyclopedia of Data Science and Machine Learning, Eds. John Wang, IGI Global, 2023, DOI: 10.4018/978-1-7998-9220-5.ch187.

[25] R. Anand, J. Singh, D. Pandey, B. K.Pandey, V. K.Nassa and S. Pramanik, "Modern Technique for Interactive Communication in LEACH-Based Ad Hoc Wireless Sensor Network", in Software Defined Networking for Ad Hoc Networks, M. M. Ghonge, S. Pramanik and A. D. Potgantwar, Eds, Springer, 2022, https://doi.org/10.1007/978-3-030-91149-2_3

[26] B. K. Pandey, D. Pandey, S. Wairya, G. Agarwal, P. Dadeech, S. R. Dogiwal and S. Pramanik, "Application of Integrated Steganography and Image Compressing Techniques for Confidential Information Transmission", in Cyber Security and Network Security, https://doi.org/10.1002/9781119812555.ch8, Eds, Wiley, 2022.

[27] A. Mandal, S. Dutta, S. Pramanik, "Machine Intelligence of Pi from Geometrical Figures with Variable Parameters using SCILab", in Methodologies and Applications of Computational Statistics for Machine Learning, D. Samanta, R. R. Althar, S. Pramanik and S. Dutta, Eds, IGI Global, 2021, pp. 38-63, DOI: 10.4018/978-1-7998-7701-1.ch003

[28] Y. Meslie, W. Enbeyle, B. K. Pandey, S. Pramanik, D. Pandey, P. Dadeech, A. Belay, A. Saini, "Machine Intelligence-based Trend Analysis of COVID-19 for Total Daily Confirmed Cases in Asia and Africa", in Methodologies and Applications of Computational Statistics

for Machine Learning, D. Samanta, R. R. Althar, S. Pramanik and S. Dutta, Eds, IGI Global, 2021, pp. 164–185, DOI: 10.4018/978-1-7998-7701-1.ch009

[29] A. Bhattacharya, A. Ghosal, A. J. Obaid, S. Krit, V. K. Shukla, K. Mandal and S. Pramanik, "Unsupervised Summarization Approach with Computational Statistics of Microblog Data", in Methodologies and Applications of Computational Statistics for Machine Learning, D. Samanta, R. R. Althar, S. Pramanik and S. Dutta, Eds, IGI Global, 2021, pp. 23–37, DOI: 10.4018/978-1-7998-7701-1.ch002

[30] S. Pramanik, K. M. Sagayam and O. P. Jena, Machine Learning Frameworks in Cancer Detection, ICCSRE 2021, Morocco, 2021.

[31] S. Pramanik, M. G. Galety, D. Samanta, N. P Joseph, Data Mining Approaches for Decision Support Systems, 3rd International Conference on Emerging Technologies in Data Mining and Information Security, 2022.

[32] D. Samanta, S. Dutta, M. G. Galety and S. Pramanik, A Novel Approach for Web Mining Taxonomy for High-Performance Computing, The 4th International Conference of Computer Science and Renewable Energies (ICCSRE'2021), 2021, DOI:10.1051/e3sconf/202129701073.

[33] R. Bansal, A. J. Obaid, A. Gupta, R. Singh, S. Pramanik "Impact of Big Data on Digital Transformation in 5G Era", 2nd International Conference on Physics and Applied Sciences (ICPAS 2021), doi:10.1088/1742-6596/1963/1/012170, 2021.

[34] S. Pramanik, Niranjanamurthy M. and S. N. Panda, Using Green Energy Prediction in Data Centers for Scheduling Service Jobs, ICRITCSA 2022, Bengaluru, 2022.

[35] V. Vidya Chellam, V. Veeraiah, A. Khanna, T. H. Sheikh, S. Pramanik and D. Dhabliya, A Machine Vision-based Approach for Tuberculosis Identification in Chest X-Rays Images of Patients, ICICC 2023, Springer

[36] S. Praveenkumar, V. Veeraiah, S. Pramanik, S. M. Basha, A. V. Lira Neto, V. H. C. De Albuquerque and A. Gupta, Prediction of Patients' Incurable Diseases Utilizing Deep Learning Approaches, ICICC 2023, Springer

[37] S. Pramanik, An Adaptive Image Steganography Approach depending on Integer Wavelet Transform and Genetic Algorithm, Multimedia Tools and Applications, 2023, https://doi.org/10.1007/s11042-023-14505-y.

[38] P. T. Khanh, T.H.Ngọc, and S. Pramanik, Future of Smart Agriculture Techniques and Applications, in Advanced Technologies and AI-Equipped IoT Applications in High Tech Agriculture, Eds, A.Khang, IGI Global, 2023.

12

Using the VIKOR Model: How UAVs Help Precision Agriculture in Agri-Food 4.0

Kaushal Kishor[1], R. Murugan[2], Sunil Kumar[3], Parth Sharma[4], Prashant Dhage[4], Dharmesh Dhabliya[5], and Sabyasachi Pramanik[6]

[1]Department of Agriculture, Sanskriti University, India
[2]Department of Computer Science and IT, School of CS and IT, Jain (Deemed to be) University, India
[3]College of Agriculture Science, Teerthanker Mahaveer University, India
[4]Symbiosis Law School, Nagpur Campus, Symbiosis International (Deemed University), India
[5]Department of Information Technology, Vishwakarma Institute of Information Technology, India
[6]Haldia Institute of Technology, India

Email: kaushal.ag@sanskriti.edu.in; murugan@jainuniversity.ac.in; sunilagro.chaudhary@gmail.com; parthsharma@slsnagpur.edu.in; prashantdhage@slsnagpur; dharmesh.dhabliya@viit.ac.in; sabyalnt@gmail.com

Abstract

The Fourth Industrial Revolution is a potential innovative avenue for individual existence that may see robots and artificial intelligence take the place of humans and physical labor. All facets of human existence are adopting and using the Fourth Industrial Revolution (4IR) principle. While business leaders prepare for the inevitable and quick adoption in the energy, automotive, communications, facilities, security, medical, and various organization divisions, academicians are doing rigorous study into the revolution. As the largest employer of human resources, the agriculture and food sector—also known as "Food 4.0"—is a significant industry which is anticipated to greatly aid from the idea and implementation of 4IR as it moves into a new age of growth.

12.1 Introduction

12.1.1 The First to Fourth Industrial Revolutions

IR refers to the displacement of one or other existing technologies by newer creations whose adoption and spread should result in an abrupt shift in human growth during the course of a condensed era of technological advancement. Three prior industrial revolutions have been emphasized in various writings. With the introduction of mechanization using steam and waterpower, the First Industrial Revolution (Industry 1.0) [1] began to emerge at the end of the 18th century. The use of mechanized steam engine systems increased the output capacity of several industrial sectors. The Second Industrial Revolution, often known as Industrial 2.0 [2], took place during the late eighteenth and earlier twentieth centuries. Through the utilization of electrical energy, conveyor belt technology, the invention of the telegraph, and other methods, this period gave rise to mass manufacturing. Automation, the usage of electronic equipment, and the development of computer and information technology were all brought about by the Third IR, or Industry 3.0. The events in the middle of the twentieth century culminated in the 1970s. Digital development was rapidly advancing at this time. In the extractive, manufacturing, processing, and engineering facilities industries, automation, and IT were major economic drivers. Between the beginning of the First IR and the commencement of the Third IR, there were around 25 decades. The Fourth IR, also known as Industry 4.0 or 4IR, or the age of cyber–physical systems, is now in existence. It is the age of technological integration of the physical and virtual worlds. Figure 12.1 shows how the IR progressed from Industry 1.0 to Industry 4.0 and the characteristics of every age.

The Fourth IR (4IR) is a newly coined term that refers to a period of fast social and economic change brought about by disruptive supply- and demand-driven technology advancements. In 2011, the German federal government unveiled its “Industry 4.0” high-tech plan, which subsequently inspired the 2016 Davos conference’s “4th IR (4IR)” concept. The creator and chairman of 4IR [3], Klaus Schwab, stated that the Fourth Industrial Revolution (4IR) has come as an uprising with an altogether new realm and effect from the past IRs during the World Economic Forum (WEF) annual conference in 2016. He is entitled to introduce the 4IR since he used the phrase “Fourth Industrial Revolution” in his study. The word, according to the University of

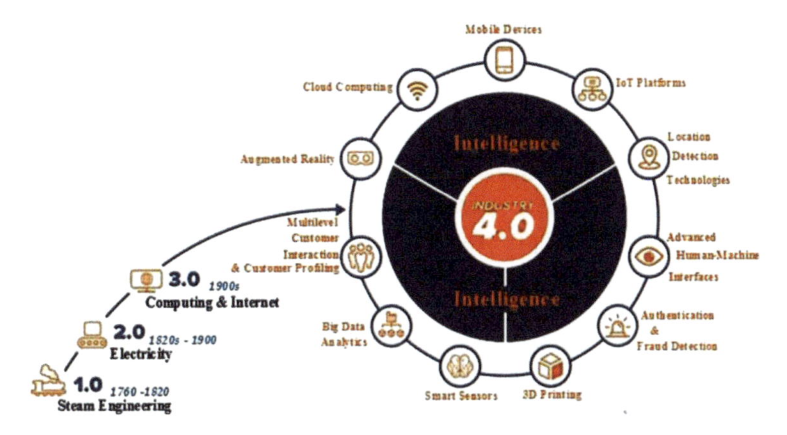

Figure 12.1 The features of the First through Fourth IRs.

California, may be traced back to Rostow (1983), a research article published by W. Rostow.

The breadth of 4IR's coverage goes well beyond the simple integration of technology into numerous facets of social advancement. It is a crucial part of our near future. The notion and defining terms are continuously evolving and are not yet well defined. The notion of the Fourth Industrial Revolution (4IR) is still developing, but individuals, businesses, and policymakers are already looking at its economic prospects, as was previously said about prior IRs at an early stage. The 4IR has unique characteristics that set it apart from prior revolutions, despite the American House of Commons' assessment in which the word is poorly interpreted with highly dramatic impacts on society and the economy. The simultaneous, real-time, and shifting advance in the technological uprising across several industries is proof that 4IR is in operation. Gene sequencing is being replaced by nanotechnology. Renewable energy is being replaced by quantum computing. The fusion of these technologies across several domains has given rise to 4IR. By linking large populations online, 4IR technologies have the potential to significantly enhance the environment, corporate productivity, asset management, organizational uptime, and other areas. The world is moving away from fundamental framework automation, the use of electronic devices, computer and IT infrastructures, and toward a fusion of the real and virtual worlds thanks to 4IR. Interoperability, autonomy, and cutting-edge artificial intelligence will make fusion possible in the 4IR age. According to [4], the distinctive characteristics of 4IR are to combine breakthroughs in digital, physical, and biological technology, integrate those facilities over the Internet for a quick and low-cost global technological revolution, and produce a wide range of influences on human existence.

The technologies that will alter human growth and change society over the next 10 years are still in the early stages of development and are mostly archetypal experiments. The catalyst for progress in the 4IR is the confluence of various technologies. Among these technologies are ML, the IoT, latest AI, intelligent materials, 3-D printing, transportation technology, bio-informatics and bio-fabrication, robotics, and blockchain. Through technology advancement, 4IR has now brought about tangible improvements to the economic and social aspects of human existence. According to [5], cyber–physical systems have enabled people to work together to use hardware, software, and virtual technologies for all aspects of production and maintenance. The permitted technologies were grouped under the distinctive characteristics of 4IR by Klaus Schwab. Robotics, 3-D printing, autonomous cars, novel materials, and other physical technology are examples. Digital technologies consist of, but are unlimited, Bitcoin, the IoTs, and blockchain. Genetic sequencing, 3-D printing, synthetic biology, and genetic modification are examples of biological technologies. 4IR will open up a lot of scopes, but with the unavoidable difficulties of calculated complexity brought on by the unreliability which evolves with newer technological advancements. Three main categories have been determined by studying the 4IR's potential and problems. First, the development of digital technology has defined mankind. Second, the development of digital technology has greatly benefited mankind. Third, despite the advantages that digital technology has brought to mankind, it also comes with problems and drawbacks.

The 4IR is a potential innovative path that might replace or supplement human existence, robotics and artificial intelligence combined with human intellect and physical labor. All facets of human existence are adopting and using the 4IR principle. While business leaders prepare for the inevitable and quick deployment in the energy, automotive, communications, services, security, and various industrial sectors, academics are doing rigorous study into the revolution. Known as "food 4.0," the agriculture and food industry is the largest employer of human resources, a significant industry which is anticipated to gain much from the idea and use of 4IR in bringing the industry into the next age of progress.

12.2 Food and Industrial Revolutions

The food business has probably been the most important sector for the advancement of the world economy and the survival of the populace since the industrial revolution. Concern about food supply is linked with the ongoing rise in global population. Despite the fact that food production must increase to meet the needs of nine billion people by the middle of the twenty-first century, there has been tremendous food growth in the last 50 years. According to a recent research, in order to meet the rising demand for food without seeing a large price hike, production levels would need to expand by between 75% and 100%. According to [6], the proportion of the making improved and GDP that is dependent on the food industry is quite higher in the majority of nations. The sector's degree of significance has made the problems it faces a crucial and worldwide problem.

According to [7], the difficulties are mainly connected to rising food needs, ecological protection, food security, climate change, governmental policies, market and trade concerns, and other things. To stay afloat and compete in the market, the food industry is continually introducing new strategies to fight these problems. These strategies have continuously focused on enhancing technical knowledge in conjunction with the contemporaneous industrial revolution. However, a significant issue in the industry is the need of high costs for small- and medium-scale farmers in underdeveloped nations to embrace innovative technology. The fundamentals of early man's success in evolving from a hunter to a hunted via his sense for food production may be used to characterize the components of food revolutions. Agriculture and the food sector have been serving three fundamental purposes across industrialization eras: increasing production and manufacture to feed the burgeoning non-agricultural population, creating an excess of the expert labors needed for firms and cities, and generating capital assemblage to be used in various current spheres of the economy.

12.2.1 1.0 Food in firms

Through its technical advancement, the manufacturing sector has always been a catalyst for the progress of the food business throughout history. Advances in the industrialization of food processing occurred during Industry 1.0 that is also frequently called as the "Agricultural Revolution," notably due to system mechanization and steam engine uses. In the beginning, food processing and farming were made possible by horsepower, as seen in Figure 12.2. The revolution was made possible by the significant developments in Great Britain's agricultural industry starting in the mid-1600s. Up until the mid-1840s, when agricultural science and engineering reached their peak of maturity, advancements made possible by technology and capital investment remained limited throughout this time. In comparison to other European nations, America had

(a) (b)

Figure 12.2 Industry 1.0's food revolution in horsepower: (a) cultivation and (b) processing.

a less than 2/3 farming share via the functional population engaged in agriculture earlier in the mid-1750s. The British government passed Enclosure Acts, which gave affluent landowners ownership and tenancy over small-scale farmers' agricultural holdings. The one-time occurrence caused a large exodus of some former farmers to the metropolis and overseas, producing a sharp decline in the number of farmers and the amount of food produced in rural areas, but a growth in the industrial operating class in urban areas. As a result, the time period benefited from the new commercial farming techniques by using better crop management, commonly called as four-field system. In order to avoid leaving any fields fallow while keeping a higher deposition of minerals and nutrients in an area, crops are rotated on farms every season.

Additionally, there was a vigorous introduction of agricultural equipment for more productive and efficient farming between 1800 and 1850. The equipment covered every facet of food production, with grain threshers, automated seed planters, reapers, and harvesters. In addition, fertilizer usage was introduced to boost agricultural productivity, which significantly expanded the amount of land available for food production. Animal food output also increased steadily throughout this time period. New breeds were introduced and improved methods of raising animals were arranged. As a result, more robust, bigger, and higher-quality animals were produced.

12.2.2 2.0 Food in firms

The food industry underwent mass production during the Industry 2.0 era thanks to electricity-based technology. In order to implement the new systems, new agricultural techniques that are based on the production of fodder crops and stall-fed animals have to be used. It was also necessary to employ new tools and implements, which presented the traditional difficulty

(a) (b)

Figure 12.3 Using steam engine technology, the food revolution is built on (a) cultivation and (b) processing.

of the expense of implementing new entities. Industry 2.0 brought forth application-specific technology advancements that were mostly reliant on the kind of crop and the climate. During this time, the yield of food crops was enhanced by the use of commercial quantities of chemically manufactured fertilizers. Through a development in research and technology, fungicide and insecticides were also introduced in addition to fertilizer additions. A French botanist named Pierre-Marrie Millardet created the Bordeaux blend of fungicide in 1885 to assist Ireland fight off an epidemic of a terrible potato disease. However, there was a problem with the food supply bcing negatively impacted by outside influence from other industries. Although the first negative impact was severe, food output and availability steadily increased over time.

Agriculture and the food revolution advanced more slowly during Industry 2.0 than most other industries. The industrial and engineering industries, as opposed to agriculture and the food sector, have benefited from localization. The sorts of technology produced for food production were moreover heavily reliant on animals for a power source despite the development of electricity and the steam engine. Due to the requirement for the power sources to be located on the production site, the advantages of the technical breakthrough of the steam engine were comparatively absent on the agricultural and food industry. At first, just a small portion of the food production was mechanized. According to reports, the winnowing machine was developed by Oliver Baker in the United States in 1977, and the threshing machine by Anthony Kepp in the United Kingdom in 1984. According to [8], the first tractors having full internal combustion engines were created and used in the Atlantic region in 1786. Figure 12.3 shows an automated food production system with the help of a steam engine.

12.2.3 3.0 Food in industry

In Industry 3.0, the agricultural and food industries saw a change near the end of the twentieth century. It was the Green Revolution, sometimes referred to as the agricultural revolution. The time was characterized by a revolution in digital technology, including computers, electronics, semiconductors, robotics, and automation that saw the introduction of specialized equipment into the many operational sectors of agricultural and food production. Crop methods, storage procedures, and animal production all showed significant advancements and enhanced productivity. The agricultural and food industries in the American countries were the leader to undergo a technological revolution, which led to a dramatic rise in yield via mechanized agriculture and extremely huge livestock output by indoor techniques of animal raising, having a hidden cost. With the supply of fresh seedlings, irrigation and chemical inputs, lenient loans, and expanded extension facilities, governments aided farmers. To satisfy the growing population's increased need for food, there was a concerted worldwide push to end hunger and raise food indices across nations. Some positive outcomes came from the initiatives. Having a cumulative of $82.36 billion in sales for the year, the United States had the biggest food sector sales in 1966.

Agriculture and food had a significant technical improvement during Industry 3.0. In 1982, Monsanto Company launched the well-known genetic alteration of plant and agricultural cells. The potato crop, a variety which was defiant to pests and herbicides, served as the first experimental specimen. Additionally, the development of software, IT, and mobile devices at this time had a significant influence on the food industry. In order to keep coworkers in the loop of communication, farmers and extension workers started using mobile devices in 2004. Mobile devices were also used as part of the enabling systems in the fields of marketing, transportation, and food storage. With proper real-time access to data on the seedlings and chemical inputs, the gadgets assist employees in staying connected both within and outside of the field. These aided farmers in increasing food yields and harvests. Additionally, novel pest-resistant crop varieties, inventive irrigation technology, and improved communication channels were all made possible by advancements in ICT and the usage of mobile devices. The year 2015 saw the advent of big data analytics, which has subsequently transformed the food industry. Data is the king, and data is a lovely bride. Now, agriculturists and others who work in the food industry might utilize data to make educated choices and prevent losses via the sustainable usage of resources. A prominent instance is a digital platform created by Climate FieldView

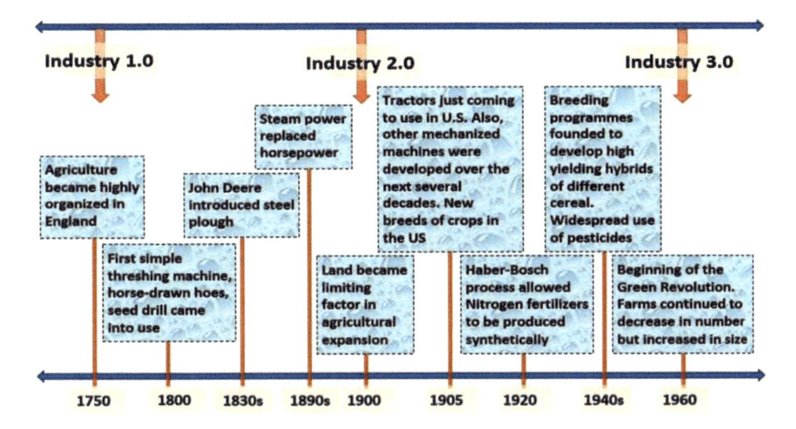

Figure 12.4 First to third food revolution timeline.

that combines data collecting, agronomic modeling, weather monitoring, and improved field expertise for farmers. The platform helps the farmer prepare for a healthy crop and encourages environmentally sound actions. In conclusion, the food industry's knowledge and information deficit was closed thanks to the linking of knowledge and innovation through ICT infrastructures. Mobile phone apps significantly helped information sharing for pest monitoring and disease surveillance in the same year that big data was introduced.

Industry 3.0 saw a lot of success in the food industry, but there were also a lot of difficulties throughout the time. The unequal distribution of accomplishments from one location to another was the first setback. Food production increased, although in some locations it did so slowly while in others it increased quickly. Famine still existed despite the rise in incentives that encouraged a decrease in hunger. The increasing food output also resulted in greater production costs and a worsening of the environment, which contributed to climate change. The disappearance of arable land and the development of pests and illnesses that were resistant to traditional agricultural methods presented another problem for rural communities. In reality, the sector's technical development did not provide a guarantee for the availability of food, food security, a fair distribution of income, or a healthy ecology. As a result, the emergence of the later food revolution, known as Food 4.0, was inevitable given the interaction of the problems and triumphs. Figure 12.4 illustrates the history of the food revolutions from Industry 1.0 to Industry 3.0.

12.2.4 Food in Industry 4.0

Industry 4.0's methodologies, methods, technologies, and strategies are presently what drive the agricultural and food industries. The food industry now makes use of cutting-edge ICTs that take the capabilities of the IoT, contemporary equipment, and tools. The implementations of the 4IR have ushered in a newer age of food production known as Food 4.0, where automation, connectivity, digitization, the use of renewable energies, and the effective usage of resources are principal.

4IR is the age of cyber–physical systems. With helpful technologies like intelligent sensors, robotics, smart greenhouses, drones, vertical farms, WSNs, 4IR has huge advantages for precision and food production systems. Other positives of the 4IR include digitization and contemporary technology, with a focus on the economic, social, and ecological problems in the food production industry. The quarter's goals are to maximize the use of renewable energy, water, fuel, and chemicals while reducing costs.

Before the invention of 4IR, chance had played a significant role in the growth of agriculture and food. The evolution of the sector has been an example of art based on changes in civilization and generation. Although relatively modest, the sector's technological development was clearly defined and gradually developed on a succession basis. Both in terms of idea and applications, the 4IR are distinctive. Food 4.0 is therefore already in a position to greatly profit from the revolution's elements. The 4IR's planning, direction, and emphasis on effort and financial components are the areas on which the food sector is most focused. Food 4.0 is a result of the necessity to prevent famine and poverty and provide access to inexpensive food for the anticipated increase in the global population.

According to projections, the food business will develop into a high-tech industry with all of its segments connected with technologies like AI [9–13], the IoT, robots [14–18], big data [19–23], and others. With the help of farm equipment, agricultural systems will be combined into a single entity that combines farm management, irrigation, and production forecasts. The development will affect all facets of society, including social, economic, and business areas.

Business models are an expression of ethical principles alongside many sectors. As shown in Figure 12.5, the combination of 4IR automations will have three significant effects on the food industry.

The answers for many problems will be improved by precise optimization, particularly in the sphere of food waste. The agricultural sector's corresponding character points to instability in the inputs and products. According

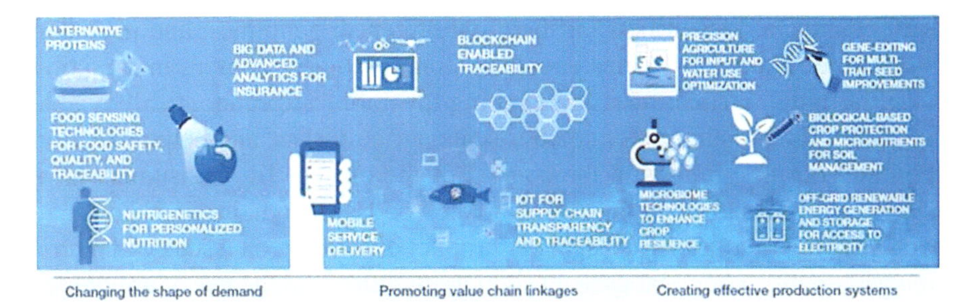

Figure 12.5 4IR technologies combined to allow advancements in the food industry.

to the FAO white report, there is sufficient food produced worldwide to feed everyone. Nevertheless, a large number of people continue to perish from hunger as a result of food production waste. About 75% of the cumulative water generated is used for agriculture, 35% of the viable crop is grown, and the rest water and 35%–50% of the food produced are squandered. The reversal of human resources and rural production components will also have an influence on farming and food production. Prior IRs saw significant worker movement from rural to urban areas (particularly Industry 2.0). Due to the room for leisure, the workforce in cities would choose rural labor during 4IR. Therefore, throughout the present revolution, labor, money, and technology resources that were formerly abandoned by agricultural communities will return. In addition, the Food 4.0 processes will have a stronger rapport with our environment. The 4IR technologies will make it possible for the food production technique to gain from precise and definite weather forecast and control. Since the improvements in 4IR would allow a higher degree of decision-making and implementations, the food systems won't be subject to alterations of nature and human intelligence. Climate change-related pest and disease difficulties, as well as the exorbitant costs of processing and producing cattle in the immediate aftermath of the last revolution, will no longer exist in the present. In summary, the Fourth Industrial Revolution (4IR), which focuses on sustainability and agriculture, has begun. All of the sustainable development objectives listed in Table 12.1 has a lot in common with the uprising in food systems that are presented in this chapter.

12.3 The Need for Food and Its Adaptation 4.0

The main takeaway from Food 4.0 is that 4IR offers the food business a great potential to develop into the ideal condition of a "Food Renaissance." This is a phase in which the key stakeholders embrace new technology, such as big data analytics, robots, artificial intelligence, genomics, and precision agriculture. The technologies enable customers to access more information about farming practices, assist in creating favorable and economically viable functional circumstances for smaller-scale agriculturists, and help reposition food and farming brands in the context of the present IR. Consumers have access to information about the complete food value chain, and they may even alter their requests for a specific product depending on issues with nutrition and quality, chemical usage, origin tractability, and the use of antibiotics in the processing of animals. An investment in new technology is necessary to meet the needs. The motivation for smallholder farmers comes from a gestation period to concentrate on new entrants' development and the creation of productive mindsets; family practices and technology; skill development by mentoring; the development of business skills; and the design of funding options for the role holders. In order to recruit young people and open up professional opportunities, agribusiness and the food sector also require deliberate branding.

As AI, personality-embedded machines, rejuvenated and regrown tissue, enhanced musculoskeletal systems, and brains that respond to artificial stimuli increase in the modern period, perspectives of humanity are predicted to change. The obvious finding is that when disruptive technology is used to create disruptive business models, they may spread widely. The new models bring about the necessary quick innovative shift in type, value of employment, and location. However, the various nations and regions are not reacting as rapidly as necessary to the new revolution while simultaneously adapting. The regions of Australia are far behind North America and Europe. The Australian food production industry is not very informed of how 4IR may affect its many enterprises. Middle-level company activities often have low levels of digitization, which raises concerns about sluggishness in the food sector. The Industry 3.0 activities in digitization in labor savings, waste minimization, increased immediate visibility, data warehousing, and quality management have mostly been the center of attention. The majority of food ingredient producers use SCADA systems to collect vast amounts of process data, with little to no capability for analysis. Consequently, there were fewer possibilities for machine learning.

Food 4.0 requires concurrent maturation for the individuals, technology, approaches, and structure in order to generate the necessary environment. A board-level commitment, the creation of a company plan for innovation, agility, and flexibility to the rapidly altering market dynamics, and newer techniques that may capitalize on evolving technology should all be required for maturity level lifting. Establishing digitalized planning and control systems

Table 12.1 Food systems' relevance to the sustainable evolution targets.

	SDG goals	Description
1	No poverty	Almost 75% of the worlds impoverished population work in agriculture and live in rural places
2	There is no hunger	The amount of food produced may feed the whole globe's population, yet roughly 750 million people are in a life-threatening state of malnutrition.
3	Overall health and wellbeing	Malnutrition is the primary cause of the worldwide sickness, which results in around 4 billion individuals being overweight or lacking in certain micronutrients.
4	Quality education	25% of children under the age of five throughout the globe suffer from malnutrition, which hinders brain development and lowers academic achievement.
5	Gender equality	Although women make up 43% of the agricultural workforce, they do not have equal access to markets, technology, or land.
6	Clean water and sanitation	According to [24], 72% of freshwater extractions are now covered by food systems.
7	Accessible and clean energy	Current food production relies significantly on fossil fuel and uses around 25% of the world's energy supply.
8	Reasonable economic and employment growth	The greatest global workforce is in agriculture. In underdeveloped nations, it employs around 60% of the workforce.
9	Industry, innovation, and infrastructure	According to [25], most of the 850 million populations living in village areas operate in farming and lack access to power.
10	Reduced inequalities	In the past 30 years, 70% of the world's population has moved to nations that either have little or no care for equality. Healthy food accessibility results from equality.
11	Sustainable cities and communities	By 2055, about 55% of the globe's people will dwell in city areas, changing consumer demand and placing more strain on resources.

(*Continued*)

Table 12.1 *Continued.*

	SDG goals	Description
12	Liable consumption and production	Approximately 40%, or 1.3 billion tons, of the world's food output is lost or squandered.
13	Climate change mitigation	Food systems are responsible for 20%–30% of world emissions. Food yields are predicted to decrease by 25% as a result of climate change.
14	Under water marine life	17% of the world's animal protein consumption and 45% of fish piles have been destroyed.
15	Life on land farming	Accounts for 51% of deforestation as of 2015, which is a big contribution.
16	Strong institutions	The impact of war on food security may be seen in the rise in the quantity of underprivileged person from 800 million to 900 million in 2018 and 2020, respectively.
17	Collaborations for the objectives	By forming collaborations, it is possible to overhaul the food system and open up $4.5 trillion in yearly potential for the private sector by 2025.

with fixed capacity scheduling should be part of the harder job in operations management. Additionally, it makes sure that the best possible amount of manufacturing resources are available and that production processes are motivated by consumer demand. The formulation, handling, and configuration of a well-controlled, market-responsive innovation should be well-controlled and standardized for food manufacturing to ensure repeatability and to remove dependency on implicit knowledge. As a result, the successful Food 4.0 processes and recipes may not be exposed to the cultivated operator use.

As a result, the Food 4.0 initiative has made it clear that higher institutions must update their curriculum to include the technologies, theories, and abilities of 4IR as well as participate in food training, education, and technology. It is done with the express purpose of generating competing benefits for early adopters. The success of every country relies on its capacity to drive technological progress and enhance the demand for and skills in science, innovation, and technology, according to South African President Cyril Ramaphosa. To build the future that is envisioned, leadership is required. In Industry 4.0, this chapter seeks to provide a broad and calculated vision of the upcoming and prospects of the food system. It doesn't concentrate on creating any new theories, but rather on igniting interest in the possibilities for the food sector in the age of cyber–physical systems. Transparency in knowledge and processes, data management, and food safety all serve to illustrate the broad scope of the food revolution.

12.4 Transparency of Information and Process

The management of information and process transparency is a very important endeavor that is needed in the field of the food revolution. Consumers have recently become much more aware of how the different food chain levels affect their health. Clients nowadays want for information on the origin, sustainability, traceability, and safety of the processes used to make and transport their food. They increasingly also consider food quality, ecological compliance, and protection while making purchases. In the meanwhile, several studies are being conducted to ascertain customers' willingness to pay a greater price for sustainable food items. In order to maintain openness and sustainability of food information and procedures, the food business must now take into account evolving customer demands.

In the food sector, sharing product information is a highly important and essential responsibility. Global compliance mandates that food organizations and producers make critical product information accessible to all stakeholders. As a result, Food 4.0 enables the integration of cyber–physical systems and technologies, enabling producers to efficiently communicate the necessary data about their goods. The confluence and cooperation of technologies and systems is a key characteristic of Food 4.0. Stakeholders in the food chain will have access to all product information on different virtual platforms, from the stage of raw material specifications through the final completed product. Analytics are used to collect large amounts of high-quality data and create a baseline performance that is clear and consistent.

The digital onset may be utilized to construct steady development exercises and forecast future execution via decision-making using the data.

12.4.1 Transparency prompted by tractability

The term "transparency of the food chain" is the availability of knowledge on goods upon request from interested parties, without undue delay, noise, or obfuscation. The technique of having visible what is "invisible" in the food systems is called traceability. Thus, by aggregating the amount of information available to customers, traceability efforts are a critical point of differentiation among food sector participants. Traceability assists in facilitating a vast-ranging monitoring on the health, economic, social, and ecological effects of different food production techniques as well as estimating the true cost of food. Traceability aims to satisfy the customers' desire for transparency. Food manufacturers may integrate potential efficiencies, such as fresh value sources and cost reductions, with traceability-based transparency. The impact of traceability in having client orders for food information and techniques, improving the likelihood of recognizing, reacting to, and stopping protective goals in the food system, assisting food chain optimization and minimization of food losses, and verifying declarations in backing of the sustainability aims are all ways that transparency in the food chain may be enhanced. Customers might be given information such as the price of production and the amount of water used to produce a crop, for example. Additionally, businesses might label their goods with verified claims of quality and flavor. Because of traceability, customer happiness is raised and brand loyalty is ensured. In order to overcome the present issues in food systems and significantly advance the sustainable development objectives, traceability draws on the many disruptive technologies.

12.4.2 Openness promotes sustainability

Transparency in data suggests that information must be correct, applicable, readily accessible, and well-timed in the required amount. Additionally, legible information should be sent, and the arrangement and logic of the information transmission must be appropriate. The communication methods and channels of the food supply chain must now strictly adhere to transparency requirements in order to reach stakeholders and consumers [26–30]. The scope to fulfill the needs of the current era without compromising the happiness of future era is explained by sustainability in the food sector. Environmental concerns, social concerns for the safety and health of people, and predicted financial gains are all included under sustainability. Consumers' decisions and purchasing patterns should be guided by a thorough understanding of the components of the food chain's sustainability, including the products, businesses, and procedures. Therefore, it means that the whole food supply chain should be sufficiently transparent.

However, for huge companies with intricate food supply networks, sustainability is a significant concern. For instance, Unilever Plc sought to produce goods sustainably by using a "triple-bottom-line method" to find raw materials that would benefit people, the environment, and corporate profits. Although consumer demand preferences are an important aspect of sustainability, they are not the primary determining factor. Government has a significant role in environmental issues and legislation, thus other players like university researchers and supply chain actors also have an impact on sustainable practices. Therefore, transparency for sustainability has been divided into vertical and horizontal dimensional issues in literary works. While the horizontal dimension alludes to the rules and regulations that are relevant to the vertical dimension pertains to the regulations and demands placed on firms operating within a particular level of the food supply chain. In order to ensure a sustainable future for food, transparency provision is thus a given in the food sectors.

12.5 Management of Data

Food 4.0 includes data as a decision-making tool. Data is used by humans in many facets of endeavor to make decisions and advance growth. Data collecting in the agricultural and food industries often calls for the deployment of intelligent and autonomous equipment. The processing, transport, storage, and exchange of data outputs rely heavily on the Internet and low-cost computer systems. The procedures for retaining the necessary product information at all points along the use and manufacturing chain are crucial to the food revolution of the present.

The analytics management and intellect implementation of the food industry will be crucial to the current industrialization of the food sector. The 4IR technologies' integration with cyber–physical systems will enable the food industry's needed revolution and growth by providing large data access. By using current data on demand needs and food supply histories, future forecasts of food demand and supply may be generated successfully. Incorporating networked technologies that are able to create compound touch points might make it easier to share data. As a result, the development of business insights and data analysis will be made more efficient thanks to the enhanced information accessibility.

However, there are significant and current difficulties in managing data properly in the digital sphere. The component organizations of the 4IR and Food 4.0 are now facing both strategic and operational issues. The main issue facing the deployment of Food 4.0 at the moment is data security. The ability to safeguard proprietary operational information is a difficulty. There are substantial dangers, but the benefits are bigger. By limiting staff access to certain data, data security may be improved. Figure 12.6 depicts the usual data flow management in the food industry.

12.6 Food Security

How to guarantee food safety is a crucial problem in Food 4.0, even as the rapid technological progress in 4IR is also presenting new complications and

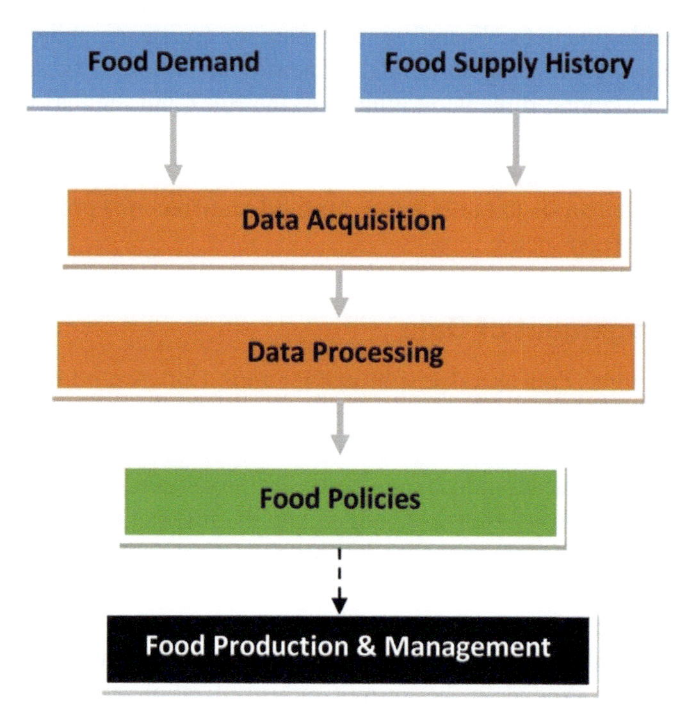

Figure 12.6 Data management in the food industry.

obstacles. According to the WHO, tainted food causes 1 in 9 persons to get ill and 560,250 deaths annually. About one-third of all instances worldwide – or a considerably greater extent and popularity – of fatality owing to food safety issues occur in emerging countries. Food safety is now a major problem on a worldwide scale as a result of this. For instance, it is estimated that 50 million US citizens get sick every year from food-borne diseases, and the International Business Agency's trade restrictions for "mad-cow disease" cost US beef producers between $2.3 billion and $1.8 billion.

Between 2004 and 2007, $2.7 was spent. The cost of the food-borne diseases in the UK is predicted at $20.4 billion, which is less of a burden when compared to the $992 billion that the food industry contributes to the US economy overall, but the safety danger it also poses to the economy and public health is significant.

Technology is not an irregularity to the potential for issues brought on by everything that creates opportunities. The food sector must adapt to several changes in its operating strata in the modern food period. For instance, there will be a drastic difference in how components are mixed. Because they make it possible to produce edible and secure synthetic materials, cyber–physical systems and technology have made synthetic meals more prominent. In order to fully profit from Food 4.0, it is not possible to rely only on technology to ensure that food is produced, sourced, made, and transported to customers in a safe manner. Technology to guarantee food safety may be supported by sustainable manufacturing practices. Reduced incidence of food pollutants will be achieved by effective preventative measures. An effective tracking system will guarantee accurate and straightforward tracking of incoming materials and a recall of defective items. Additionally, by considering the value of their consumers with the help of sustainable practices, businesses and organizations will be able to comply with regulations more easily.

12.6.1 Food safety via traceability

Effective case identification, source separation, and isolation for food safety issues depend on traceability. The technique becomes highly advantageous since it improves inspection productivity, lowers prevalence, and lowers the cost of product recalls. Traceability will not eliminate food-borne illnesses on its own, but it will assist the sector and governments in addressing food safety concerns accurately. Although product recalls because of food safety problems are uncommon, they have a significant financial effect on the food sector. A recent poll found that 58% of respondents having experienced a product recall in the previous five years, 23% of people said that financial risks from product recalls ranged from "significant to disastrous" sustained losses in sales and direct recalls that were expected to be around $30 million.

The use of traceability will make source sorting for food contamination more effective, quick, and practical, which will help to lower the risks associated with food safety. The procedures are now expensive and time-consuming. The epidemic of E coli disease linked to romaine lettuce in the Yuma area in 2018 lasted for three months, resulting in 210 illnesses, 96 hospitalizations, and 5 fatalities. The traceability for food safety is depicted in Figure 12.7, which demonstrates how 4IR technologies like distributed ledger and IoT will improve consumer safety along the entire food value chain while lowering financial risk.

12.6.2 Analytics for food safety prediction

The 4IR's artificial intelligence technology has allowed data measurement, storage, and manipulation to advance significantly – far beyond the capabilities of humans. Future consequences of a process may be reliably and

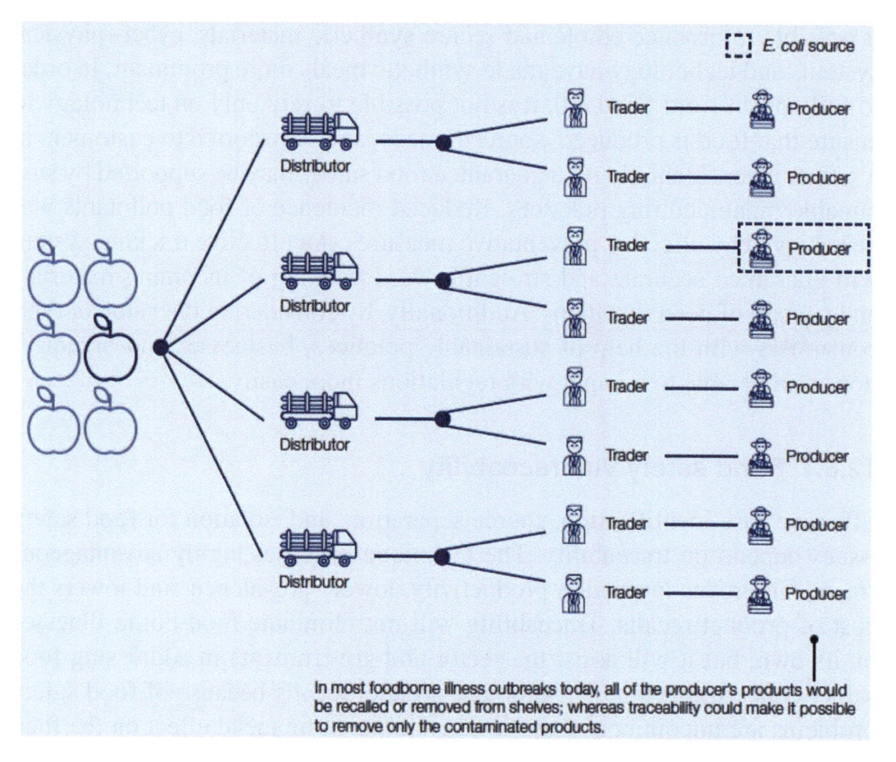

Figure 12.7 Spotting of a contaminated source with traceability.

precisely anticipated with adequate and accessible data. It has been said that the food industry fell behind other industries during the industrial revolutions in implementing new technology. Predictive analytics is already widely used in fields like fleet management, insurance, and public policy; the food business is just now becoming familiar with it, similar to other wearable technologies. The food industry is rapidly expanding in established countries, as well as in developing ones, with a roughly equal rise in consumer objections and a striking decline in the number of food hygiene regulators. Predictive analytics are the workable strategy. Agencies in New York, Illinois, have shown the effectiveness of predictive analytics to identify serious danger in the food industry with accuracy.

12.7 Methodology

The promotion of sustainable practices and technical advancements has led to a surge in interest in smart agriculture. Unmanned aerial vehicles (UAVs) are essential to smart agriculture since they support several stages of the farming process. It is imperative that action be done on the crucial and difficult topic of UAVs' contribution to precision and sustainable agriculture taking into consideration, especially for smallholder farmers to save money and time, and enhance their knowledge about agriculture. Therefore, the purpose of this work is to provide an integrated group framework for making decisions to choose the best agricultural unmanned aerial vehicle. Prior research on UAV assessment (i) could not effectively represent uncertainty, and (ii) expert weights are not carefully ascertained; (iii) the significance of experts and criterion kinds are ignored when calculating the weight of the criterion, and (iv) there is a lack of individualized rating for UAVs keeping dual weight entities in mind. Using the relevant research and expert views as a guide, nine crucial selection criteria are identified here, and five current UAVs are regarded for assessment. In order to bridge the gaps, a new integrated framework for appropriate UAVs is proposed in this study, taking into account q-rung orthopair fuzzy numbers (q-ROFNs) choosing. Specifically, by presenting the regret measure, the experts' weights may be methodically estimated. Additionally, an aim driven by a weighted logarithmic percentage change for the purpose of calculating criterion weight, the weighting (LOPCOW) approach is developed, and an algorithm with VIKOR (visekriterijumska optimizacija i kompromisno resenje) method in conjunction with the Copeland technique. The results demonstrate that the terms "camera," "power system," and "radar system" are the most important when choosing an agricultural unmanned aerial vehicle. It may also be deduced that the most promising UAV is the T30 DJ AGRAS. Given that UAV use in agriculture is only going to increase, this framework may be a useful tool for managers, farmers, and other decision-makers, legislators as well as other interested parties.

12.7.1 An algorithm for personalized ranking

A crucial stage in the decision-making process is ranking, which entails giving the alternatives a priority based on their rank values. The ranking is influenced by the alternative's performance or preference over the criteria. The preference matrix and the criterion weights are used to calculate the rank values of the alternatives. It is evident from [31] that past ranking methods simply took preference information into account when rating alternatives; they did not take into account the experts' a-priori selections. In real-world scenarios, expert views on each option should be taken into consideration as possible information for the decision-making process.

Based on the Lp norm concept, VIKOR is a visually appealing and widely used ranking system. It essentially uses distance standards to rate options and chooses a compromise solution. According to [32], the VIKOR approach performs better than its near relative TOPSIS and may be used to describe outranking approaches as special instances. Additionally, it can be deduced that VIKOR is straightforward and elegant, and it closely mimics decision-making that is driven by humans. By incorporating individual preferences into VIKOR and compiling expert-generated customized rank values, a potent ranking algorithm is provided that respects the human decision-making process while simultaneously logically ranking options. In order to arrive at a fair and logical ranking of alternatives, it must also be assumed that the ranking process considers the kind of criteria for rank estimation as well as the subjective decisions made by experts. The suggested method is shown below to get the alternatives' individual and cumulative rankings using expert rating data.

12.7.2 Algorithm: q-ROFN–VIKOR personalized ranking

Input: a criterion weight vector of $1 \times B$ and decision matrices $A \times B$

Results: $1 \times B$ vectors and a $1 \times B$ cumulative rank vector

Method:

Start

Step 1. Based on each expert's evaluation data, identify the ideal and anti-ideal solution criteria.

Step 2. Determine the $1 \times B$ group utility vector for every choice matrix.

Step 3. For every decision matrix, get the unique regret vector of $1 \times B$.

Step 4. To get the merit function of $1 \times B$, use the linear combination.

Step 5. Apply the Copeland technique to get the final fused rank vector of $1 \times B$.

End

In Figure 12.8, the suggested model's framework is shown. Experts are observed to rate UAVs according to certain criteria. The rating is first provided in qualitative words, which are then translated into q-ROFNs using tabular data. The meticulous determination of expert weights results in an

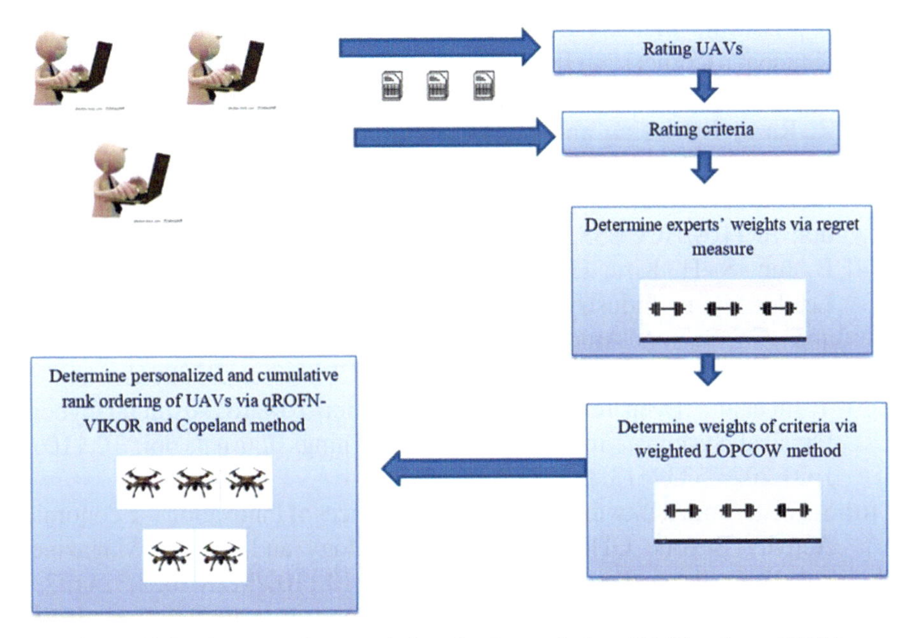

Figure 12.8 A proposed approach for selecting agricultural UAVs using q-ROFNs.

expert weight vector that is used to rank thc UAVs and compute the weights of the criteria. Specifically, both phases use the experts' significance ratings to provide intuitive justification for the decision-making process. Experts are implied to play a significant function in decision-making, therefore it makes sense to include their significance into the process. The stakeholders are assisted in understanding the ordering of UAVs from both viewpoints by both the cumulative ranking based on the Copeland approach and the individualized ranking of UAVs based on each expert's data. Experts' weights and choice matrix are taken into account while determining the criterion weight vector. Finally, UAVs are ranked using the weight vectors of both experts and criteria.

References

[1] F. -Y. Wang, J. Yang, X. Wang, J. Li and Q. -L. Han, "Chat with ChatGPT on Industry 5.0: Learning and Decision-Making for Intelligent Industries," in IEEE/CAA Journal of Automatica Sinica, vol. 10, no. 4, pp. 831–834, April 2023, doi: 10.1109/JAS.2023.123552.

[2] F. Missiroli, "Industry 4.0: Opinion of a Roboticist on Machine Learning [Student's Corner]," in IEEE Robotics & Automation Magazine, vol. 30, no. 2, pp. 124–126, June 2023, doi: 10.1109/MRA.2022.3214395.

[3] Z. Raza, I. U. Haq and M. Muneeb, "Agri-4-All: A Framework for Blockchain Based Agricultural Food Supply Chains in the Era of Fourth Industrial Revolution," in IEEE Access, vol. 11, pp. 29851–29867, 2023, doi: 10.1109/ACCESS.2023.3259962.

[4] B. Jun, S. H. Kim, H. Choi, J. H. Jeon and D. Yu, "Technological Leadership in Industry 4.0: A Comparison Between Manufacturing and ICT Sectors Among Korean Firms," in IEEE Access, vol. 11, pp. 28490–28505, 2023, doi: 10.1109/ACCESS.2023.3259065.

[5] Y. Jin et al., "Deep Temporal State Perception Towards Artificial Cyber-Physical Systems," in IEEE Internet of Things Journal, doi: 10.1109/JIOT.2023.3239413.

[6] S. Fölster and A. Litwin, "Hidden Consequences of Innovation—Economic Activity Beyond GDP," in IEEE Technology and Society Magazine, vol. 42, no. 1, pp. 37–44, March 2023, doi: 10.1109/MTS.2023.3241323.

[7] W. Min et al., "Large Scale Visual Food Recognition," in IEEE Transactions on Pattern Analysis and Machine Intelligence, doi: 10.1109/TPAMI.2023.3237871.

[8] A. R. Sagor, T. M. Mridul, M. M. H. Bhuiyan, F. A. Mridha, M. R. Hazari and M. A. Rahman, "Design and Implementation of Solar PV Operated E-Power Tiller," 2023 3rd International Conference on Robotics, Electrical and Signal Processing Techniques (ICREST), Dhaka, Bangladesh, 2023, pp. 368-373, doi: 10.1109/ICREST57604.2023.10070045.

[9] R. Jayasingh, J. Kumar R.J.S, D. B. Telagathoti, K. M. Sagayam and S. Pramanik, "Speckle noise removal by SORAMA segmentation in Digital Image Processing to facilitate precise robotic surgery", International Journal of Reliable and Quality E-Healthcare, vol. 11, issue 1, DOI: 10.4018/IJRQEH.295083, 2022

[10] S. Pramanik, K. M. Sagayam and O. P. Jena, Machine Learning Frameworks in Cancer Detection, ICCSRE 2021, Morocco, 2021.

[11] D. Samanta, S. Dutta, M. G. Galety and S. Pramanik, A Novel Approach for Web Mining Taxonomy for High-Performance Computing, The 4th International Conference of Computer Science and Renewable Energies (ICCSRE'2021), 2021, DOI:10.1051/e3sconf/202129701073

[12] R. Bansal, A. J. Obaid, A. Gupta, R. Singh, S. Pramanik "Impact of Big Data on Digital Transformation in 5G Era", 2nd International Conference on Physics and Applied Sciences (ICPAS 2021), doi:10.1088/1742-6596/1963/1/012170, 2021.

[13] S. Pramanik, AJ Obaid, N M, and SK Bandyopadhyay "Applications of Big Data in Clinical Applications", Al-Kadhum 2nd International Conference on Modern Applications of Information and Communication Technology, *AIP Conference Proceedings* 2591, 030086 (2023), https://doi.org/10.1063/5.0119414

[14] V. Vidya Chellam, V. Veeraiah, A. Khanna, T. H. Sheikh, S. Pramanik and D. Dhabliya, A Machine Vision-based Approach for Tuberculosis Identification in Chest X-Rays Images of Patients, ICICC 2023, Springer

[15] S. Praveenkumar, V. Veeraiah, S. Pramanik, S. M. Basha, A. V. Lira Neto, V. H. C. De Albuquerque and A. Gupta, Prediction of Patients' Incurable Diseases Utilizing Deep Learning Approaches, ICICC 2023, Springer

[16] S. Pramanik, Niranjanamurthy M. and S. N. Panda, Using Green Energy Prediction in Data Centers for Scheduling Service Jobs, ICRITCSA 2022, Bengaluru, 2022.

[17] P. T. Khanh, T. H. Ngọc, and S. Pramanik, Future of Smart Agriculture Techniques and Applications, in Advanced Technologies and AI-Equipped IoT Applications in High Tech Agriculture, Eds, A. Khang, IGI Global, 2023.

[18] T. H. Ngọc, P. T. Khanh and S. Pramanik, Smart Agriculture using a Soil Monitoring System, in Advanced Technologies and AI-Equipped IoT Applications in High Tech Agriculture, Eds, A. Khang, IGI Global, 2023.

[19] Reepu, S. Kumar, M. G. Chaudhary, K. G. Gupta, S. Pramanik and A. Gupta, Information Security and Privacy in IoT, in Handbook of Research in Advancements in AI and IoT Convergence Technologies, Eds, J. Zhao, V. V. Kumar, R. Natarajan and T. R. Mahesh, IGI Global, 2023.

[20] V. Veeraiah, V. Talukdar, Manikandan K., S. B. Talukdar, V. D. Solavande, S. Pramanik, A. Gupta, Machine Learning Frameworks in Carpooling, in Handbook of Research on AI and Machine Learning Applications in Customer Support and Analytics, Eds, Md S. Hossain, R. C. Ho and G. Trajkovski, IGI Global, 2023.

[21] S. Pramanik, "An Effective Secured Privacy-Protecting Data Aggregation Method in IoT", in Achieving Full Realization and Mitigating the Challenges of the Internet of Things, Eds, M. O. Odhiambo and W. Mwashita, IGI Global, 2022, DOI: 10.4018/978-1-7998-9312-7.ch008

[22] S. Pramanik and S. Bandyopadhyay, Analysis of Big Data, in Encyclopedia of Data Science and Machine Learning, Eds. John Wang, IGI Global, 2023, DOI: 10.4018/978-1-7998-9220-5.ch006

[23] S. Pramanik and S. Bandyopadhyay, Identifying Disease and Diagnosis in Females using Machine Learning, in Encyclopedia of Data Science and Machine Learning, Eds. John Wang, IGI Global, 2023, DOI: 10.4018/978-1-7998-9220-5.ch187.

[24] Y. Han et al., “Dynamic Mapping of Inland Freshwater Aquaculture Areas in Jianghan Plain, China,” in IEEE Journal of Selected Topics in Applied Earth Observations and Remote Sensing, vol. 16, pp. 4349–4361, 2023, doi: 10.1109/JSTARS.2023.3269430.

[25] I. Froiz-Míguez, P. Fraga-Lamas and T. M. Fernández-CaraméS, “Design, Implementation, and Practical Evaluation of a Voice Recognition Based IoT Home Automation System for Low-Resource Languages and Resource-Constrained Edge IoT Devices: A System for Galician and Mobile Opportunistic Scenarios,” in IEEE Access, vol. 11, pp. 63623–63649, 2023, doi: 10.1109/ACCESS.2023.3286391.

[26] S. Pramanik, M. G. Galety, D. Samanta, N. P Joseph, Data Mining Approaches for Decision Support Systems, 3rd International Conference on Emerging Technologies in Data Mining and Information Security, 2022.

[27] R. Bansal, B. Jenipher, V. Nisha, Jain, Makhan R., Dilip, Kumbhkar, S. Pramanik, S. Roy and A. Gupta, “Big Data Architecture for Network Security”, in Cyber Security and Network Security, Eds, Wiley, 2022, https://doi.org/10.1002/9781119812555.ch11

[28] R. Anand, J. Singh, D. Pandey, B. K.Pandey, V. K. Nassa and S. Pramanik, “Modern Technique for Interactive Communication in LEACH-Based Ad Hoc Wireless Sensor Network”, in Software Defined Networking for Ad Hoc Networks, M. M. Ghonge, S. Pramanik and A. D. Potgantwar, Eds, Springer, 2022, https://doi.org/10.1007/978-3-030-91149-2_3

[29] K. Dushyant, G. Muskan, A. Gupta and S. Pramanik, “Utilizing Machine Learning and Deep Learning in Cyber security: An Innovative Approach”, in Cyber security and Digital Forensics, M. M. Ghonge, S. Pramanik, R. Mangrulkar, D. N. Le, Eds, Wiley, 2022, https://doi.org/10.1002/9781119795667.ch12

[30] D. K.aushik, M. Garg, Annu, A. Gupta and S. Pramanik, Application of Machine Learning and Deep Learning in Cyber security: An Innovative Approach, in Cybersecurity and Digital Forensics: Challenges and Future Trends, M. Ghonge, S. Pramanik, R. Mangrulkar and D. N. Le, Eds, Wiley, 2022, DOI: 10.1002/9781119795667.ch12

[31] Y. Zheng and D. X. Wang, “A Comparative Study of Preference Ordering Methods for Multi-Criteria Ranking,” 2023 10th IEEE

Swiss Conference on Data Science (SDS), Zurich, Switzerland, 2023, pp. 108–111, doi: 10.1109/SDS57534.2023.00023.

[32] X. Li, C. Guo, L. Gupta and R. Jain, "Efficient and Secure 5G Core Network Slice Provisioning Based on VIKOR Approach," in IEEE Access, vol. 7, pp. 150517–150529, 2019, doi: 10.1109/ACCESS.2019.2947454.

13

Crop Monitoring in Real Time in Agriculture

Umesh Kumar Mishra[1], M.P. Karthikeyan[2], Ajay Rastogi[3], Ayesha Khatun[4], Rakesh Nair[4], Ankur Gupta[5], and Khaled M. Rabie[6]

[1]Department of Agriculture, Sanskriti University, India
[2]Department of Computer Science and IT, School of CS and IT, Jain (Deemed to be) University, India
[3]College of Computing Science and Information Technology, Teerthanker Mahaveer University, India
[4]Symbiosis Law School, Nagpur Campus, Symbiosis International (Deemed University), India
[5]Department of Computer Science and Engineering, Vaish College of Engineering, India
[6]Department of Engineering, Manchester Metropolitan University (MMU), UK

Email: umeshm.ag@sanskriti.edu.in; karthikeyan.mp@jainuniversity.ac.in; ajayrahi@gmail.com; ayeshakhatun@slsnagpur.edu.in; rakeshnair@slsnagpur.edu.in; ankurdujana@gmail.com; k.rabie@mmu.ac.uk

Abstract

Many nations, including India, still use traditional agricultural techniques. Our farmers' lack of adequate information makes the status of the agriculture industry even direr. Since agricultural methods heavily depend on weather forecasts and projections, which may not always be accurate, farmers often suffer enormous losses that result in debt and multiple agriculturist suicides. Ample soil moisture, good soil, clean air, and efficient irrigation all significantly influence crop output and cannot be ignored. Currently, the world's population is growing at an alarming rate, and agricultural production cannot keep up with the rising demand. By 2050, the population of the globe is predicted to reach nine billion, necessitating an increase in the agricultural supply of at least 70%. Monitoring plant development at all phases, from seeding through culture, is required to accomplish this.

13.1 Introduction

Indigenous agricultural practices are used in many nations. Farmers' lack of basic understanding is contributing to the agriculture sector's worsening condition. The ability to harvest crops successfully depends on the availability of sufficient soil moisture, good soil, clean air, and effective irrigation, all of which cannot be ignored. The rapidly growing population, which is causing the agricultural according to [1] supply, is not keeping up with the rising demand.

By 2045, the globe's people will reach to 9.7 billion. As a result, the agricultural supply must expand by at least 70% to fulfill demand. In order to do this, it is vital to keep an eye on the plant's development during each step, from seeding through cultivation.

When agricultural specialists consider the issues with global food production, "Data driven agriculture" is seen as the most viable solution. The capacity to map each field and describe it with a ton of data is known as data-driven farming. Data-driven farming refers to the effective use of large amounts of data to improve on-farm precision agriculture. Data-driven agriculture entails taking the appropriate soil data at the appropriate time to make the correct decisions that will increase longer-term usefulness. For instance, one might want to know how much moisture is still in the soil four inches below the surface or how much nutrient content is present throughout the farm. If such a map can be created, it could make it possible to use precision agriculture techniques. The capacity to do site-dependent applications is precision agriculture. An illustration makes precision farming easier to grasp. While farmers may evenly spray pesticides or water throughout their whole land, precision farming allows them to selectively administer it where it is necessary. Agriculturists will now utilize lesser water or pesticides, which will reduce water waste and pesticide usage, both of which are good for the environment. Figure 13.1 depicts the many elements that affect the crops. A few years ago, the application of this technology was costly. However, by means of improvement of technology, the cost has considerably decreased.

As a result of technological development, "Real time crop tracking in farming" may one day become a reality. Various field tasks, such as moisture monitoring and water level monitoring, this system may be used to monitor a variety of things, including temperature, crop status, weeds, and natural activities.

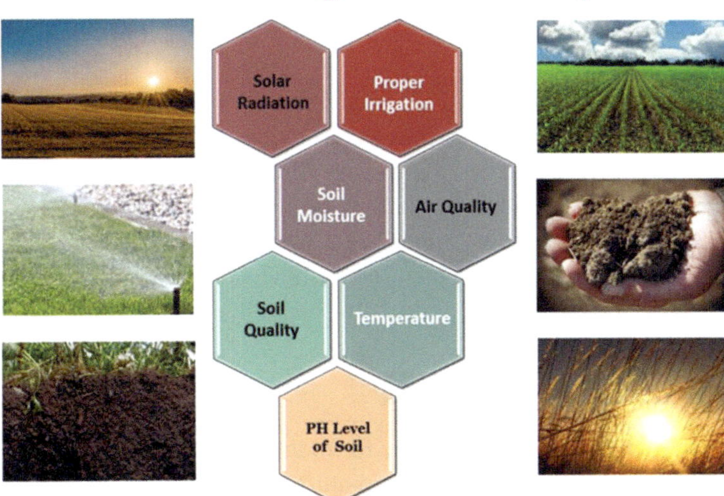

Figure 13.1 The crop's influencing elements.

To permit agriculturists to frequently track the plant development parameters and take the appropriate steps to optimize production, agriculture professionals are working to digitalize agriculture and associated farming operations. Due to its remote monitoring capabilities and excellent reliability, this novel invention will benefit the agricultural sector. Agriculture will become a more dependable and lucrative industry thanks to this idea. However, how widely it is used mostly relies on how well-known it is among farmers and how well it affects the agricultural economy.

This concept will not only replace antiquated agricultural methods, but it will also benefit society by generating new jobs. Its main objective is to use computerized approaches to track the expansion of yield. Accurate values of the many factors that affect the development of the crop may be readily obtained.

13.1.1 Monitoring system field measures

13.1.2 Field record and crops

These devices record every specific detail pertaining to the farms and crops. It looks at the alteration in data over time and explains why it changed. The system primarily records the date of planting crops, the kind of crops planted jointly, the field's history, and other pertinent information on the caliber of the soil and the crops cultivated.

13.1.3 The indexes of vegetation

By offering a variety of the various plants found on the farm, the real-time tracking framework evaluates the quality of the crops. This technique provides a good picture of how a land is doing in relation to various fields that are growing the similar crop. Given knowledge on plants includes the typical farm plants, historical plant cover, and the farm's underperforming areas.

13.1.4 Situations with the soils

Because of the excess of water in the farm, the real-time tracking framework primarily captures soil moisture, temperature, and soil drainage. Additionally, it keeps track of soil optimization outcomes.

13.1.5 Test of climate and precipitation

The most common meteorological tests are those for dew points, snow cover, air and soil temperatures, and dew points.

In the current digital age, individuals are attempting to combine the human brain with the artificial brain as technology advances. The ongoing research attempts have given rise to a newer area of AI. The development of technology is the sole reason why this real-time monitoring is now feasible.

13.2 Real-time Crop Monitoring Is Required

Since agricultural methods heavily depend on weather forecasts and projections, which may not always be accurate, farmers often suffer enormous losses that result in debt and mass farmer suicides. Consequently, a crop monitoring system is required. The following are additional causes:

- Since the majority of the monitoring is done remotely, the farmer may acquire information that can help them better their company.
- Farmers could get information quickly and with less work.
- Inspecting the water extent in farms and crops is a constant necessity.

- By keeping an eye on fields remotely, you may avoid risky situations like being shocked by electrical circuits; traveling back to examine on farms at odd hours, wasting energy, managing sources inconveniently and physically, etc.
- Even a person with physical disabilities may readily utilize these automated devices since they are so simple to operate.
- It helps in handling situations where there is too much water gushing in the farm or when the pump is running when the source is dry.
- Despite having a favorable position in the globe for agricultural methods due to the sharp rise in population, many nations still struggle to fulfill the need for food supply. Therefore, there is a dire need to advance food production methods in order to expand output.
- Why utilizing a real-time tracking framework by agriculturists will have a significant economic effect. This allows farmers to assess the state of their crops throughout the growing season, a crucial factor in the support of agricultural programs.
- Various natural disasters, such as floods, hailstorms, cyclones, and earthquakes, seriously damage the infrastructure and the crops in the afflicted regions. Many nations on the planet are vulnerable to various types of natural disasters. For instance, tropical cyclones often affect coastal regions in India. In 2018, three powerful cyclones – Gaja on November 16, Titli on October 11, and Chaki on November 17 – impaired cyclic crops like date, potato, and papaya trees in Tamil Nadu and Andhra Pradesh's east coast regions. The traditional study-based methodologies utilized to compute loss value are labor-and time-intensive. The same technologies that are used for real-time crop tracking may be used to address this flaw in the conventional survey.

13.3 System for Real-time Crop Tracking

As was already said, the people of the globe have been rapidly growing recently and are projected to come to 10 billion people in 2035. Consequently, as the population has grown, so has the need for supplies.

Farmers use a variety of methods to meet demand, but this is insufficient; as a result, they use toxic pesticides in ways that are out of control and injure the land. This causes the field to be infertile. Therefore, there is a clear

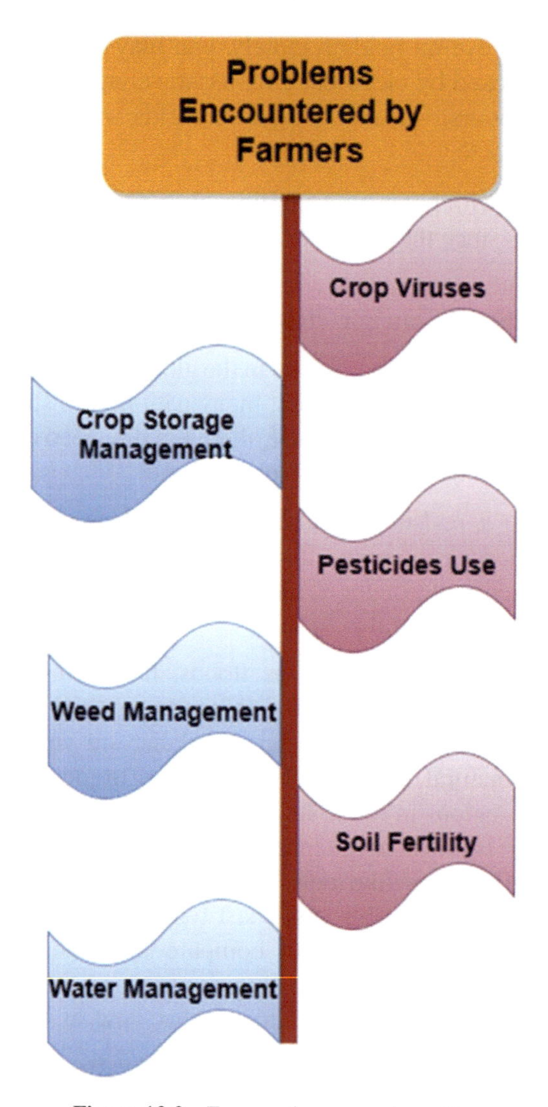

Figure 13.2 Farmers face a number of issues.

necessity for various automation approaches. As seen in Figure 13.2, farmers deal with issues including crop viruses, crop storage management, pesticide usage, weed control, soil fertility, and water management. These issues can all be managed by different methods with less effort like automated methods such as IoT [2–6], ML [7–11], and AI [12–16], etc. The productivity may be increased and soil fertility increased by adopting automation in agricultural

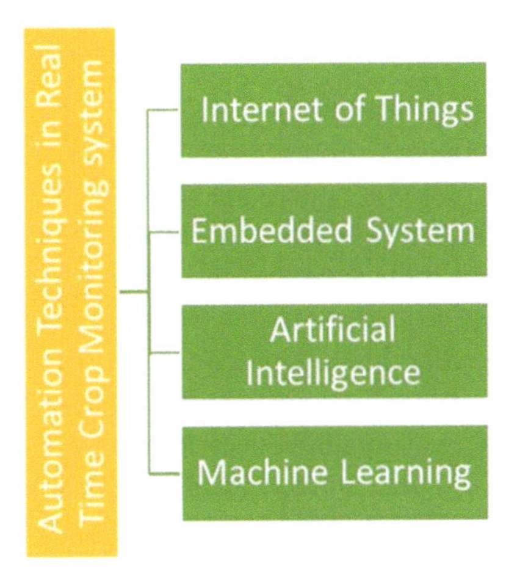

Figure 13.3 Automated crop monitoring systems using automation techniques.

methods. Because of this, the automation of farming is evolved and has attracted the attention of many nations.

In agriculture, technology has advanced that boosts fertility and productivity. The ability to capture speedy responses to fleeting occurrences is now made possible by systems for real-time crop monitoring. These approaches entail examining satellite photographs of various farms and crops to ascertain both the negative and positive development of the crops (users often utilize satellite images, but there may be alternative ways as well). The farmer may determine if the crops are thriving or deteriorating by examining the several vegetation indices of a company over time. The farmer may use this information to determine what corrective actions might be needed on the farm. The satellite-based crop monitoring system enables a user to observe the crops that are growing in different fields, places, and geographies. Following data collection, the system analyzes various fields and provides various users with interpretations of the data.

Depending on which technologies are used, real-time agricultural monitoring systems may be classed as follows:

- A crop monitoring system based on artificial intelligence and machine learning
- Crop tracking framework based on embedded systems
- Crop monitoring system powered by IoT

The automation approaches used in real-time crop tracking systems are depicted in Figure 13.3. A real-time crop tracking framework may include more than one of the technologies indicated above.

13.4 Monitoring System for Crops using AI and ML

People must become more inventive and efficient in their agricultural practices as a result of the growing global population and the scarcity of available land. They must learn how to grow more crops on less space while also improving the fertility of their farms. Humans have created clever robots that carry out a variety of tasks by studying their surroundings in this age of sophisticated technology. AI is a field of computer science which may develop a sense of the environment and provide the best outcomes. An artificial intelligence (AI) system allows a computer to perform tasks based on existing knowledge. Today, AI is used in various fields, including security, the finance industry, the medical sector, agriculture, and the education field, among others. Some domains improve the job done by a machine and enhance the development of more advanced technology, like DL [17–21], convolutional neural network [22–26], artificial neural network, and machine learning.

Currently, businesses are using AI approaches to aid in a variety of domains, including:

- to produce more powerful crops on the farm;
- to keep an eye on growth and soil conditions in fields;
- pest management in agriculture; and
- to create a pattern out of farmer data, devise a variety of various agricultural activities for the food supply chain, and help with workload management.

Numerous causes, including climate change, worries about food security, and population increase, have prompted different businesses to search for more cutting-edge techniques to safeguard and enhance harvest development. As a result, AI is progressively becoming a component of the technological development of the sector. “AI and ML” are essential to the farming industry’s ability to monitor crops in real time. The following categories seem to be where artificial intelligence in farming is most prevalent:

13.4.1 Robots for farming

Numerous organizations are concentrating on the creation of the autonomous robots that might undertake fundamental farming tasks efficiently than human employees, such as collecting crops at a higher rate and volume.

13.4.2 Plant and soil tracking

Companies use deep learning algorithms and AI (artificial intelligence) to analyze statistics collected by a variety of drones. Businesses use software-based technologies to monitor the health of the soil and crops. A few years ago, field data were manually entered, which wasted time while also being imprecise. With today's cutting-edge technology, the whole field is geo-tagged in order to identify the real field area. The system aids in weather forecasting, remote sensing, tracking and planning field operations (for total traceability), teaching farmers how to choose the best approaches, monitoring entire crop health and crop estimation, and providing indications for insect activity in the crops. Artificial intelligence and other cutting-edge technologies can interpret and comprehend data to power real-time visuals.

Because they lack the education necessary to comprehend the technology underlying crop lifecycles, quality measures, pests, and the most recent micro fertilizers, small farmers all around the globe continue to conduct traditional farming. The automated frameworks depend on image-dependent resolutions to provide information on both the final quality of the harvested crops and the general health of the crops throughout the budding stage. It offers the following advantages:

- Farming product grading: Since the color, form, and size of fresh items like fruits, vegetables, grains, etc. make them stand out; their analysis and grading are dependent on the photographs taken by the drones. An agriculturist uploads a photo that he took with his phone, which the technology interprets and uses to assess the commodity's grade in real time and having no effort.
- Alarms on harvest infestation: An agriculturist may learn whether pests, viruses, or alien plants (weeds) are growing alongside their crops with only a click, and technology also provides the cure. The system (which makes use of DL, image processing [27–31], and artificial intelligence) can detect crop viruses or insect infestation in the harvests. Along with the criteria, it also offers advice on how the condition may be treated and prevented from spreading further.

13.4.3 Machine learning for predictive analytics

In order to track and predict numerous environmental factors, such as changes in weather that affect agricultural productivity, developed ML concepts are utilized. The optimal time to plant seeds may be predicted by a number of AI and ML systems, which may also alert users anytime there is a danger of pest invasion. For farmers, determining when to plant crops may be a difficult task. Excessive rainfall and drought can also be problematic. For this, a business intelligence and machine learning-based app may be created. The app has the ability to send linked farmers guidance messages with subject lines like "ideal time to sow." Farmers may save money by not having to install any sensors. To get this message of counsel or alarm, all they need is a cell phone. The following are a few applications as examples:

- AQUATEK: This app enables farmers to use their mobile devices to monitor the health of their fields and take corrective action before a problem arises. Farmers may measure water consumption and make exact, precise irrigation plans by using an app alone, rather than using an expensive irrigation management system.
- Kisan Yojana: "Kisan Yojana" is a well-known agricultural software that Android users may download for free. Using this app, farmers may learn about all government initiatives. It closes the information gap between the rural population and government. Using an application, many state government initiatives may be provided with ease.
- IFFCO Kisan: "IFFCO Kisan" is among the top agricultural applications available to farmers. This Android application may deliver weather prediction information, agriculture warnings to farmers in 10 different Indian languages, the most recent agricultural advice, and many farming recommendations. Through the use of this program, they may easily enlist the assistance of agricultural specialists. There are other further initiatives, "Haritha" being one of them. According to [32], it is helpful for automated irrigation systems for home gardens. To determine the best time to plant crops in a particular place, one may gather historical climatic data (for a few previous years) and do analysis using artificial intelligence. Moisture Adequacy Index, or MAI, is used to determine when to plant. It is a common metric used to determine if rainfall and soil moisture levels are sufficient to fulfill the fundamental water needs of crops. Prior to planting seeds, the nearby agriculturists will wait for a text message and automated alert voice calls informing them of the harvest phase and whether or not their field's harvests are at danger of a pest outbreak (depending on climate circumstances).

13.4.4 Efficiency in the supply chain

Companies use real-time data from numerous data sections to create efficient and clever supply chains. These data streams formerly originated from several sources. And artificial intelligence and similar technologies are helpful to guarantee an efficient and trustworthy supply chain at different steps of farming. Here are a handful of them:

- Transition detection: Using crowd-sourced data from buyer/maker dealings and shipment together with a transaction detection algorithm, data examination on several data lines is done to find higher-margin transactions.
- Quality maintenance: Computer vision and AI-dependent self-grading and sorting of fruits and vegetables is done in order to set up agricultural commodity's worldwide standards to effectively sell across nations.
- Credit risk management: Crowd-sourced data, techniques, and analytics are used to address the credit delinquency, one of the most difficult problems in the present supply chain, to assure a lower risk operation.
- Agriculture mapping: A map of agricultural resources and products with a 1 sq. kilometers may be determined utilizing crowd-sourced fusion and satellite-based image analysis.

Provided the vast efficiency of farming, it is crucial to ensure that technology is used as effectively as possible to benefit both consumers and farmers. With the aid of technical advancement and supportive governmental legislation, several companies are sprouting in the nation to make advanced technology like AI broadly accessible in agriculture. AI is a huge benefit for the agriculture sector, which is highly reliant on climatic circumstances that are often unexpected. In reality, since the agriculture economy is heavily dependent on meteorological circumstances, which are sometimes unexpected, AI may be quite beneficial. Given its advantages, AI is expected to change the agriculture sector. With its processing capacity rising and adoption costs decreasing, artificial intelligence is becoming more practical, particularly for smallholder farmers.

13.4.5 Real-time crop tracking advantages of AI and ML

Figure 13.4 illustrates some of the advantages of AI and ML in real-time crop tracking. Following is a list of some of them:

Figure 13.4 Advantages of a real-time crop tracking system with artificial intelligence.

Artificial intelligence is assisting farmers in the analysis of field data: Every day, farms produce hundreds of data points in the farms. By examining temperature, weather, water use, and soil conditions, AI may help agriculturists in having better choices. Technologies based on AI may assist agriculturists take knowledgeable decisions regarding the crops, seeds, and resource use.

13.4.5.1 Artificial intelligence makes precision farming happen

AI technology is used in precision agriculture to help discover pests on farms, nutritional deficiencies in plants, and illnesses in plants. AI allows for the detection and targeting of weeds, assisting in the selection of the appropriate herbicide and preventing the use of unneeded herbicides to protect our food. Precision agriculture is what is meant by this.

13.4.5.2 Utilizing AI to boost productivity and improve farming accuracy

Agriculturists may develop forecasting systems employing AI to boost productivity, precision, or farming accuracy. These algorithms can forecast weather patterns months in advance, which can help farmers, make decisions. Small farms in some of the developing nations might benefit greatly from this kind of forecasting since their knowledge and data may be constrained in a variety of ways. These days, farmers are now looking to the skies to examine data from their operations underneath the surface of the earth. DL techniques and computer vision are used to analyze data obtained from drone flights over fields. Drones equipped with AI-based cameras may take photographs of the whole farm and analyze them in almost real-time to spot trouble spots and possible improvements. In comparison to utilizing humans to monitor farms, utilizing these drones may address a big region of a farm in a relatively short amount of time.

Due to a decline in interest in farming, a scarcity of labor is one of the major problems that most farmers are now experiencing. Farms historically require a significant number of people, generally seasonal, to harvest crops and maintain productivity. Since humans have moved away from agricultural communities and are now living in cities, there are fewer individuals who are able and ready to relocate to rural regions. Artificial intelligence farm robots are one of the answers to this labor scarcity situation. These machines maximize the human labor to provide the greatest results in the shortest amount of time. Robots can harvest crops more quickly and in greater quantities than a human worker could. Due of their propensity for continuous labor, these robots can locate and remove weeds more effectively while also lowering agricultural expenses.

13.4.6 Crop monitoring system based on an embedded system

Additionally, the main components of AI and ML are theories and data. Algorithms and programming are the two main concepts of AI and ML. For the aim of implementing these logic-based ideas and techniques, there must be a software-hardware interface. The structure of the embedded system is shown in Figure 13.5.

"Embedded systems" are the method through which this work may be completed. Embedded systems are hardware devices that have chips with unique software programming.

In order to implement hardware, we employ a variety of sensors, including light, humidity, temperature, and soil moisture sensors.

13.4.7 Crop monitoring system depending on the IoTs

IoT refers to a common set of items or devices that may communicate with one another via an Internet connection. It is really beneficial in farming.

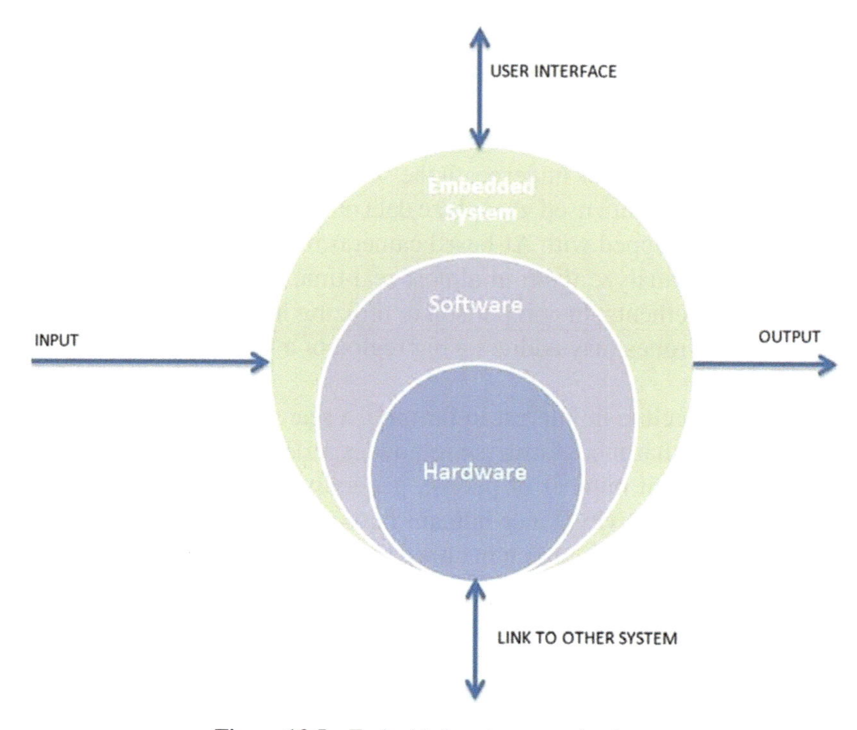

Figure 13.5 Embedded system organization.

Using an Internet connection, data obtained from sensors may be transferred to a database on a web server. Data may be encoded using a database server, and the procedure will begin. For instance, irrigation would begin automatically based on information obtained from moisture and temperature sensors, and farmers' mobile phones will get notification messages on a regular basis. According to Mahalakshmi S. D. (2016), field may now be observed from any location.

13.5 A Real-time Crop Tracking System's Benefits

Real-time crop tracking system may be used to monitor and gather data in a farm, improving crop productivity. Here are a few other advantages:

- Proper planting and watering will boost production rate.
- With the aid of weather predictions and different types of soil moisture sensors, water will only be used when necessary in the proper location.

- Farmers may see numerous factors in real time, including harvest production rate, soil moisture, light intensity, etc., which helps to speed up the decision-making process.
- Farmers may save costs overall and resource usage by utilizing real-time crop tracking system.
- Farm-related equipment may be monitored and maintained in accordance with labor efficiency, output rates, and failure prediction.
- New technology will enhance the quality of the production.

13.6 The Use of Real-time Crop Tracking Systems by Organizations and People

Following are some other individuals or groups who, in addition to farmers, may benefit from a real-time crop tracking system:

- Agronomists and farming managers: The crop tracking system saves time and is more accurate. Agronomists, who are experts in soil management and the production of field crops, employ this method often in order to make the most use of available resources.
- Business owners: Since their projections are based on extensive data, business owners should trust them when making investment decisions.
- State organizations: The approach is highly helpful for state firms since they may depend on it when making judgments about ecological issues and food security.

13.7 Implementations of Real-time Crop Tracking Systems and Their Prospects

The real-time crop monitoring system has several uses that might benefit farmers in the future.

- Younger farmers are investing wisely compared to their more experienced counterparts in order to use artificial intelligence in farming.
- In order to identify the kind of plant, farmers are now adopting a variety of innovative AI approaches, such as deep learning, which may also assist them in fostering an atmosphere conducive to long-term development which unquestionably has accelerated production and improved its quality.

- By convolutional neural network, recurrent neural network [33–36], or various computational networks, the application of AI methods in agriculture to identify illness and weeds in fields is expanding extremely quickly.
- Utilizing Internet of Things, wireless technology, communication protocols, and sensors – that are necessary for weather management and tracking even without human existence in the farms – can be used to deliver the specific needed environment to the plants utilizing greenhouse farming without human interaction.
- An instance of real-time tracking is the employment of robots, which can be used to all facets of farming, from planting to harvesting, which typically requires 25 to 30 employees to do the same task.
- Drones with cameras that are utilized for continuous real-time farm tracking and data collecting for future analysis are another intriguing usage of artificial intelligence technology in agriculture.

IoT, ML, sensors, and various advanced technologies will be required to create a real-time crop tracking system or to self-drive the various applications in farming.

13.8 Conclusion

Farming technologies based on ML and AI is enabling farmers to collect relevant data. The efficiency of smart technology for agriculture to improve sustainability and competing in their farm products must be understood by both small farmers and big landowners. If small farmers use digital technologies effectively in their fields, the need may be successfully satisfied despite the population's fast growth.

References

[1] V. Jayakumar, A. B. K. Mohideen, M. H. Saeed, H. Alsulami, A. Hussain and M. Saeed, "Development of Complex Linear Diophantine Fuzzy Soft Set in Determining a Suitable Agri-Drone for Spraying Fertilizers and Pesticides," in IEEE Access, vol. 11, pp. 9031-9041, 2023, doi: 10.1109/ACCESS.2023.3239675.

[2] R. Jayasingh, J. Kumar R.J.S, D. B. Telagathoti, K. M. Sagayam and S. Pramanik, "Speckle noise removal by SORAMA segmentation in Digital Image Processing to facilitate precise robotic surgery",

International Journal of Reliable and Quality E-Healthcare, vol. 11, issue 1, DOI: 10.4018/IJRQEH.295083, 2022

[3] S. Pramanik, M. G. Galety, D. Samanta, N. P Joseph, Data Mining Approaches for Decision Support Systems, 3rd International Conference on Emerging Technologies in Data Mining and Information Security, 2022.

[4] D. Samanta, S. Dutta, M. G. Galety and S. Pramanik, A Novel Approach for Web Mining Taxonomy for High-Performance Computing, The 4th International Conference of Computer Science and Renewable Energies (ICCSRE'2021), 2021, DOI:10.1051/e3sconf/202129701073.

[5] R. Bansal, A. J. Obaid, A. Gupta, R. Singh, S. Pramanik "Impact of Big Data on Digital Transformation in 5G Era", 2nd International Conference on Physics and Applied Sciences (ICPAS 2021), doi:10.1088/1742-6596/1963/1/012170, 2021.

[6] S. Pramanik, Niranjanamurthy M. and S. N. Panda, Using Green Energy Prediction in Data Centers for Scheduling Service Jobs, ICRITCSA 2022, Bengaluru, 2022.

[7] V. Vidya Chellam, V. Veeraiah, A. Khanna, T. H. Sheikh, S. Pramanik and D. Dhabliya, A Machine Vision-based Approach for Tuberculosis Identification in Chest X-Rays Images of Patients, ICICC 2023, Springer

[8] S. Praveenkumar, V. Veeraiah, S. Pramanik, S. M. Basha, A. V. Lira Neto, V. H. C. De Albuquerque and A. Gupta, Prediction of Patients' Incurable Diseases Utilizing Deep Learning Approaches, ICICC 2023, Springer

[9] S. Pramanik, AJ Obaid, N M, and SK Bandyopadhyay "Applications of Big Data in Clinical Applications", Al-Kadhum 2nd International Conference on Modern Applications of Information and Communication Technology, *AIP Conference Proceedings* 2591, 030086 (2023), https://doi.org/10.1063/5.0119414

[10] P. T. Khanh, T. H. Ngọc, and S. Pramanik, Future of Smart Agriculture Techniques and Applications, in Advanced Technologies and AI-Equipped IoT Applications in High Tech Agriculture, Eds, A. Khang, IGI Global, 2023.

[11] T. H. Ngọc, P. T. Khanh and S. Pramanik, Smart Agriculture using a Soil Monitoring System, in Advanced Technologies and AI-Equipped IoT Applications in High Tech Agriculture, Eds, A. Khang, IGI Global, 2023.

[12] Reepu, S. Kumar, M. G. Chaudhary, K. G. Gupta, S. Pramanik and A. Gupta, Information Security and Privacy in IoT, in Handbook of Research in Advancements in AI and IoT Convergence Technologies,

Eds, J. Zhao, V. V. Kumar, R. Natarajan and T. R. Mahesh, IGI Global, 2023.

[13] V. Veeraiah, V. Talukdar, Manikandan K., S. B. Talukdar, V. D. Solavande, S. Pramanik, A. Gupta, Machine Learning Frameworks in Carpooling, in Handbook of Research on AI and Machine Learning Applications in Customer Support and Analytics, Eds, Md S. Hossain, R. C. Ho and G. Trajkovski, IGI Global, 2023

[14] S. Pramanik, "An Effective Secured Privacy-Protecting Data Aggregation Method in IoT", in Achieving Full Realization and Mitigating the Challenges of the Internet of Things, Eds, M. O. Odhiambo and W. Mwashita, IGI Global, 2022, DOI: 10.4018/978-1-7998-9312-7.ch008

[15] S. Pramanik and S. Bandyopadhyay, Identifying Disease and Diagnosis in Females using Machine Learning, in Encyclopedia of Data Science and Machine Learning, Eds. John Wang, IGI Global, 2023, DOI: 10.4018/978-1-7998-9220-5.ch187.

[16] S. Pramanik and S. Bandyopadhyay, Analysis of Big Data, in Encyclopedia of Data Science and Machine Learning, Eds. John Wang, IGI Global, 2023, DOI: 10.4018/978-1-7998-9220-5.ch006

[17] R. Bansal, B. Jenipher, V. Nisha, Jain, Makhan R., Dilip, Kumbhkar, S. Pramanik, S. Roy and A. Gupta, "Big Data Architecture for Network Security", in Cyber Security and Network Security, Eds, Wiley, 2022, https://doi.org/10.1002/9781119812555.ch11

[18] K. Dushyant, G. Muskan, A. Gupta and S. Pramanik, "Utilizing Machine Learning and Deep Learning in Cyber security: An Innovative Approach", in Cyber security and Digital Forensics, M. M. Ghonge, S. Pramanik, R. Mangrulkar, D. N. Le, Eds, Wiley, 2022, https://doi.org/10.1002/9781119795667.ch12

[19] B. K. Pandey, D. Pandey, S. Wairya, G. Agarwal, P. Dadeech, S. R. Dogiwal and S. Pramanik, "Application of Integrated Steganography and Image Compressing Techniques for Confidential Information Transmission", in Cyber Security and Network Security, https://doi.org/10.1002/9781119812555.ch8, Eds, Wiley, 2022.

[20] D. K.aushik, M. Garg, Annu, A. Gupta and S. Pramanik, Application of Machine Learning and Deep Learning in Cyber security: An Innovative Approach, in Cybersecurity and Digital Forensics: Challenges and Future Trends, M. Ghonge, S. Pramanik, R. Mangrulkar and D. N. Le, Eds, Wiley, 2022, DOI: 10.1002/9781119795667.ch12

[21] M. Sinha, E. Chacko, P. Makhija and S. Pramanik, Energy Efficient Smart Cities with Green IoT, in Green Technological Innovation for

Sustainable Smart Societies: Post Pandemic Era, C. Chakrabarty, Eds, Springer, 2021, DOI: 10.1007/978-3-030-73295-0_16

[22] A. Bhattacharya, A. Ghosal, A. J. Obaid, S. Krit, V. K. Shukla, K. Mandal and S. Pramanik, "Unsupervised Summarization Approach with Computational Statistics of Microblog Data", in Methodologies and Applications of Computational Statistics for Machine Learning, D. Samanta, R. R. Althar, S. Pramanik and S. Dutta, Eds, IGI Global, 2021, pp. 23–37, DOI: 10.4018/978-1-7998-7701-1.ch002

[23] A. Mandal, S. Dutta, S. Pramanik, "Machine Intelligence of Pi from Geometrical Figures with Variable Parameters using SCILab", in Methodologies and Applications of Computational Statistics for Machine Learning, D. Samanta, R. R. Althar, S. Pramanik and S. Dutta, Eds, IGI Global, 2021, pp. 38–63, DOI: 10.4018/978-1-7998-7701-1.ch003

[24] Y. Meslie, W. Enbeyle, B. K. Pandey, S. Pramanik, D. Pandey, P. Dadeech, A. Belay, A. Saini, "Machine Intelligence-based Trend Analysis of COVID-19 for Total Daily Confirmed Cases in Asia and Africa", in Methodologies and Applications of Computational Statistics for Machine Learning, D. Samanta, R. R. Althar, S. Pramanik and S. Dutta, Eds, IGI Global, 2021, pp. 164–185, DOI: 10.4018/978-1-7998-7701-1.ch009

[25] S. Pramanik, "Carpooling Solutions using Machine Learning Tools", in Handbook of Research on Evolving Designs and Innovation in ICT and Intelligent Systems for Real-World Applications, K. K. Sarma, N. Saikia and M. Sharma, IGI Global, 2022, DOI: 10.4018/978-1-7998-9795-8.ch002.

[26] G. Taviti Naidu, KVB Ganesh, V. Vidya Chellam, S Praveenkumar, D. Dhabliya, S. Pramanik, A. Gupta, Technological Innovation Driven by Big Data, in "Advanced Bioinspiration Methods for Healthcare Standards, Policies, and Reform", Hadj Ahmed Bouarara IGI Global, 2023, DOI: 10.4018/978-1-6684-5656-9

[27] R. R. Chandan, S. Soni, A. Raj, V. Veeraiah, D. Dhabliya, S. Pramanik, Ankur Gupta , Genetic Algorithm and Machine Learning, in "Advanced Bioinspiration Methods for Healthcare Standards, Policies, and Reform", Hadj Ahmed Bouarara, IGI Global, 2023, DOI: 10.4018/978-1-6684-5656-9

[28] D. Mondal, A. Ratnaparkhi, A. Deshpande, V. Deshpande, A. P. Kshirsagar and S. Pramanik, Applications, Modern Trends and Challenges of Multiscale Modelling in Smart Cities, in Data-Driven Mathematical Modeling in Smart Cities, IGI Global, 2023, DOI: 10.4018/978-1-6684-6408-3.ch001

[29] P. K. Mall, S. Pramanik, S. Srivastava, M. Faiz, S. Sriramulu and M. N. Kumar, FuzztNet-Based Modelling Smart Traffic System in Smart Cities Using Deep Learning Models, in Data-Driven Mathematical Modeling in Smart Cities, IGI Global, 2023, DOI: 10.4018/978-1-6684-6408-3.ch005

[30] B. K. Pandey, D. Pandey, V. K. Nassa, A. S. Hameed, A. S. George, P. Dadheech and S. Pramanik, A Review of Various Text Extraction Algorithms for Images, in The Impact of Thrust Technologies on Image Processing, Nova Publishers, 2023.

[31] B. K. Pandey, D. Pandey, V. K. Nassa, A. S. George, S. Pramanik and P. Dadheech, Applications for the Text Extraction Method of Complex Degraded Images, in The Impact of Thrust Technologies on Image Processing, Nova Publishers, 2023.

[32] A. A. Desai, R. A. Metri, S. R. Patil, A. A. Nagargoje and D. S. Desai, "Automated Irrigation System for Efficient and Portable Farming," 2023 International Conference on Power, Instrumentation, Control and Computing (PICC), Thrissur, India, 2023, pp. 1–6, doi: 10.1109/PICC57976.2023.10142376.

[33] R. Ghosh, R. Bhunia, S. Pramanik, S. Mohanty and P. K. Patnaik, Smart City Healthcare System for Survival Forecast for Cardiac Attack Situations using Machine Learning Techniques, in Data-Driven Mathematical Modeling in Smart Cities, IGI Global, 2023, DOI: 10.4018/978-1-6684-6408-3.ch019

[34] A. Gupta, A. Verma, S. Pramanik, "Advanced Security System in Video Surveillance for COVID-19", in An Interdisciplinary Approach to Modern Network Security, S. Pramanik, A. Sharma, S. Bhatia and D. N. Le, CRC Press, 2022.

[35] S. Pramanik, An Adaptive Image Steganography Approach depending on Integer Wavelet Transform and Genetic Algorithm, Multimedia Tools and Applications, 2023, https://doi.org/10.1007/s11042-023-14505-y.

[36] S. Pramanik, D. Samanta, R. Ghosh, S. K. Bandyopadhyay, "A New Combinational Technique in Image Steganography", International Journal of Information Security and Privacy, vol. 15, issue 3, article 4, IGI Global, 2021. DOI: 10.4018/IJISP.2021070104, ESCI, Scopus.

14

Smart Farming Utilizing a Wireless Sensor Network and the Internet of Things

Dinesh Singh[1], D. Janet Ramya[2], Manish Joshi[3], Sukhvinder Singh Dari[4], Ahmar Afaq[4], Dharmesh Dhabliya[5], and Sabyasachi Pramanik[6]

[1]Department of Agriculture, Sanskriti University, India
[2]Department of Computer Science and IT, School of CS and IT, Jain (Deemed to be) University, India
[3]College of Computing Science and Information Technology, Teerthanker Mahaveer University, India
[4]Symbiosis Law School, Nagpur Campus, Symbiosis International (Deemed University), India
[5]Department of Information Technology, Vishwakarma Institute of Information Technology, India
[6]Haldia Institute of Technology, India

Email: dinesh.ag@sanskriti.edu.in; d.janet@jainuniversity.ac.in; gothroughmanish@gmail.com; director@slsnagpur.edu.in; ahmar@slsnagpur.edu.in; dharmesh.dhabliya@viit.ac.in; sabyalnt@gmail.com

Abstract

The agricultural industry in India is negatively impacted, and crop output is declining daily. Therefore, in order to enhance productivity, it is crucial to find and put into practice a solution to the issue. To advance the agricultural sector, smart technologies are being implemented. The value of technologies like Internet of Things, big data, cloud-dependent facilities, and GPS is growing in the agricultural sector. Due to the need for increased crop analysis accuracy, real-time field data processing, and automated farming approaches for future advancement, there is an increasing demand. With the use of these intelligent processes, a smart agricultural business is what is anticipated. The pros and cons of the Internet of Things (IoT) and different kinds of sensors for data collection have been covered by the writers in this chapter.

14.1 Introduction

The smart gadgets are employed in smart agriculture, but their utilization is limited by their expensive cost and farmers' general lack of awareness of emerging technology. The cost rise is a result of the widespread use of smart gadgets, which raises the final product's price overall and prevents farmers from adopting new technology to boost the worldwide market's productivity for food items. Farmers may simply transition to Internet of Things in order to minimize the workload, time, and expense of human labor. The traditional method of manually measuring the meteorological parameters is carried out and is continually monitored. IoT [1–3] technology, which is extremely effective and trustworthy, aids in obtaining data on key environmental factors, including climate, humidity, temperature, and soil fertility. Internet of Things offers online crop tracking systems that enable an agriculturist to link with their field whenever they're away from it. Microcontrollers [4–6] and wireless sensor networks are combined to monitor, automate, and regulate agricultural activities. Today's enormous number of improved technology, together with a variety of equipment and procedures, enable the agricultural industry to operate effectively. IoT combines a variety of devices to automatically convert the acquired data into information. From the time crops are first planted on a farm until they are distributed to consumers at the greatest price, agriculture is fraught with difficulties. The most crucial area for a nation's growth is agriculture.

Since the 1970s, climate change and unpredictable rainfall have had an impact on the agricultural cycle. The output of raw materials continues to steadily decline when conventional techniques are used. Due to these results, a strategy known as "smart agriculture" has lately been embraced by Indian farmers. This approach combines a number of smart technologies. IoT technologies use WSNs [7–10] for remote tracking to assist several applications in providing their output more rapidly and precisely. India and other developing nations are in a position to adopt innovative technology and methods for the rapid production of food supplies. Numerous initiatives use wireless sensor networks to collect data from various sensor devices linked to numerous nodes and transmit it back using wireless protocol. The information gathered is used to forecast environmental conditions. Although measuring and tracking these environmental variables may aid in climate prediction, they do not provide a complete answer for enhancing agricultural productivity. There are also more factors that might significantly reduce output. The area of agriculture should build an automated method to solve the issue. Therefore, it is crucial to design an aggregated model that provides every climatic parameter impacting the production at all phases in order to give a clear solution to such a difficult issue. However, as the data acquired are dynamic and susceptible to alter, it is challenging to deploy a fully automated model in agriculture owing to a number of technological challenges. Despite much study, no product has yet been created to assist farmers in their efforts. According to [11], IoT has been integrated with the Internet's next big thing. This unique feature includes previously undiscovered facilities, Internet of Things framework, protocol, and applications. IoT enables a variety of sensors and items that can communicate with one another without the need for human intervention. The Internet of Things includes physical devices like sensor devices that watch and collect data from many Things. The IoT has consistently connected people, sensors, objects, and services on a worldwide scale. Since the dawn of time, humans have used a variety of traditional farming methods to cultivate food in accordance with their needs and ability to pay. Numerous factors, including weather, soil, temperature, and moisture, influence various agricultural practices. Traditional farming employs two key methods: shifting agriculture and nomadic herding. Rice-dependent, root crop-dependent, grain legume-dependent, agrisilvicultural, silvopastoral, and agrisilvopastoral approaches make up semi-commercial systems. Learning about smart farming and integrating various techniques to increase agricultural output through time has helped in the growth of the nation.

Plantations, agroforestry, and ranching are other commercial system practices. Automation systems have a high degree of precision and efficiency, which is why the project is incorporating IoT. Farmers still face issues due to the changing times, such as crop damage by animals, water logging, and a lack of knowledge about air and soil characteristics that cannot be foreseen or prevented. Few IoT-based research projects are being proposed in the agriculture sector for monitoring and maintaining the supply chain management system for farming goods. Internet of Things offers a WSN tracking framework by integrating GSM and GPRS characteristics, which is a crucial component of the productivity of e-agriculture. The pH sensor, humidity sensor, GPRS technology, moisture sensor, relay, water level sensor, Arduino Uno microcontroller output, and electric motor are just a few of the many components that make up smart e-agriculture. According to [12], the primary goals of farm management information systems (FMIS) are to comply with regulations, cut production costs, adhere to farming standards, compute higher product quality, and check for safety precautions. According to Michael J. O'Grady, technical improvement drives radical modernization in agricultural care and provides means for putting agriculture-based models into practice. The goal of IoT is to support high output, reduce costs, maintain quality over time, meet agricultural standards, and ultimately satisfy demand via FMIS.

Internet of Things technologies rely on hardware such as WSNs, network-linked smart phones, weather stations, and webcams to collect copious amounts of ecological and agricultural data for performance tracking. It includes human observations obtained and recorded by mobile smart phone apps, time series data gathered by sensors, spatial data collected from cameras, and more. For every sort of particular farm, these data are examined to remove outliers and forecast unique crop suggestions. Energy efficiency is a crucial factor in the growth of green IoT approaches.

14.2 Intelligent Farming Technologies Background

14.2.1 IoTs

According to [13], the Internet of Things is a network that uses devices like radio frequency identification (RFID) [14–16], device sensors, GPS [17–19], device scanners, and different data-sharing devices to integrate all gadgets into real-time applications. The IoTs are already used to develop applications for monitoring and managing a variety of objects, from household appliances to large structural domains like multinational corporations. Due to the effect of the supply chain operations reference model [20], IoT applications for agricultural commodities have been expanded. The greenhouse monitoring automation system is an application utilized often in planting that is accomplished by IoT gateway.

14.2.2 Cloud computing

Both consumers and businesses employ cloud computation in the realm of IT infrastructures. It may be utilized to a variety of domains, including the environment, medicine, and maintenance. The agriculturists in corporate sector will make it easier for rural communities to have affordable access to the agricultural services they need.

14.2.3 Massive data analysis

It comprises a significant amount of data, including vital information gathered from social networking sites, sensor data, remote sensing data, weather forecasting, and the agriculture sector. For world of farming [21, 22], big data analytics [23–25] aids in predicting the weather conditions that are optimal for growing at a certain moment. Using supply chain management, this lowers the price of the agricultural goods.

14.2.4 Mobile technology

A technique that facilitates communication between devices on multiple ends is mobile computing. It is utilized for providing prompt and dependable communication facilities in applications that are utilized on a regular basis. Mobile computing is mostly utilized in the agriculture industry to quickly transport information in off-the-grid locations. Farmers are primarily served by daily climate updates and news on the prices of agricultural products through mobile devices, which enable them to market their goods in an effective way.

14.2.5 Sensors and wireless sensor networks

It is significant to know how we go about our daily business. This has sensors that monitor a variety of ecological alterations in the farming region, including humidity, pressure, temperature, and other aspects which aid agriculturists in timely cultivation forecast.

14.2.6 Data analysis

Knowledgeable data is retrieved from the data repository through data mining. Approaches such as clustering and classification are used to identify crops and types of seeds, and association rule mining aids in the selection of acceptable crops for cultivation. Using classification methods in the agricultural sector aids in classifying the different types of soil and crops.

14.3 The Agricultural Industry's Challenges

- A lack of knowledge about manufacturing and its variables
- The weather prediction offers little information
- Unaware of sales distribution information

- Illiteracy and inadequate ICT infrastructure
- Agriculturists are unaware of the advantages that ICT in farming offers
- The Center for Research and Marketing's lack of research abilities
- Significant changes in the weather
- Less educated and younger individuals are interested in careers in agriculture
- The cost of machines is greater
- Added manual labor

14.4 IoT

For properly transmitting data, the IoTs link physical objects with electronics, communication, and sensor technologies. Data is collected and examined utilizing IoT-dependent devices. It is transmitted to the user, who can thereby use it for performing remote monitoring and command in the farms.

14.4.1 IoT characteristics

- Transmission: Wireless protocols are utilized. To send and receive the data collected which can be used in tracking and regulating, every devices interact with one another via transmission.
- Connectivity: Sensors and different devices are linked to a secured network.
- Things: These may be actuators, but they are the entities which collect data and exchange communications throughout an IoT system. Heart monitoring, biochip transponders, vehicles having built-in sensors, farm operation devices, and rescue functions are just a few of the many gadgets that are available.
- Data: Data is the glue that holds the Internet of Things together and is what drives action and insight.
- Intelligence: By utilizing preprocessing and data analytics, sensors make intelligent IoT devices.
- Action: Actuators may be utilized to do it either manually or automatically.

- Ecosystem: The IoT's surroundings or domain.
- IoT features have a major consequence on the supply chain for farming. The farming logistics are also provided.

14.4.2 IoT's advantages for farming

Agriculturists may reduce yield expenditures and optimize inputs because of the implementation of Internet of Things and its usage in farming. IoT in farming enhances agriculture and facilitates the mentioned benefits:

- Fruitful input minimizes the cost.
- Avoids crop loss or makes preventative measures by continuously monitoring the crops and assessing the weather and illness.
- IoT-dependent optimization of water usage by the help of sensors
- More fruitful design of agricultural operations
- Time and manpower minimization
- Keeping an eye on crop growth
- Utilizing sensors to direct water-content and PH levels in optimal crop growth
- Crop sales may enhance in the global market.
- The major hurdles to apply IoT are the cost of the equipment and Internet obtainability.

14.4.3 Internet of Things in farming

According to how it performs, IoT applications are categorized as tracking, precision farming, tracking and tracing, greenhouse production, and agricultural equipment. Internet of Things has a significant influence on agriculture, particularly in the areas of crop and livestock irrigation, water quality tracking, soil tracking, disease and pest management, equipment weather tracking, automation, and precision farming.

14.4.4 Monitoring

In agriculture, there are several things to keep an eye on. Crop farming, aquaponics, forestry, livestock farming, as well as other industries including water, fuel, and animal feeds, are a few of them. Farmers may plan ahead and save costs by using the tracked and archived data.

14.4.5 Monitoring and detection

It collects data throughout the supply chain so that the whole history of the product is realized, achieving the confidence of the consumer. It collects data on I/P, O/P, storage, communication, and processing.

14.4.6 Farming equipment

It may aid enhance crop yield and reduce grain loss. It may map sufficiently to run the equipment in autopilot mode with the help of GPS and GNSS. The equipment will collect information on irrigation, fertilization, and nutrition to help agriculturists in mapping the farms in this manner.

14.4.7 Precision farming

The usage of predictive analytics and real-time data accumulating to make decisions that would provide an effective yield lessened the influence of environmental variables and cost. Big data analytics, GPS, and sensor nodes [26] are utilized for this.

14.4.8 Climate production

The advantage of this technique is that it may grow any variety of plant at anytime, anywhere, by having the correct climatic situations. It is also called "glasshouse technology." WSN is utilized to have a look on the environment. IoT is being used to reduce the need for human resources, save energy, monitor activity effectively, and connect with consumers.

14.5 Selection of Crop

14.5.1 Internet of Things (IoT) crop selection: Suitability of the land

It is a burden for agriculturists to detect farms and different variables, yet crops require various atmospheric circumstances and environments to grow rapidly. UAVs [27–29], which are utilized for in-flight surveillance and are mini airplanes that collect data about the surroundings such as visual, temperature, humidity, weather, thermal, and air pressure, are capable of using artificial intelligence in conjunction with IoT to provide surveys. Further, it proposes particulars on crops, with their height, count, health, weed detection, and seasonal consequences.

14.5.2 Level of nutrients in the crops

The right balance of nutrients is crucial for high-quality meals. By utilizing intelligent farming and the Internet of Things, the production level of nutrients may be controlled. Utilizing the nutrition analyzer sensor, monitor, manage, and regulate the nutrient level. By relating certain ions in the side of the membrane, the sensor determines the quantity and quality of ions present as water travels across it. The existence of the six ions is concurrently detected and sent to the smartphone. With this knowledge, a farmer may alter the production maturity rate and give the right balance of nutrients to improve plant output.

14.5.3 Crop surveillance

Crop monitoring is crucial in IoT-based agriculture. Understanding crop development in a small-scale context is essential. As the environment changes globally, research and study on crop development are required. The growth pattern and environmental conditions provide scientific advice and solutions for enhancing agricultural productivity. It may simulate how a crop grows in various settings across different locations to enhance overall performance. By enabling networking to convey information, two different kinds of sensor nodes are employed to track the crop. The platform for environmental parameter acquisition collects data on soil and weather conditions. The picture capturing device offers photographs of the expanding crop, allowing us to immediately track its development. We may obtain the data from the sensor nodes in the sensor network for monitoring agricultural condition through the Internet.

14.6 Agriculture with Precision (Pa)

PA demands an agricultural Decision Support System [30–32]. It is an important phrase in the world of agriculture. Precision agriculture first evolved in Texas in 1995, and since then, it has been a focus of survey and application all around the globe. This kind of agriculture is used for enhancing decision-making in crop yield. According to standard definitions, PA is a "IT dependent farm management framework to locate, survey, and direct variability in different farms for optimum profit, property, and protection of the farm resource." The main intention of PA design is to develop a DSS for entire farm management and to increase input return while preserving resources.

It is required to manage and disseminate the I/P information on a website at certain time gaps in order to reduce operating costs and enhance performance. Farmers are looking for more effective ways to improve power and reduce costs due to the rise in input prices for the production of commodities. A single technology or a number of them are employed for various agricultural sector models. Multiple technologies will be used to provide different features. PA is the major aspect of precision farming. It suggests a fresh perspective on issues in the agricultural sector dependent on environmental concerns to stabilize production.

This summarizes the precise irrigation methods now in use, along with specific crops and areas for open agricultural and greenhouse systems. It also makes reference to the Android apps created for PA is discussed more. Data collection is done for field environmental tracking using a variety of sensors in accordance with system requirements. Data gathering, reporting, collecting GPS data into GIS, tractability systems like animal tracking identification, and web cams to see the farm are all examples of PA applications.

Developing mobile technology and mobile-based data services provide a means of addressing the information imbalance that exists in a variety of industries, including agriculture, health care, and education. Between the supply and distribution of agricultural supplies and agricultural infrastructure, there is a huge chasm that will be bridged by mobile technology. The open source development platforms for Android apps allow for the creation of very sophisticated applications by developers. There are various frameworks available for the implementation of PA ranging from simple ones to many highly developed ones. The classification of PA is dependent on the needs of the agriculturists.

Different techniques have been used for automated irrigation, including scheduling irrigation based on canopy temperature distribution, irrigation systems depending on soil moisture constituent, calculated with the aid of dielectric moisture sensors to control actuators and to reduce water use, scheduling irrigation of crops using crop water stress index (CWSI), estimating plant evapotranspiration (ET), and irrigation using infrared thermometers. To determine the plants' water needs, the DSS is connected with the ET. For agricultural crops, authors [33] presented an automated irrigation design. It has sensors to measure temperature and soil moisture. Techniques for controlling plant watering utilizing a microcontroller-dependent gateway were presented depending on the data supplied by the sensors and the threshold values.

The following is a list of PA technology benefits:

- The effective use of manure and chemical spraying to protect ground water.
- By reducing freshwater withdrawals via precision irrigation.
- By quickly and skillfully reacting to insect and fungal infections, crop harm may be avoided.
- By using novel polyculture techniques.

By immediately signaling a difficulty in the farm to offer online assistance for future investigation, information on PA diagnostic procedures may be sent to smart phones or tablets and made accessible to smaller farms.

14.7 Technologies for Sensing

Sensing technologies are used to gather information that may be utilized to enhance agricultural productivity and reduce environmental impact. There are several precision farming techniques that make use of this data.

14.7.1 Information on topography and boundaries

GPS is utilized to capture a precise topographic map of any field, which may then be used to anticipate the location of weeds and yield maps. One may locate the wetlands, field borders, and roads that are obtainable to assist agriculturists in planning their farms with the use of the data.

14.8 Systems for Monitoring Yield

In order to provide information regarding crop weight production depending on the distance, time, and location recorded utilizing GPS and kept for analysis, they are mounted to the crop harvesting trucks.

14.8.1 The yield map

By putting GPS sensors on harvesting equipment to capture geographical coordinates and fusing that with yield monitoring, it creates yield maps.

14.8.2 A weed map

With the use of a data recorder and GPS to determine the position, it creates maps based on the operator's judgment. This may overlap the yield, fertilizer, and spray maps. To get the information, the device might be equipped with a vision recognition system.

14.8.3 Mapping of salinity

It is dragged through the fields that are impacted by salinity using a sled and a salinity meter. Salinity mapping interprets the changing conditions and variations in salinity throughout time.

14.9 Tools for Applying Fertilizer at a Variable Rate

It controls liquid, granular, and gaseous fertilizer ingredients using a yield map and a pattern regarding the health of the plant that was discovered by optical surveys. It may be automated utilizing an on-board GPS computer or controlled manually.

14.9.1 Spraying controllers that are flexible

After locating and mapping the locations of the weeds, it regulates the amount and mixture of the spray administered by turning on and off the herbicide boom spray.

14.10 A Range of Sensors

When selecting a sensor, several considerations must be taken into account. They are as follows:

- Accuracy – How closely the result corresponds to the right value or a standard
- Environmental circumstances – Examines meteorological variables including temperature and moisture content
- Range measures the separation between the sensors
- Resolution – It detects even minute alterations
- Expenditure
- Repeatability – Repeatedly measuring the data in the same setting

14.11 Different Sensor Configurations in E-Farming

Below is a list of the many sensors used in e-farming.

S.NO	Types of sensors	Function
1.	Airflow	To measure the soil permeability
2.	Location	To find latitude, longitude and altitude using GPS
3.	Camera	To get the picture of the crop, environment features and monitoring
4.	Microphone	Helps with predictive maintenance of machinery
5.	Accelerometer	To find the agriculture activity, leaf Angle Index etc
6.	Gyroscope	To detect rotation of equipment
7.	Electrochemical	To find the ions in the soil using electrodes
8.	Dielectric soil moisture	To measure the dielectric constant in the soil using electrodes
9.	Optical	light is used to measure the soil properties
10.	Mechanical	To measure resistance and soil compaction using probes

Figure 14.1 Various sensors.

14.11.1 (Lm35) Temperature sensor

This sensor aids in temperature measurement through electrical output. Additionally, it measures the ideal temperature for crops. Additionally, it provides data about the previous day's temperature. Compared to other devices like thermostats, temperature sensors provide findings that are more accurate.

14.11.2 Hr 202 humidity sensor

A humidity sensor is used to measure the amount of water on the ground. This calculates how much water the crop needs. The HR202 is an organic macromolecule-based humidity-sensitive resistor. Hospitals, storage facilities, workshops, the textile industry, etc., all utilize it. The functional temperature range is between 20% and 95%.

14.11.3 Water level sensor in Rh

The spherical and cylindrical float balls are sometimes referred to as water level floats sensors. They might be used in switch mechanisms as well. These are constructed of pliable material that floats in liquids like water.

14.12 AIT Stands for Agricultural Information Technology

AIT is a powerful instrument used to efficiently use resources in all facets of agriculture to increase output. AIT's agriculture information management subtechnology is used to manage agricultural information and make wise decisions that will increase production. Data gathering, monitoring, and automation – what we refer to as "smart agriculture" – will undoubtedly enhance output while lowering costs. These activities are made possible by data insights, inexpensive sensors, and the use of IoT technology. The use of databases in cloud computing enables administration of data connected to the output performance of plants and analysis of them for decision-making. Scalable network architecture may be used to monitor and manage farming operations and rural produce.

14.13 Smart Agriculture Benefits

- Increasing agricultural production – This may be done by using the right water, pesticides, planting techniques, and harvesting techniques, which will raise productivity.
- Water conservation – Apply water correctly by evaluating the need using sensors that measure soil moisture and forecast the weather.
- Real-time data and production insight – Farmers may use Internet of Things to remotely track production levels, soil moisture, and sunshine intensity.
- Reduced operation costs – By automating the process, it is possible to reduce costs, efficiently use resources, and human mistake.
- Improved production quality – Agriculturists may comprehend the procedure by evaluating the quality and may change the care to improve the quality.
- Accurate farm and field evaluation – Periodic production rate study will provide an accurate forecast for future output.
- Better livestock farming – Livestock may be managed and monitored effectively using location tracking and geofencing.
- Lesser environmental footprint – The use of water and increased productivity per acre have a quick and beneficial effect on the environmental footprint.
- Remote monitoring – Using an Internet connection, farmers may remotely monitor their crops from various locations.

- Equipment tracking – The use of equipment for upkeep and tracking depending on output, workforce costs, and deficiency predictions.

They may enhance agriculture by using IoT apps by

- Smart IoT platform's daily supply chain transformation
- IoT for transforming agriculture

Here are the first five strategies for smart agriculture:

1. Agricultural parameter sensing
2. Gathering of data
3. Field-based information transmission for decision-making
4. Data analysis for early warning and decision assistance
5. Following warnings, corrective action must be implemented

User, server, and data side are present.

The information is obtained from many sensors that are connected to the framework, including

- sensors for measuring temperature, humidity, gas, solar, and soil moisture;
- to boost production rate, decisions may be made using this data.

14.13.1 In the server

The information sent from the sensor to the server is examined for numerous characteristics, such as the soil's pH level, humidity, and light, and then decisions are made and the system is monitored as a result.

The user may get the info on their mobile device. He may review the data from a distance and use the updated information to determine the required actions to boost crop development.

14.13.2 Use of the agricultural information cloud

- Controlling and tracking of agricultural produce security
- Effective production and management techniques
- Plant growth monitoring

14.14 Agriculture Information Management Systems

Data processing and storage are necessary for the massive amounts of complex, dynamic, and geographical data needed for smart agriculture.

To handle and do data analytics on the many forms of data, several MIS [34] for farming are created. Almost no commercially available platforms are:

14.14.1 OnFarm

OnFarm is a free, industry-standard, and enterprise farm management solution that analyzes data and presents information from many sources.

14.14.2 Phytech

Direct sensing, data analytics, plant status, and suggestions are utilized by the plant IoT platform Phytech to make decisions that will boost production and improve watering.

14.14.3 Semios

Semios focuses on orchard irrigation, network coverage, frost, pests, and diseases. Using real-time monitoring services, it offers notice of field events.

14.14.4 EZFarm

The IBM project EZfarm manages water resources and tracks soil quality and plant health.

14.14.5 KAA

KAA offers remote crop monitoring, resource mapping, animal tracking, statistics on feeding and product for livestock and humans, smart logistics, and storage.

14.14.6 CropX

By keeping an eye on the soil, CropX increases crop productivity while managing water and reducing energy costs.

The procedure for acquiring sensor data involves digitizing the data such that it may be gathered, saved, and used for decision-making.

14.14.7 PV solar panel

The sensor is attached to a small solar panel that, when exposed to sunlight, produces DC electricity. The output is transformed by the Arduino into digital values that may be saved for analysis.

14.14.8 Gas detector

The threshold value of CO_2 is set using a potentiometer and a linked sensor, such as the MG-811. Based on the CO_2 concentration, a digital signal (ON/OFF) will be generated. Using the data gathered by the sensor, the CO_2 constituent may be estimated. CO_2 is crucial for the development of plants. The farmer may utilize a sprinkler depending on the situation.

14.14.9 pH sensors for soil

The crop's production may be impacted by the acidity of the soil. A pH meter with a reference and conducting electrode may be used to measure the hydrogen ion concentration of the soil. They were placed into the ground to bring about conduction, which changes depending on the amount of hydrogen ions present.

14.14.10 Megapixel-rich cameras

High resolution cameras are used to keep an eye on the plants for disease signs. For detecting the symptoms and determine the severity of harm, the captured photos are compared to the data that has been recorded in the control center.

14.14.11 Board for Arduino

The Arduino board that serves as a link between the sensors and Zigbee, performs ADC. It is the control hub for the sensor panel. It exchanges data with Zigbee and sends it to the center hub. In a centralized hub, the data is processed and stored for use in decision-making.

14.14.12 Zigbee

Having Arduino and Raspberry Pi serving as the hosts on each side, Zigbee may function as a transmitter and receiver. The Zigbee communication protocol is a good fit since it is portable and uses little power.

14.14.13 Raspberry Pi

It is a tiny computer that fits in your pocket that may be used for networking and computing. It has an Internet connection so that a remote control device may automate the system.

It is a key element of the Internet of Things, a potent controller, and a tool for making decisions. It has a Zigbee connection that allows it to broadcast and accept data from various sensor modules. Versions of the Raspberry Pi are varied.

14.14.14 PIR device

It is utilized to track the movement of people, animals, and entities by detecting the infrared radiation reflected and released by an item. When it senses a change in temperature, it transforms that change into voltage, which may then be utilized to make decisions.

14.15 Softwares

14.15.1 Simulator for PROTEUS

It is a commonly used simulation program for many microcontrollers, which include all electronic components. In order to prevent harm from poor design, this program may be used for testing before actual hardware testing.

14.15.2 Version 4 of AVR Studio

It creates a hex file that can be burnt into the microcontroller to do the desired actions. This file may then be utilized to compile the embedded c code, examine for faults, and make it.

14.15.3 Trace Dip

It is an EDA/CAD program featuring a schematic capture editor, PCB [35] layout editor, component editor, pattern editor, and export features for making printed circuit boards and schematic diagrams.

14.15.4 SinaProg

It is an application for programming AVR-based microcontrollers that includes a hex downloader and a fuse bit calculator.

14.15.5 OS Raspbian

This is an open source, Debian-based computer OS [36, 37] designed specifically for Raspberry Pi, complete with programs and tools. It comes with a lot of precompiled programs for Raspberry Pi installation.

14.16 Conclusion

The chapter talks about the difficulties facing the agricultural industry. In the realm of agriculture, crop output is directly impacted by environmental factors and climatic conditions, which makes it challenging for farmers to make decisions. The IoT technologies have greatly benefited farmers by enhancing agricultural production rates and lowering costs. It is detailed how various data collecting sensors are used in agriculture. The development of new IoT and WSN technologies has had a significant influence on agriculture. These will be combined with AI and other data mining techniques, resulting in the development, invention, and application of fresh concepts for effective manufacturing and marketing.

References

[1] Reepu, S. Kumar, M. G. Chaudhary, K. G. Gupta, S. Pramanik and A. Gupta, Information Security and Privacy in IoT, in Handbook of Research in Advancements in AI and IoT Convergence Technologies, Eds, J. Zhao, V. V. Kumar, R. Natarajan and T. R. Mahesh, IGI Global, 2023.

[2] S. Pramanik, "An Effective Secured Privacy-Protecting Data Aggregation Method in IoT", in Achieving Full Realization and Mitigating the Challenges of the Internet of Things, Eds, M. O. Odhiambo and W. Mwashita, IGI Global, 2022, DOI: 10.4018/978-1-7998-9312-7.ch008

[3] Dhamodaran S, S. Ahamad, J.V.N. Ramesh, G. Muthugurunathan, Manikandan K, S. Pramanik, D. Pandey, Thrust Technologies' Effect on Image Processing, IGI Global, 2023

[4] Dhamodaran S, S. Ahamad, J.V.N. Ramesh, S. Sathappan, A. Namdev, R. R. Kanse, S. Pramanik, Fire Detection System Utilizing an Aggregate Technique in UAV and Cloud Computing, Thrust Technologies' Effect on Image Processing, IGI Global, 2023

[5] V. Veeraiah, D.J. Shiju, J.V.N. Ramesh, Ganesh Kumar R, S. Pramanik, A. Gupta, D. Pandey, Healthcare Cloud Services in Image Processing, Thrust Technologies' Effect on Image Processing, IGI Global, 2023

[6] S. Ahamad, V. Veeraiah, J.V.N. Ramesh, R. Rajadevi, Reeja S R, S. Pramanik and A. Gupta, Deep Learning based Cancer Detection Technique, Thrust Technologies' Effect on Image Processing, IGI Global, 2023

[7] R. Ghosh and S. Pramanik, "A Novel Performance Evaluation of Resourceful Energy Saving Protocols of Heterogeneous WSN to Maximize Network Stability and Lifetime, International Journal of Interdisciplinary Telecommunications and Networking, vol. 13, issue 2, 2021. (ESCI, DBLP)

[8] R. Ghosh, S. Mohanty, P. Pattnaik and S. Pramanik, "A Performance Assessment of Power-Efficient Variants of Distributed Energy-Efficient Clustering Protocols in WSN", International Journal of Interactive Communication Systems and Technologies, vol. 10, issue 2, article 1, 2020, DOI: 10.4018/IJICST.2020070101.

[9] R. Ghosh, S. Mohanty, P. K. Pattnaik and S. Pramanik, "Performance Analysis Based on Probablity of False Alarm and Miss Detection in Cognitive Radio Network", International Journal of Wireless and Mobile Computing, vol. 20, no. 4, https://doi.org/10.1504/IJWMC.2021.117530, 2020.

[10] R. Jayasingh, J. Kumar R.J.S, D. B. Telagathoti, K. M. Sagayam and S. Pramanik, "Speckle noise removal by SORAMA segmentation in Digital Image Processing to facilitate precise robotic surgery", International Journal of Reliable and Quality E-Healthcare, vol. 11, issue 1, DOI: 10.4018/IJRQEH.295083, 2022

[11] W. Shi, W. Xu, X. You, C. Zhao and K. Wei, "Intelligent Reflection Enabling Technologies for Integrated and Green Internet-of-Everything Beyond 5G: Communication, Sensing, and Security," in IEEE Wireless Communications, vol. 30, no. 2, pp. 147–154, April 2023, doi: 10.1109/MWC.018.2100717.

[12] V. K, T. P, K. I. J, K. M and K. P. Sambrekar, "Design and Development of an Efficient Agriculture Management System in Cloud Computing using Machine Learning," 2023 IEEE 8th International Conference for Convergence in Technology (I2CT), Lonavla, India, 2023, pp. 1–7, doi: 10.1109/I2CT57861.2023.10126486.

[13] Q. -V. Pham, M. Zeng, O. A. Dobre, Z. Ding and L. Song, "Guest Editorial Special Issue on Aerial Computing for the Internet of Things (IoT)," in IEEE Internet of Things Journal, vol. 10, no. 7, pp. 5623–5625, 1 Aprill, 2023, doi: 10.1109/JIOT.2023.3248970.

[14] D. Samanta, S. Dutta, M. G. Galety and S. Pramanik, A Novel Approach for Web Mining Taxonomy for High-Performance Computing, The

4th International Conference of Computer Science and Renewable Energies (ICCSRE'2021), 2021, DOI:10.1051/e3sconf/202129701073.
[15] S. Pramanik, M. G. Galety, D. Samanta, N. P Joseph, Data Mining Approaches for Decision Support Systems, 3rd International Conference on Emerging Technologies in Data Mining and Information Security, 2022
[16] R. Bansal, A. J. Obaid, A. Gupta, R. Singh, S. Pramanik "Impact of Big Data on Digital Transformation in 5G Era", 2nd International Conference on Physics and Applied Sciences (ICPAS 2021), doi:10.1088/1742-6596/1963/1/012170, 2021.
[17] B. K. Pandey, D. Pandey, V. K. Nassa, A. S. George, S. Pramanik and P. Dadheech, Applications for the Text Extraction Method of Complex Degraded Images, in The Impact of Thrust Technologies on Image Processing, Nova Publishers, 2023.
[18] B. K. Pandey, D. Pandey, V. K. Nassa, A. S. Hameed, A. S. George, P. Dadheech and S. Pramanik, A Review of Various Text Extraction Algorithms for Images, in The Impact of Thrust Technologies on Image Processing, Nova Publishers, 2023.
[19] P. K. Mall, S. Pramanik, S. Srivastava, M. Faiz, S. Sriramulu and M. N. Kumar, FuzztNet-Based Modelling Smart Traffic System in Smart Cities Using Deep Learning Models, in Data-Driven Mathematical Modeling in Smart Cities, IGI Global, 2023, DOI: 10.4018/978-1-6684-6408-3.ch005
[20] D. Mondal, A. Ratnaparkhi, A. Deshpande, V. Deshpande, A. P. Kshirsagar and S. Pramanik, Applications, Modern Trends and Challenges of Multiscale Modelling in Smart Cities, in Data-Driven Mathematical Modeling in Smart Cities, IGI Global, 2023, DOI: 10.4018/978-1-6684-6408-3.ch001
[21] P. T. Khanh, T. H. Ngọc, and S. Pramanik, Future of Smart Agriculture Techniques and Applications, in Advanced Technologies and AI-Equipped IoT Applications in High Tech Agriculture, Eds, A. Khang, IGI Global, 2023.
[22] T. H. Ngọc, P. T. Khanh and S. Pramanik, Smart Agriculture using a Soil Monitoring System, in Advanced Technologies and AI-Equipped IoT Applications in High Tech Agriculture, Eds, A. Khang, IGI Global, 2023.
[23] A. Gupta, A. Verma and S. Pramanik, Security Aspects in Advanced Image Processing Techniques for COVID-19, in An Interdisciplinary Approach to Modern Network Security, S. Pramanik, A. Sharma, S. Bhatia and D. N. Le, Eds, CRC Press, 2022.

[24] K. Dushyant, G. Muskan, A. Gupta and S. Pramanik, "Utilizing Machine Learning and Deep Learning in Cyber security: An Innovative Approach", in Cyber security and Digital Forensics, M. M. Ghonge, S. Pramanik, R. Mangrulkar, D. N. Le, Eds, Wiley, 2022, https://doi.org/10.1002/9781119795667.ch12

[25] D. K.aushik, M. Garg, Annu, A. Gupta and S. Pramanik, Application of Machine Learning and Deep Learning in Cyber security: An Innovative Approach, in Cybersecurity and Digital Forensics: Challenges and Future Trends, M. Ghonge, S. Pramanik, R. Mangrulkar and D. N. Le, Eds, Wiley, 2022, DOI: 10.1002/9781119795667.ch12

[26] G. Taviti Naidu, KVB Ganesh, V. Vidya Chellam, S Praveenkumar, D. Dhabliya, S. Pramanik, A. Gupta, Technological Innovation Driven by Big Data, in "Advanced Bioinspiration Methods for Healthcare Standards, Policies, and Reform", Hadj Ahmed Bouarara IGI Global, 2023, DOI: 10.4018/978-1-6684-5656-9

[27] A. Mandal, S. Dutta, S. Pramanik, "Machine Intelligence of Pi from Geometrical Figures with Variable Parameters using SCILab", in Methodologies and Applications of Computational Statistics for Machine Learning, D. Samanta, R. R. Althar, S. Pramanik and S. Dutta, Eds, IGI Global, 2021, pp. 38–63, DOI: 10.4018/978-1-7998-7701-1.ch003

[28] S. Pramanik, "Carpooling Solutions using Machine Learning Tools", in Handbook of Research on Evolving Designs and Innovation in ICT and Intelligent Systems for Real-World Applications, K. K. Sarma, N. Saikia and M. Sharma, IGI Global, 2022, DOI: 10.4018/978-1-7998-9795-8.ch002.

[29] Y. Meslie, W. Enbeyle, B. K. Pandey, S. Pramanik, D. Pandey, P. Dadeech, A. Belay, A. Saini, "Machine Intelligence-based Trend Analysis of COVID-19 for Total Daily Confirmed Cases in Asia and Africa", in Methodologies and Applications of Computational Statistics for Machine Learning, D. Samanta, R. R. Althar, S. Pramanik and S. Dutta, Eds, IGI Global, 2021, pp. 164–185, DOI: 10.4018/978-1-7998-7701-1.ch009

[30] A. Bhattacharya, A. Ghosal, A. J. Obaid, S. Krit, V. K. Shukla, K. Mandal and S. Pramanik, "Unsupervised Summarization Approach with Computational Statistics of Microblog Data", in Methodologies and Applications of Computational Statistics for Machine Learning, D. Samanta, R. R. Althar, S. Pramanik and S. Dutta, Eds, IGI Global, 2021, pp. 23-37, DOI: 10.4018/978-1-7998-7701-1.ch002

[31] A. Gupta, A. Verma, S. Pramanik, "Advanced Security System in Video Surveillance for COVID-19", in An Interdisciplinary Approach

to Modern Network Security, S. Pramanik, A. Sharma, S. Bhatia and D. N. Le, CRC Press, 2022.

[32] S. Pramanik, AJ Obaid, N M, and SK Bandyopadhyay "Applications of Big Data in Clinical Applications", Al-Kadhum 2nd International Conference on Modern Applications of Information and Communication Technology, *AIP Conference Proceedings* 2591, 030086 (2023), https://doi.org/10.1063/5.0119414

[33] R. Barreto, J. Cornejo, J. C. Suárez Quispe, C. N. Ochoa and D. O. Tacuri, "Conceptual Design of an Automated Irrigation and Fixation System for Long-Term Crop Care of Cherry Tomatoes for Applications in Microgravity Machines," 2023 Third International Conference on Advances in Electrical, Computing, Communication and Sustainable Technologies (ICAECT), Bhilai, India, 2023, pp. 1–5, doi: 10.1109/ICAECT57570.2023.10117929.

[34] S. Pramanik, An Adaptive Image Steganography Approach depending on Integer Wavelet Transform and Genetic Algorithm, Multimedia Tools and Applications, 2023, https://doi.org/10.1007/s11042-023-14505-y.

[35] Z. Ma, S. Wang, H. Sheng and S. Lakshmikanthan, "Modeling, Analysis and Mitigation of Radiated EMI Due to PCB Ground Impedance in a 65 W High-Density Active-Clamp Flyback Converter," in IEEE Transactions on Industrial Electronics, vol. 70, no. 12, pp. 12267–12277, Dec. 2023, doi: 10.1109/TIE.2023.3239904.

[36] F. -Y. Wang, J. Li, R. Qin, J. Zhu, H. Mo and B. Hu, "ChatGPT for Computational Social Systems: From Conversational Applications to Human-Oriented Operating Systems," in IEEE Transactions on Computational Social Systems, vol. 10, no. 2, pp. 414–425, April 2023, doi: 10.1109/TCSS.2023.3252679.

[37] X. Tan and I. Hakala, "StateOS: A Memory-Efficient Hybrid Operating System for IoT Devices," in IEEE Internet of Things Journal, vol. 10, no. 11, pp. 9523–9533, 1 June1, 2023, doi: 10.1109/JIOT.2023.3234106.

15

Intelligent Agriculture using Autonomous UAVs

Manoj Kumar Mishra[1], J. Bhuvana[2], Anu Sharma[3], Nuzhat Fatima Rizvi[4], Nikhil Polke[4], Sabyasachi Pramanik[5], and Hafiz Husnain Raza Sherazi[6]

[1]Department of Agriculture, Sanskriti University, India
[2]Department of Computer Science and IT, School of CS and IT, Jain (Deemed to be) University, India
[3]College of Computing Science and Information Technology, Teerthanker Mahaveer University, India
[4]Symbiosis Law School, Nagpur Campus, Symbiosis International (Deemed University), India
[5]Department of Computer Science & Engineering, Haldia Institute of Technology, India
[6]School of Computing and Engineering, University of West London, UK

Email: manoj.ag@sanskriti.edu.in; j.bhuvana@jainuniversity.ac.in; er.anusharma18@gmail.com; nuzhatrizvi@slsnagpur.edu.in; nikhilpolke@slsnagpur.edu.in; sabyalnt@gmail.com; sherazi@uwl.ac.uk

Abstract

The evolution of farming robots and their impact on the sector are all topics covered in this chapter, along with intelligent farming and its elements, robotic frameworks, and ML approaches. By minimizing the target's time, it is intended to apply the gathering of autonomous unmanned aerial aircraft and UGVs that are communicated with one another. Utilizing particle swarm optimization, it sought to explore several strategies for the ideal number of stops. The best time to harvest the apples distributed to the stalls was calculated using deterministic, binary mixed (0-1) integer modeling and the *k*-means approach. With the help of this modeling, it is established in which

UAV may be used to gather the apple and how to determine using 0-1 binary modeling if the air vehicle has successfully gathered the fruit or not. The unmanned UGV's path was created using the 2-opt, closest neighbor, and nearest insertion approaches.

15.1 Introduction

Robotics and sensing innovations from the twenty-first century have the ability to address persistent issues in the agriculture sector. By transitioning to an agricultural system that uses robotic technology, crops may be made more productive and sustainable. Robotic automation studies are being conducted by a large number of researchers worldwide in an effort to lower production costs and improve quality in greenhouses that grow fruits and vegetables. Robotic autonomous pickers that monitor crop development and gather production are being tested. Sensing technologies assist livestock producers in managing and regulating the health of their animals. Without utilizing the the random usage of farming pesticides, tracking and sustentation tasks are executed to eradicate hazardous pests and diseases in order to enhance soil quality.

The majority of these technologies are still in the development stage in start-up businesses, even if some of them are now accessible. Large-scale manufacturers of agricultural equipment have not yet shown a desire to charge in autonomous agricultural technology. This is due to a concern that the economic model these businesses established in the existing yield and sales markets may alter. If start-up businesses using digital farming technology are successful in putting their plans into action in the agricultural industry, it will also permanently alter how we produce food. By 2050, our current food production will have doubled and will continue to grow at this pace, meeting the needs of 70% of the world's population.

Engineers and scientists have been developing answers by developing remarkable technologies and systems for the major issues that mankind has faced for millennia as a result of the industrialization technique that began with the industrial revolution. The first significant issue that engineers and scientists will have to deal with in the first half of the twenty-first century is Agriculture 5.0. According to the idea of "Agriculture 5.0," farms utilize delicate agricultural techniques and apply digital agriculture by using tools like unmanned functioning and automated decision support system. In light of this, Wang et al. [1] define Agriculture 5.0 as the utilization of automated robotics and certain AI technologies.

To satisfy the annual need for agricultural labor, seasonal laborers in traditional farming throughout the globe go to agricultural jobs in various nations and areas of the world. Seasonal workers are unable to leave because of the worldwide COVID-19 outbreak, putting several nations in risk of a decline in the field of agriculture. In addition, every year sees a rise in the pace of migration from rural to cities. Due to this, the world is increasingly eschewing an agrarian civilization, and urban populations are expanding quickly. These issues indicate that there will be significant labor issues in the agriculture industry in the next years, according to [2]. Chen et al. [3] state that automated robotic systems in farming yield produce by operating 30 times as much as a labor in the industry. We must accept that additional outbreaks similar to COVID-19 will occur, abandon the traditional use of human labor, and quicken the switch to robotic farming yield.

A study by [34] found that farming robots may gather crops in the farm more quickly and in greater quantities than human labor. In most industries today, robots are not as quick and effective as people, but this is not the case in the agriculture industry. Daily monotonous chores on farms are quickly completed by agricultural robots owing to the fast evolving autonomous robotics and computer programs that use artificial intelligence. These innovations heralded the beginning of Agriculture 5.0, a new age in agriculture.

In the research by Nelson et al. [4], farming robots are demonstrated to significantly boost production and lower farm running expenses in the nations with higher rates of employing digital farming. As was already noted, the use of robotics in agriculture is growing quickly. For agriculturists, particularly those having smaller field, plenty of farming technologies are quite costly. Agriculturists who have tiny plots of land and smaller economies in close proximity to this are currently unable to access robots and frameworks in this field of technology. But, as output grows yearly, technology becomes less expensive, and agriculture is backed by the government, small farmers will also utilize this technology as part of growth plans.

The year 2015 saw a global decline in crop yields and agricultural productivity. The idea of an agricultural robot was presented to the globe in an effort to solve these issues and efficiently satisfy the growing demand. The use of robots in both the stock market and the world's agriculture industry is on the rise. When compared to farmers, agricultural robots can execute tasks far more quickly and with superior quality, according to a research by Verified Market Intelligence.

According to [6], agriculture technology start-up companies' market worth rose by almost $800 million between 2013 and 2017. Parallel to the advancement of artificial intelligence, start-up businesses that utilize robots and ML to tackle challenges in farming are growing since 2020. In the previous five years, the amount of venture capital invested in the area of AI has surged by 380%. "Less resources, more production" is another way to describe this new approach to agriculture. Because according to the Food and Agriculture Agency of the UN (FAA) estimate from 2009, 9.1 billion people will inhabit the planet by 2050. The most fundamental and important goal in the next years will be to avoid climate change in order to provide sustainable food production for the expanding global population.

Modern agricultural techniques are applied in industrialized nations, and as a result, these nations produce more than they anticipated. As a result, industrialized nations began to enjoy comparative advantages in agriculture. The global population has been growing quickly, which has raised per capita consumption but also created a number of issues. These issues include:

- Is unaware that chemical pesticides and fertilizers were utilized
- Inefficient use of groundwater
- A decline in agriculturally productive regions
- Property concerns
- Issues brought on by mechanization
- Clearing forests to make way for homes or agricultural land
- Klavuz (2019) is unaware of agricultural activity in erosion-prone locations.

15.2 Background

The agriculture business is evolving into a high-tech industry, opening up new possibilities for product creation and technology application. As the sector develops, autonomous systems and agricultural instruments with specialized uses take center stage. The agricultural [7–10] platform may be categorized widely. General purpose robots are created to carry out a range of activities in many contexts, whereas task-based robots are created to carry out a single activity on a predetermined product.

Agricultural robots can now carry out a variety of agricultural tasks, including crop imaging, pest and weed management, sprinkling water and pesticide and harvesting, thanks to technological advancements. Depending on the area and application, these procedures may be totally or partially autonomous. Although there are fast and accurate agricultural robots [11–13], there are still few uses for them in agriculture because of the challenges and unstructured surroundings. For instance, a robot capable of planting seeds could not know the distances between the seeds in order to do it as efficiently as possible. Depending on the crop variety and the soil conditions, a water or pesticide spraying robot may not be able to regulate the quantity of water or insecticide to be sprayed. Although robots are already used in certain aspects of agriculture, they are not yet intelligent enough to decide for themselves based on a variety of physical, natural, and environmental elements. By retaining past data, machine learning sub-algorithms and evolving technologies continue to improve the existing model. Therefore, autonomous robots utilized in agricultural technology will eventually become more sophisticated.

Fruit harvesting platforms are utilized to decrease time spent outdoors picking, reduce human mobility while harvesting, and provide the best working conditions. In other words, using these platforms boosts production throughout the harvest season.

Depending on a much perfect and resource-structured technique, smart agriculture has the possibility to create a more fruitful and sustainable farming yield. The producer should get additional value from smart agriculture in the format of improved decision-making or much effective business operations and management. Here, precision agriculture, agricultural automation, and agricultural robots are three technological sectors that are closely tied to smart agriculture. The use of robotic, automated control, and AI [14–17] systems in various facets of farming production, like farmbots and farm UAVs [18, 19], is known as agricultural automation and robotics. For farmers, agricultural robots improve productivity in a number of ways. Some of such include robotic arms, autonomous tractors, and unmanned aerial vehicles. Innovative and inventive applications are made use of using these technical advancements. The autonomy of agricultural robots frees farmers from tedious, repetitive, and time-consuming chores so they may concentrate on rising overall output efficiency. Electromechanical machines known as autonomous robots are able to discover a way to complete a job that has been set to them. They are also suitable for doing static jobs dynamically or to avoid potential dangers by utilizing sensors' (sensors') data collected from the surroundings to defend it.

The following is a list of the most typical applications for agricultural robots in the agriculture sector:

- Weed management
- Picking and harvesting
- Autonomous seeding, spraying, and mowing
- Packaging and separation
- Platforms for services

Robots used for harvesting and picking select items using image processing and a configuration of robotic arms. Apple can only do quality control and grading by avoiding repeated processing in a single action. Analyzing the product processing data from the harvesting robot may help to control the packaging and processing procedures as well as estimate company revenue.

The primary machine in the orchard is currently guided by an operator utilizing components that have developed and spread as a result of technological improvement; the rest fruit harvesting devices are accessible on machines which are done spontaneously. Instances of these devices include the robotic fruit picker created by robotics for apple picking.

The sensors and cameras on the device locate the fruits in the 3-D space plane when the operator positions the collector in the region to be gathered. The robotic arms proceed toward the fruit in the next step, rip it off along the axis of the fruit stem, and then drop it onto the carriage once the position information of the fruit has been assessed in the processor. Following this procedure, the conveyor delivers the fruits to a frame revolving on a tray at the rear of the device. With the help of this equipment, a harvest worker may determine if a fruit is ripe or not, tear it, and put it in a case. A harvest worker is ten times less effective than the machine, according to the manufacturing business robotics.

Fruit harvesting equipment is also available that doesn't need a human operator. The SW 8250 strawberry harvesting device created by Agrobot [20–22] is an example of this sort of device. The SW 8250 is the first automated robot on the market which can identify and gather strawberries. It does this by carefully examining each strawberry as it is harvested to avoid damaging the fruit. It includes an operating width of 6 meters, a working span of 4 meters, and the ability to alter the distance between the lines, making it suitable for use in fields and greenhouses. It accomplishes tasks like attaining the harvest ripeness and locating the fruit in the 3-D space plane by scanning the plants in the manner they were inserted using color and infrared sensors. The fruit is then removed from the handle and put into the bottom-level containers by the robotic arms.

Researchers at MIT have created a robot dubbed SWEEPER that picks peppers. The SWEEPER robot goes around agricultural area and scans peppers using a number of machine learning methods. The robot can gather peppers without harming their blooms because it uses data such as color and height to determine when a pepper is ripe. The robot was taught to choose peppers from more than 1000 pepper photos using a machine learning technique. Additionally, according to researchers, the robot's algorithm may be taught to gather fruits and vegetables like oranges and tomatoes. Having autonomous

UAVs and unmanned ground vehicles, Tevel-tech, a startup business in UK, has begun to execute the apple-growing project. They discovered a collecting technique that needed significant work and cost in a quick and less expensive way by contacting software created for farmers to utilize. In the next years, we'll see more initiatives including UAV/UGV harvest collection techniques for agriculture that collaborate with swarm formation. Having the expansion of active harvesting of these autonomous agricultural technologies, the transition to Agriculture 5.0 will quicken autonomous systems.

One of the primary unsolved issues in farm robotics is the aggregated actions of categories of heterogeneous robots and their delivery/transportation to the point of quick solution of applicable difficulties. The creation of a theoretical foundation for the autonomous operation of a mobile platform in supporting a group of UAVs under high level, mission-dependent remote control through intuitive UI in the course of agricultural functioning is the primary goal of the study in this project. The project's incorporation of several UAV parking places is a standout aspect.

In contrast to present studies, the chapter will produce an optimum combined working of groups of diverse unmanned vehicles. The chapter's primary jobs are:

1. Creation of the theoretical underpinnings for the mobile platforms used by UAVs to act autonomously.
2. Creating a theoretical framework for diverse UAV group interactions to address agricultural tasks.
3. Theoretical foundation for human-platform-UAV interaction and control, along with collaborative learning and time-sensitive interface elements via gesture, speech, etc.

The chapter's primary effect will be the development of an agriculture robotic system that can reduce labor costs and worker accidents in agricultural operations, increase the area covered by automated farming, and employ UAVs and precision agricultural practices to reduce fertilizer consumption. Applications for intelligent farming and autonomous unmanned ground and air vehicles are included in the chapter.

The literature has several researches, including those on vehicle routing problems (VRPs) [23] for various UAV use scenarios on the research of various scientists; the harvest of previously identified apples on agricultural land is intended to be collected by the time minimization of autonomous unmanned aircraft and UGVs. Because of the size of the present issue, it is divided into smaller sub-problems in order to answer it in the shortest amount of time possible. The regions of the UGVs' stops were established utilizing the *k*-means clustering approach, which was then analyzed. For computations, the Julia programming language is employed. The analysis showed that three stops were the ideal quantity for 500 apples. The Gurobi solvent was used to resolve the model created to optimize when an air vehicle should pick up apples at each stop, and the desired results were obtained. The study's findings revealed that the UGV typically waited 439 seconds at each stop.

In their investigation, researchers proposed a concept in which the truck was transported first, followed by the UAV at the final site. The "Flying Sidekick Traveling Salesman Problem" is the name of this concept. In the context of the investigation, deliveries were made using both a vehicle and a UAV. Vehicles may transport passengers at once. The UAV must return to the vehicle after delivery. Researchers conducted his analysis using heuristic techniques and mixed integer linear programming (MILP).

In their research, Olivares et al. examined the employment of quadcopters of the UAV type in a manufacturing facility's internal logistics, particularly during product assembly and customization. This study's goal is for estimating and optimizing higher operational energy consumption that is a significant drawback. To find the warehouses, clusters, and subsets, internal logistics modeling was used in the research. In addition to this procedure, other approaches for each quadcopter have been developed utilizing a genetic algorithm. Additionally, it was decided what each quadcopter will carry in terms of weight, material, and route. Finally, according to the weight carried, the number of workstations visited, and the quantity of quadcopter aerodynamic efficiency, the quantity of electrical power to be discharged from each quadcopter's battery for a particular route is also optimized .

In a research by Xie et al. [24], the monitoring capabilities of local wireless networks were used to examine the route construction possibilities for autonomous unmanned aerial vehicles. The major intention of the research is to minimize the amount of travel time to the target destination using the shortest path possible. To this end, a wireless network is used to track the unmanned aerial vehicles' whereabouts. The optimum solution could not be found at the conclusion of the investigation; however researchers proposed a suitable solution by lowering the accuracy rate of the located data to be used.

Transportation times were optimized in a system with several vehicles and UAVs in the Wang et al. paper "Vehicle Routing Problem with UAVs." In this research, a number of undesirable situations were examined, and time-saving possibilities involving the employment of vehicles and UAVs rather than only trucks were suggested.

The survey by Zhao et al. [25] is the foundation for the implementation employed in the research as it is the one that comes closest to the harvest optimization area utilized by researchers in their UAV and UGV studies.

15.3 Focus of the Chapter

15.3.1 Matters, disagreements, concerns, issues, remedies, and suggestions

We are under pressure to detect a long-term remedy to the worldwide nutritional crisis due to the increasing rise of the human population. For sustaining our present nutrition standards, FAA estimates that the farming production capacity would need to rise by 70% by the year 2035, when the globe's population is anticipated to be about 10 billion. How this capacity growth will be handled, however, is a crucial concern at this time given the decline in fertile agricultural areas as a consequence of the upcoming climatic changes?

Industrial agriculture techniques now account for 11%–15% of the world's greenhouse gas emissions, and it is undeniable that if the growth is unchecked, it may hasten climate changes. The fast rise of the urban population is another effect of the growing world population. Rural regions and labor also suffer as a result of this growth. Due to the high costs of inputs and technology in the farming sector and the corresponding rise in energy demand, the techniques that lasts until 2050, will dispense with traditional farming practices and execute digital transformation in farming when it is practical in order to boost the yield by 85%.

In the globe, farming is going through a significant shift. The production of plant and animal products having greater productivity and less ecological effect is made possible by the digitalization of agricultural applications.

In the next years, agriculture will encounter several difficulties on a worldwide basis. Some of these difficulties include:

- The world's rapid population rise
- Changes in climate
- A rise in energy demand
- Limited resources
- Cities' rapid population expansion
- Less productive agricultural land

- Increased competition on global markets
- The dearth of arable land in many emerging nations

Between 1960 and 2015, agricultural productivity rose by more than three times thanks to the utilization of Agriculture 2.0, land, water, and various natural resources. The mechanization and proliferation of food and farming both increased significantly during this time. Despite increased productivity in agriculture on a worldwide scale, the costs of increased food production and economic expansion have once again been borne by nature. The planet's once-vast forests have almost all vanished, groundwater supplies are steadily declining, and wildlife is still being lost. Nitrates, pesticides, and herbicides have poisoned our groundwater supplies. Every year the burning of fossil fuels releases millions of tons of greenhouse gases into the air, causing climate change and global warming.

A significant factor in the rise in agricultural output and economic expansion is the evolution of digital technologies and applications. The importance of rural regions and farmers in particular cannot be overstated when it comes to the digitization and modernization of agriculture. The proper handling of these difficulties will be crucial for the future growth of the agriculture industry globally. The development of digital technology and applications has been shown to be a significant driver of both economic growth and improvements in agricultural productivity, according to a number of international studies. Improvements to the automation of machinery and equipment are ongoing processes in the field. The widespread usage of increasingly advanced machinery is reflected in the agriculture industry's technical and technological sophistication in current applications.

The agricultural industry, which has a history dating back ten thousand years, has experienced a technological metamorphosis during the last century that has elevated it to an entirely new level. Production was quite high at this time, and efficiency was improved. The term "Agriculture 4.0" [26] is now used to describe a new understanding of agriculture that is much more sustainable, ecologically affectionate, and energy proficient, and that may be referred to as the "newer version" of farming. The agricultural industry is undergoing a significant digital development as a result of modern technology, and this digital evolution is currently influencing how the agricultural industry will evolve in the future. Analyzing the key phases of agricultural growth is crucial for illuminating these digital technologies.

With the IoT, efficiency in farming is to be maximized. The cost is reduced because natural resources are utilized to the necessary extent. Similar to this, smart farm systems assess and provide all the production-related variables to the manufacturer at once. By doing this, resource waste is reduced and high-quality goods are generated. Additionally, simultaneously operating and communicating devices enable the development of quick decision-making processes. By lowering the workforce, options for efficient, high-quality, and organic production are established. The builder has the chance to oversee and control the whole field from a tablet or smart phone.

Smart agriculture is considerably more precise and regulated than conventional farming when it comes to raising animals and growing crops. The usage of IT (information technologies) is a crucial part of this agriculture management strategy which successfully implements Agriculture 4.0, but other technical tools including sensors, control systems, robots, autonomous vehicles, and autonomous equipment are also utilized. Farmers' usage of mobile devices and their dependability, together with the usage of lower-cost satellites for imaging and location, are all significant technical advancements that have an impact on the adoption of smart agriculture.

The usage of UAVs in the agriculture industry is an excellent illustration of how technology has evolved through time. Agriculture is now developing into an area of integrated technology that uses UAVs. Various agricultural applications being developed in agriculture use unmanned aerial vehicles. These applications include soil and land analysis, irrigation, crop spraying, planting, and product health evaluation. The main advantages of UAVs are product healthcare mapping, imaging, simplicity of usage, time savings, and possibility for competent improvement are all in use. In the agricultural industry, UAV enables real-time data collecting, processing-dependent design and arrangement, and a higher interchange of high-tech goods.

After taking flight, unmanned aerial vehicles gather multispectral, thermal, and visual pictures. Numerous reports are generated from this flight data, including those on weeds, stock estimation, chlorophyll evaluation, wheat nitrogen constituent, discharge mapping, crop healthcare indices, crop count and yield evaluation, crop height evaluation, farm water quantity survey, and investigation reports. Autonomous agricultural robots may now readily and securely do the agricultural chores necessary for smart agriculture thanks to recent technology advancements. Today's high-tech agricultural instruments gradually compress the soil, which diminishes the soil's fertility over time. It takes more than ten years for compacted soil to become fertile again. In huge agricultural areas, autonomous imaging and data collecting tools that operate alone are inadequate. Instead, tiny vehicles that collaborate with one another rather than big, independent vehicles provide a more practical alternative. Researchers have been working on the technological studies necessary for a single farmer to manage and operate a system made up of these autonomous instruments for a very long time. To allow team members to interact, a job control hub and sophisticated spanning design techniques must be created and worked together to complete the given tasks in a safe and effective manner.

The multiple robot framework known as "Swarm Robotics" [27–29] that may cooperate to carry out a particular farming duty is one of the subjects that have been recommended by many academics for a long time. To collaborate and build an ecosystem, many robots will need to apply genetic algorithm and artificial intelligence techniques. When robots begin learning interactively from one another and improve their performance over time, this approach is even more beneficial. For instance, a robot swarm that works in tandem with the command center may gather soil samples and provide to the development of food maps. The effectiveness of the technique can initially be subpar, but over time, by using the "good behavior-bad behavior" strategy in the DL [30–33] algorithm, performances that penalize each robot's poor conduct may be enhanced. Future digital agriculture will greatly benefit from these robots.

15.3.2 The application techniques

Unsupervised learning is accomplished using the *k*-means technique. Unsupervised learning is a type of ML where the model does not require to be verified. Instead, one should let the model operate independently to find information since it is primarily interested in unlabeled data. Compared to supervised learning, unsupervised learning algorithms allow for the completion of more difficult processing tasks.

In order to do a cluster analysis, the items in the dataset must either be divided into smaller sub-datasets or grouped according to similar characteristics. The goal of this method is to create a cluster of items that are as similar to one another as feasible while still being as unlike from one another as possible. Another goal is to have large variations across clusters and low variances for the items inside each cluster. As the datasets utilized for clustering are segmented and reflect the data structure and insights that already exist, they are often employed in data mining to find significant insights.

Since there are so many clustering algorithms accessible today, they are categorized in a variety of ways. Three major categories are used to assess these classifications. These approaches, which include compartmental, hierarchical, and mixed ones, were created by variously organizing these collections. The first technique uses algorithms that separate the clustering datasets into *k* subsets. The choice of *k* that must be calculated before the present inspection of an algorithm is one of these widely explored subjects as a result. The option in this case defines how many sets were utilized to create the dataset. One might state that a good and accurate clustering procedure is necessary for the best choice of k. Positive or negative clustering will proceed to the formation of classification frameworks, regardless of k. However, no value in these clustering procedures may be taken as correct. The nearest estimate to this value is located using the clustering approach, which also yields the final result. In other words, choosing k correctly creates cluster analysis much effective and is a biggest challenge for the classification and clustering method.

The k-means algorithm was first used in 1970. Among the currently utilized clustering techniques, it is one of the most popular unsupervised learning techniques. This approach uses a sharp clustering algorithm since it only permits one cluster to be allocated to each variable. It is a technique based on the idea that the set is expressed by the variable in the center of a cluster. Clusters of identical proportions are often discovered using the approach. Sum of squared error (SSE) is the application of the k-means approach that is most often used. The best results come from clustering having the least SSE estimate. Eqn (15.1) calculates the sum of the squares of the distances between the variables and the center points of the set to that they belong.

$$SSE = \sum_{j=1}^{L} \sum_{y \in D} dis^2(n, y). \tag{15.1}$$

It is intended to disperse k-clusters strongly by itself and independently from one another in a cluster as a consequence of this split. Finding k clusters which would lower the *SSE* function is the algorithm's main goal. Using the user-determined value for the k parameter, the algorithm separates the n-dataset into k-sets. The cluster's center of gravity is calculated by the cluster similarity estimate, which is calculated using the average value of the variables in the cluster.

"Mathematical model" refers to the process of determining variables in a framework using parameters and illuminating their interrelationships using functions.

The technique of developing physical, symbolic, and abstract models of events which arise in real life is referred to as mathematical modeling. With its intricate structure, mathematical modeling is a considerably more significant activity than just simulating a condition. Mental models refer to all of the analytical techniques utilized in mathematical modeling. Generally, mathematical models provide mathematically-formed external representations of our mental concepts which allow for the explanation and resolution of issues faced in daily life. Moreover, [35] argues that mathematical modeling is a technique which calls for mental modeling.

The research by [36] breaks down mathematical modeling into seven simple phases. The primary care issue is addressed throughout the modeling phase. A mathematical model which describes the present condition is developed in the second phase. The problem's mathematical solution is then created utilizing the mathematical model. The remedy's output is interpreted, and the correctness of the findings is examined. The current model should be reexamined and altered as needed if the accuracy of the findings is questioned as a consequence of the analysis. If a problem is not noticed at the last step, when the current solution is examined in light of real-life situations, the answer is transformed into a written or spoken report.

An approach that is widely utilized to solve optimization issues is linear programming (LP). George Dantzig first suggested the simplex algorithm, which is used successfully to resolve linear programming issues. The simple method has made linear programming more widely utilized across numerous industries. Many industries, including the military, education, and finance, still employ linear programming to solve optimization issues. According to a study among Fortune 500 organizations, linear programming was employed by 85% of these businesses.

By following variables and constraints, linear programming aims to have the objective function (max-min) as optimum as possible. In general, it may be claimed that linear programming is a deterministic mathematical method that takes into account the best allocation of limited resources.

Goal function, constraints functions, and positivity (non-negative) constraints are the three fundamental components of the linear programming (LP) paradigm. A LP model comprises the linear objective function to be optimized as well as constraint equations in the form of linear equations and/or inequalities. A LP issue often looks something like this:

$$Y_{\mathrm{max/min}} = \sum_{j=1}^{m} d_i y_i. \tag{15.2}$$

The following is a list of the linear programming assumptions necessary to get reliable results.

- Severability
- Proportionality
- Summation
- Certainty

- Linearity

Although linear programming has a wide range of applications, the following are the ones that are most commonly used:

- Production schedule
- Dietary program
- Ad environment choice
- Capital planning
- A distribution strategy
- Inventory management
- Balance of the production line

Mixed integer programming challenges are those where certain variables must be integers but others are real values. In terms of the approach to solving problems, mixed integer programming and integer programming are quite similar. Similar to integer programming, the simplex technique yields the best solution without needing an integer condition. One may represent the mixed integer linear programming issue as follows.

$$Y_{max} = 6x_1 + 4x_2. \tag{15.3}$$

In a research they carried out in (1960), A. Land and A. Doing suggested a generic calculating technique for usage in integer programming. In the years that followed, this approach – known in the literature as the "Doing method" – was used to address issues with traveling salesmen.

In certain circumstances, decision variables need more than two potential (0-1) value alternatives. These variables may also display the issue as 0-1, which will take on an integer representation. Applications of 0-1 integer programming show which many divisibility assumptions are false and that certain issues are determined by "yes" or "no" choices. In such cases, the only two options are "yes" or "no." Should we, for instance, undertake this investment or ought to should we locate the plant here? Similar to this, choices with two options are depicted using decision variables whose values are limited to 0 and 1. The "yes" or "no" option is therefore represented as shown:

$$y_i = 1 \text{ if y is yes} \tag{15.4}$$

$$y_i = 0 \text{ if y is no.}$$

According to [37], 0-1 integer issues have the following general form.

$$Y_{\min} = \sum_{i=1}^{m} d_i y_i. \tag{15.5}$$

In the KNN algorithm, a stop is chosen at random from the available stations to serve as the tour's beginning point. The following stop on the present tour is then added after determining which stop is closest to the one that was originally chosen. The process of adding stops continues until every stop has been added, and the trip cannot end until every stop has been added. This approach produces variable length tours depending on which stop is chosen as the start and which stop is chosen as the finish of the current route. However, since there are lengthy returns to connect with the other stations that were left after choosing the closest stop, the remedies which are far from the finest globally are acquired.

The nearest neighbor algorithmic heuristic technique was first introduced to the literature by [38]. By using this method to provide the best possible outcome in this research, they were able to produce an excellent TSP. In general, this heuristic technique may be utilized to TSP and vehicle route problems (VRPs) using smaller-scale specimens and is relatively straightforward to utilize.

If it does not surpass the time and capacity restrictions after satisfying various requests, it may be used as an instance of the KNN algorithm, beginning with the closest clients along the path of the truck and coming back to the warehouse. Figure 15.1 is an example route created using the nearest neighbor heuristic.

The solution in the nearest insertion algorithm method begins by a visit between two stations. After then, it is made sure that the stops on the current tour that are closest to the tour are also incorporated, and the tour's inclusion is evaluated. Following this procedure, the stop that will result in the tour's expansion being as modest as feasible is identified and added to the itinerary.

Until all of the stops are included in the itinerary, this procedure is repeated. The addition order rule and the choice of the starting point might alter the answer when using this approach. Figure 15.2 provides a detailed illustration of the tour's solution phases beginning at the warehouse.

A beginning solution is often produced using the nearest insertion and neighboring approaches. However, in their paper, [39] provide a metaheuristic approach which may be utilized to resolve TSP and other comparable combinatorial problems.

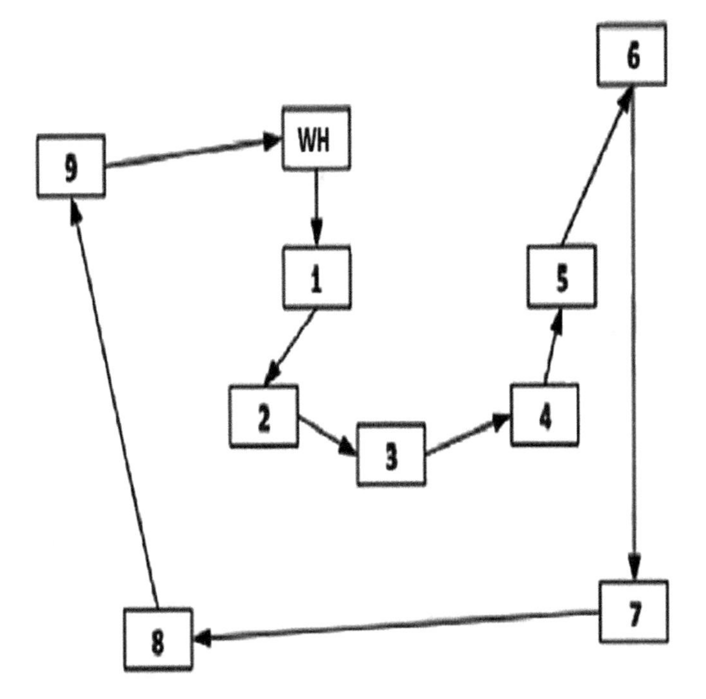

Figure 15.1 Heuristic sample 1 for nearest neighbor.

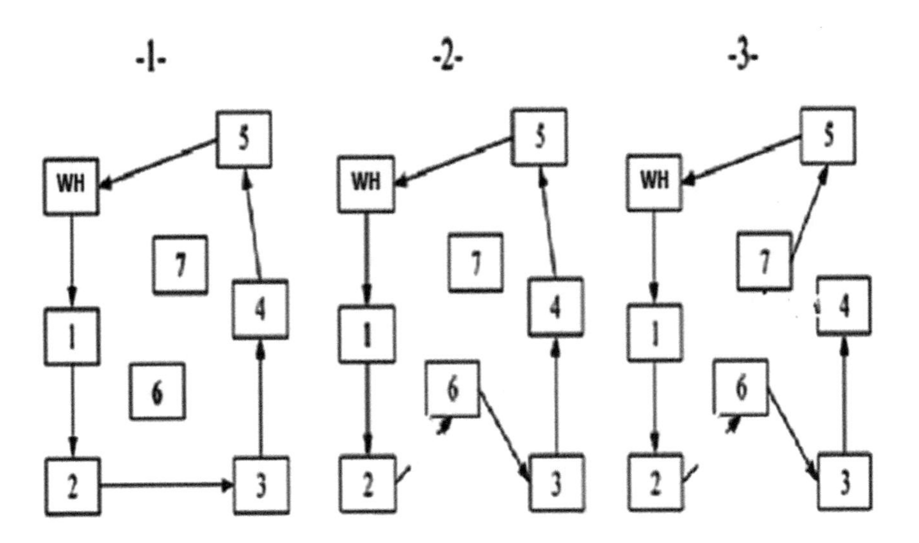

Figure 15.2 Sample 2 of the nearest insertion heuristic.

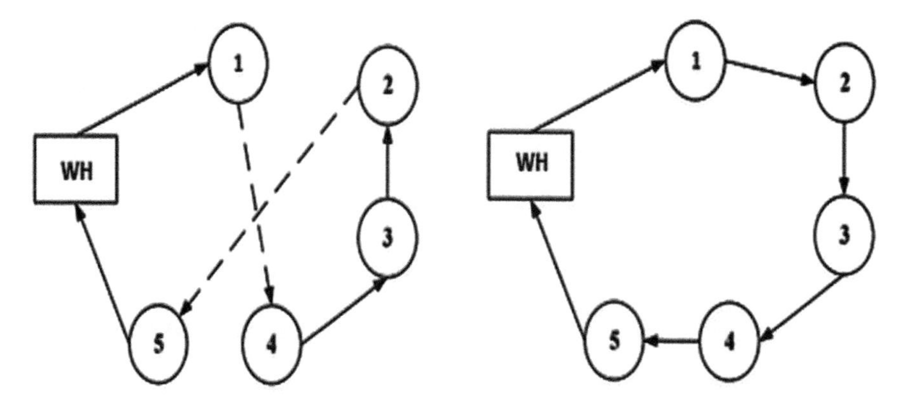

Figure 15.3 Sample 2 of the 2-opt heuristic method.

[40] introduced the in-route modification strategy using the 2-opt algorithm method.

As noted in the routes depicted as an example in Figure 15.3, the node positions were altered based on the assumption that the actions in the routes (1, 4) and (2, 5) are not the most advantageous in matters of cumulative distance and expenditure. The route was reduced and the new route produced results that were closer to the ideal ones than the old one thanks to the newer actions in directions (1, 2) and (4, 5) on the modified newer route.

A population-dependent optimization technique called PSO is based on how bird swarms look for food. Kennedy and Eberhart recommended this approach. The social particles that make up the swarm are thought to be represented by every solution in the particle swarm optimization approach. When the particles move and search independently in the solution space, [41] found that they are further influenced by the search behavior of various particles in the swarm.

In order to study the PSO technique and TSP problem-solving techniques, Pang and his colleagues targeted the transition between continuous-discrete solution spaces. When local search operations are carried out in discrete space in this research, location updating and speed computation are conducted in continuous space. By utilizing the chaotic operator, the issue of jamming to the local optimum solution is eliminated also it randomly alters the location and velocity vectors that are positioned in space. [42] investigated four distinct versions of the approach in various comparison tasks for evaluating the efficacy of the employment of chaotic operators.

The model used for this research involves breaking the issue down into two smaller challenges.

First step: UGV should pause at a certain location in order to determine the ideal moment. Researchers suggested using the *k*-means technique to identify the best stop points. The model put forward in this research employs this technique. A conditional loop controls this procedure.

The best time to select the apples distributed throughout the stops was determined in the second stage utilizing the *k*-means method and deterministic, binary mixed (0-1) integer modeling. By utilizing 0-1 binary modeling, it is identified that UAV would gather and the way it is decided only if the UAV collects the apple. Regardless of which apples are picked, the five UAVs on the UGV framework are computed to arrive at the best moment.

Utilizing the nearest neighbor and nearest insertion techniques that are part of the heuristic solution routes for the TSP, the UGV's route was first created in the third stage. Second, [43] employed 2-opt once again, a heuristic TSP remedy technique. This technique, known as tour developer, was employed to enhance the UGV's present path. The object we produced utilizing these three phases in the final stage offers us the overall harvest time of UGV as well as the time taken by unmanned aerial vehicles for harvesting in a fixed quantity of stops. We may choose the best stop from among the possible stops using the PSO algorithm, which makes use of this object, to get the shortest harvest time. The appropriate number of stops, locations, and information regarding which apples and which UAV will be gathered at these checkpoints will all be determined as a result of this suggested model may be reached with ease.

Finding these stops' positions with the fewest possible stops; while utilizing the PSO technique, it is required to identify the fewest possible stops. The amount of stops should be in line with the range of unmanned aerial vehicles for the object generated in the suggested model to function effectively, and the lower limit of the amount of stops should be discovered at the point.

In order to locate the stops with the least amount of stops, the algorithm operates as follows. The initial assumption was that there would be three stops. The apples' coordinates were discovered utilizing the *k*-means clustering technique, and they were divided into three clusters for determining where each cluster's center was located. The cycle is completed if the distance between this center point and the furthest apple in the cluster is in the collecting area (largest flying span of unmanned aerial vehicles); otherwise, the cycle is resumed, with the number of stops being increased by "1."

Table 15.1 A set of indicators, descriptions, and definitions.

Indices	Details	Definition sets
A	Apple indices	$A = \{1, 2, \ldots, a_{max}\} a_{max}$: Cumulative amount of apples on the farm
B	Stop indices	$B = \{1, 2, \ldots, b_{max}\} b_{max}$: Cumulative amount of stops on the farm
C	Unmanned aerial vehicles indices	$C = \{1, 2, \ldots, c_{max}\} c_{max}$: Cumulative amount of unmanned aerial vehicles on unmanned ground vehicles

Table 15.2 Descriptions and parameters.

Parameter	Details
E_{ab}	E is a matrix that depicts at which stop every apple must be gathered. If the quantity is "0," apple a will not be gathered from the stop b, if it is "1" it would be gathered.
F_{ab}	U is the distance of every apple in minutes to the stops.

15.3.3 Parameter description

The matrix A_{ab} indicates which stop each apple should be picked up at. Apple a won't be gathered from the stop b if the value is "0," but it would be if the value is "1."

The distance from every fruit to the stops, measured in minutes, is F_{ab}.

In accordance with the amount of stops discovered using the k-means approach and the UAV's harvesting schedule, the coordinates of the stops are determined by the data attributed to them and the distance between the stops and the apples they represent.

Table 15.1 lists the model's indices, Table 15.2 lists the parameters and definition sets, and Table 15.3 lists the decision variables. After explaining constraints and objective function, an optimization model was developed.

Table 15.3 presents the decision factors and justifications for the suggested model. The following are the restrictions in the suggested model for the issue:

Every apple is guaranteed to be gathered at any stop by any unmanned aerial vehicle thanks to the restriction stated in eqn (15.6).

$$\sum_{i=1}^{i_{max}} \sum_{l=1}^{l_{max}} L_{ijk} = 1. \tag{15.6}$$

The unmanned aerial vehicle is allowed to pick apples only in their range at every stop due to the restriction in eqn (15.7). The UAV spends as much time

Table 15.3 Decision parameters and justifications.

Decision variables	Details
G_{abc}	This variable has a value of "1" if the UAV *k* collects the apple *i* at the stop *j* and a value of "0" otherwise.
H_{abc}	It is a continuous type variable which displays how many minutes the UAV *c* needs to charge in order to retrieve the apple *i* from the Stop *b*.
I_{bc}	The continuous type variable displays the entire amount of time, in minutes, that the UAV *c* needed to gather and charge apples at the stop *b*.
J_b	The continuous type variable displays the entire amount of time in minutes that the UGV spent at the stop *b*. The UAV at the stop *b* that had the longest collecting and charging duration was used to determine this timeframe.

gathering apples as it does charging. The unmanned aerial vehicle uses 125% of the charging time when it returns with the apple. As a result, the apple's entire flight time, which must be less than 30 minutes, must be 2.2 times the distance from the station.

$$2, 2 * A_{ij} * U_{ij} * D_{ij} \leq \forall\, 30, i \in I, j \in J, k \in K. \tag{15.7}$$

The result of eqn (15.7) and (15.8) guarantees that the UAV in the model needs to charge for the number of minutes it has to pick up the apple at the stop *c*.

$$2, 2 * A_{ij} * U_{ij} * D_{ij} = \forall\, H_{ij}, \forall\, i \in I, j \in J, k \in K. \tag{15.8}$$

Eqn (15.6) denotes the overall amount of time, in minutes, that the unmanned aerial vehicle c needs to charge after collecting apples.

The UAV with the longest duration of gathering apples in stop *j*'s total time in minutes is determined by eqn (15.7).

The goal function in eqn (15.8) is to reduce the cumulative amount of time in minutes that the unmanned aerial vehicle spends at each stop having the maximum picking and charging time.

UGV harvest time; analysis is carried out utilizing the stop coordinates discovered using the *k*-means technique and the heuristic approaches of the traveling salesman problem.

The nearest neighbor and nearest insertion methods that are part of the heuristic solution approaches for the TSP, were used to create the UGV's route. Second, the heuristic TSP solution technique 2-opt was utilized once again. The technique, known as tour developer, was employed to enhance the unmanned aerial vehicle's present path.

The object the authors produced by applying three phases in an optimization model tells us the time needed by unmanned aerial vehicles for harvest in a fixed amount of stops and the overall harvest time of unmanned ground vehicle optimal number of stops for shortest harvest time. We may choose the best stop from among the potential stops using the PSO algorithm, which makes use of this object and provides us the shortest harvest time. As a consequence, this suggested model makes it simple to determine the appropriate number of pauses and places for UGV to halt, as well as the details of which apples and whose UAV would be gathered at the stops.

On a PC with 2.5 GHz CPU and 8 GB RAM, the study's suggested model was examined. The study of the apple coordinates clustering, distance calculations, and software for the mathematical model for analysis all employed the Julia programming language. The following is a list of the Julia package programs used:

- Dunning, JuMP package
- JuliaStats/Clustering.jlpackage
- JuliaStats/Distances.jlpackage
- The Field of the Traveling Salesman Heuristics
- JuliaData/DataFrames.jlpackage
- Giovineltalia/Gadfly.jlpackage
- JuliaGraphics/Cairo.jlpackage
- JuliaGraphics/Fontconfig.jlpackage
- Giovineltalia/Compose.jlpackage
- JuliaLang/Random.jlpackage
- JuliaPlots/Plots.jlpackage
- Sglyon/PlotlyJS.jlpackage
- Stdlib/LinearAlgebra.jlpackage
- Felipenoris/XLSX.jlpackage
- Jump-dev/cbc.jlpackage

The amount of apples (a_maximum = 600) and the amount of unmanned aerial vehicles on unmanned ground vehicles (c_maximum = 5) were counted

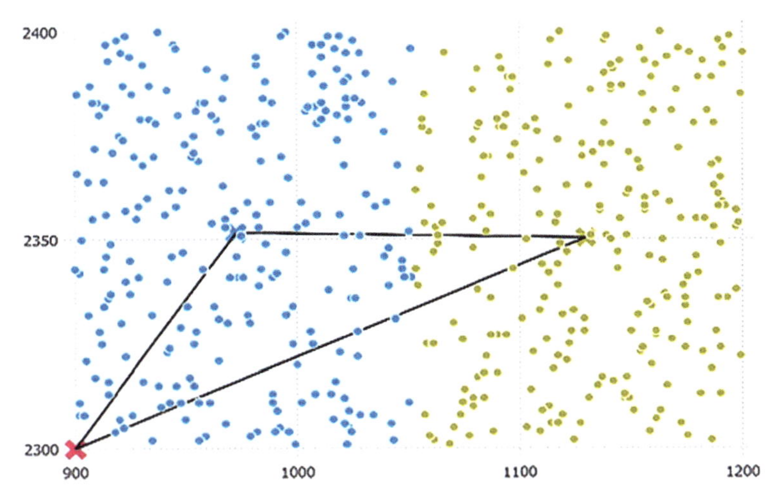

Figure 15.4 Cluster organization and stops.

during application. The "rand" function and *k*-mean clustering method were used to the JuliaStats/Clustering.jl package to determine the locations of the apples. In the flying sidekick traveling salesman problem (FSTSP), application is inefficient because the UGV is in motion during distribution or collection. This is so that the halt may be made at the best location to gather a certain amount of apples. The UGV's path was designed using the heuristic TSP approach.

Simulations with many stops are not favored because they might cause issues with black-and-white printing; instead, the scenario with UGV stop number 3 is preferable, as shown in Figure 15.4.

The minimal numbers of stops that fulfill this requirement have been utilized for allocating UAVs to stops within their range. The best values for the overall harvest collection times at the chosen stations are shown in Figure 15.5 The distance between the stops derived from apples was calculated using the JuliaStats/Distances.jl package. The number of pauses (j_max = 3) was discovered during the iterations used to determine the number of clusters and their placements. The clustering structure and pauses for 500 apples are shown in Figure 15.4. The "X" shown in Figure 15.4 symbolize the stops.

The last step of application uses the PSO algorithm to optimize the use of the object produced from the prior three phases.

It is intended to determine the ideal number of crops to use by applying the algorithm created for harvest optimization stops that offers the shortest time. Accordingly, imax-2 times *k*-means, (imax* imax+1)/2 times mathematical model, and imax times analysis of traveling salesman algorithm are

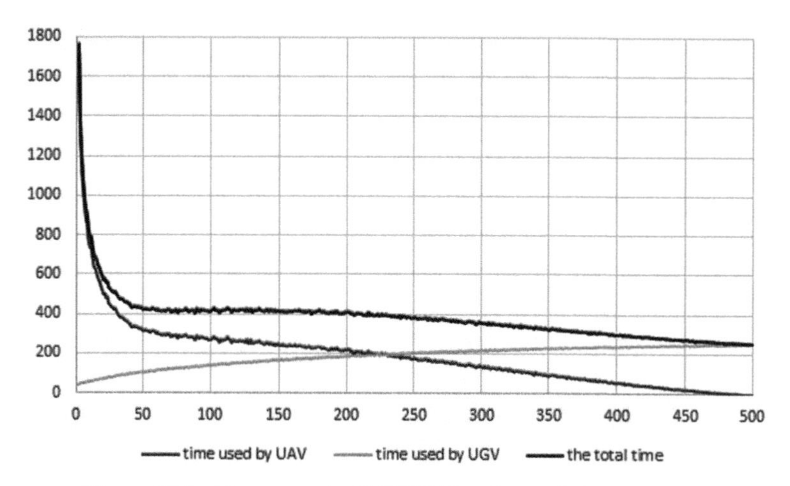

Figure 15.5 Time spent by UAV/UGV and overall charging time.

needed when all potential stops are examined. When it is essential to compare every stop to the quantity of 500 apples, it needs the study of 500 times the traveling salesman algorithm, 125,250 times the mathematical model, and 498 times *k*-means.

Even if it takes 1 second to solve each of these 125,250 mathematical models, it still takes more than 36 hours to get the final result. For this reason, the particle swarm optimization approach is anticipated to be utilized in this research to determine the ideal number of stops, under the supposition that the harvest time has a function that lowers up to a specific number of stops and grows after this number of pauses.

As a consequence, Julia programming was used to create the code, and outcomes were attained. The ideal answer was discovered in all tests as the 500th halt after a maximum of five repetitions. The findings indicate that, in relation to the number of pauses, the harvest time has a monotonously declining tendency. The method was tested for every halt that may occur, and the results are shown in Figure 15.5 for a more thorough examination.

The harvest time is a decreasing function, as shown in Figure 15.5, although it is stationary between 50 and 250 stops. As a result, it is evident that utilizing any method will result in the best stop.

The 500th stop will be a mathematical model. Analysis of the process revealed that UGV was functioning without repeating, returning between pauses, or devoting time to recharging. UAVs, on the other hand, take longer to charge than they do to fly. They must track at the same pace as the UGV while moving back and forth between each target and the stop.

Because of everything mentioned above, the UAV/UGV method is less effective than the UGV procedure. It was suggested to have a spare battery

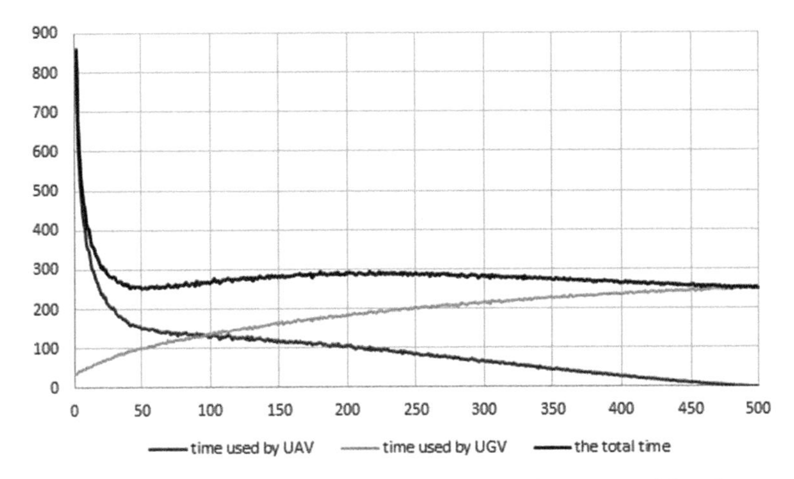

Figure 15.6 Time spent by UAV/UGV and the entire amount of free time.

for each UAV in the research done by Ferrandez et al. (2016). As a result, it is intended to cut down on charging time for UAVs to zero. Figure 15.6 will change when the situation utilized in this research is updated appropriately was attained. Figure 15.6 shows that the harvest time is a function that grows between stops 50 and 200, drops between stops 200 and 500, and declines fast to the 50th stop. The best stop is discovered to be the 500th stop, with very little variation, despite the times surrounding the 50th and 500th stops being quite near to one another. Finally, when a 20% increase in UAV speed is taken into account, Figure 15.7 was attained.

The twenty-fifth and seventy-fifth stops show a notable improvement as a result of all these changes. This range clearly includes the minimum point of the function for the harvest time. The particle swarm optimization algorithm may therefore be used to determine the ideal number of stops without having to calculate for each stop individually.

The 42nd stop was determined to be the optimum stop with 228.25 minutes after the current changes were made and the PSO algorithm was conducted using the values shown in Table 15.4. The outcome is precisely the minimal point of the overall time in Figure 15.7's result.

These presumptions must be made in order to employ the PSO method for harvest optimization. If not, the ideal number of stops will be the same as the quantity of apples.

1. Using a backup battery in UAVs will eliminate the need for charging time.
2. UGVs shouldn't move as quickly as UAVs.

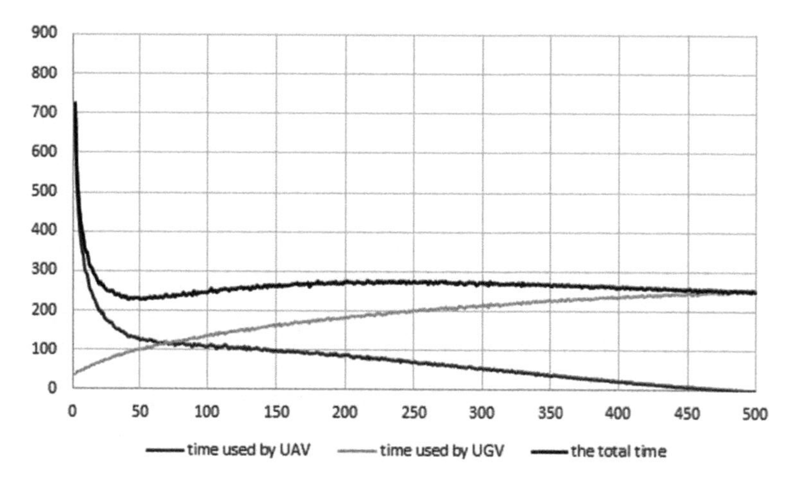

Figure 15.7 Total time without charge, UGV time, and UAV time (20% speed increases).

To ensure that the result is unaffected by the trap zone in the initial part of the graph, which is the quick decline, the method should be added to the PSO algorithm until the result generates a result that differs from the number of apples. Additionally, executing the best values of all particles until they match the global best value gave healthier outcomes in less time than repeatedly running the PSO algorithm.

15.4 Directions for Future Research

This research may serve as a model for others to follow. By maximizing inputs, smart agriculture will also greatly assist environmental issues when it is implemented with all of its components. The load-carrying capability of the UAV will be incorporated, and flying ranges and battery ratios for various weights will be simulated in our next research. Additionally, UAVs and UGV failure detection and fault tolerant control will be developed and included into the model.

15.5 Conclusion

On this research, the harvest of autonomous UAVs and UGVs on agricultural land will be collected in line with the time minimization of the apples, whose aims have already been established. Due to the size of the issue under study, it was necessary to divide it into smaller sub-problems in order to optimize the answer before offering it. According to UN statistics, by 2050, there will

Table 15.4 Parameters.

Parameter	**Value**
Variable number	1
Particle number	5
Minimum inertia weight	0.4
Maximum inertia weight	0.9
Cognitive acceleration constants C1	2
Social acceleration constants C2	2
Initial velocity	10
Maximum iteration number	50

be 10 billion people on the planet, and food consumption will rise by 70%. Food and agricultural equipment will have to be imported by nations who fail to invest in technology and take this population growth into account. There is greater space for creative agriculture and agriculture nowadays in agricultural reports 4.0 and delicate farming techniques. Farmers, manufacturers, marketers, retailers, customers, and governments that obstruct the flow of commodities and products are all impacted by this transformation. The smart agriculture business is now worth almost $2.68 billion. Smart agricultural techniques are anticipated to reach a value of $30 billion in 2030, and this will be the element that will have the most impact on the agriculture industry given the market's recent upswing. Despite the fact that there is a lot of agricultural land accessible in the globe, it is impossible to produce a productive crop because of issues like high labor costs and a lack of technology. Implementing effective and long-term digital agriculture policy has given nations with tiny surface areas, like the Netherlands and Israel, a voice in the global agricultural community. The globe will be able to satisfy the rising demand for food if agriculture completely complies with 4.0's standards and long-term plans call for switching to Agriculture 5.0.

References

[1] F.-Y. Wang, J. Yang, X. Wang, J. Li and Q. -L. Han, "Chat with ChatGPT on Industry 5.0: Learning and Decision-Making for Intelligent Industries," in IEEE/CAA Journal of Automatica Sinica, vol. 10, no. 4, pp. 831–834, April 2023, doi: 10.1109/JAS.2023.123552.

[2] G. Roggiolani, M. Sodano, T. Guadagnino, F. Magistri, J. Behley and C. Stachniss, "Hierarchical Approach for Joint Semantic, Plant Instance, and Leaf Instance Segmentation in the Agricultural Domain," 2023 IEEE International Conference on Robotics and Automation

(ICRA), London, United Kingdom, 2023, pp. 9601–9607, doi: 10.1109/ICRA48891.2023.10160918.

[3] L. -B. Chen, X. -R. Huang and W. -H. Chen, "Design and Implementation of an Artificial Intelligence of Things-Based Autonomous Mobile Robot System for Pitaya Harvesting," in IEEE Sensors Journal, vol. 23, no. 12, pp. 13220–13235, 15 June15, 2023, doi: 10.1109/JSEN.2023.3270844.

[4] H. J. Nelson, C. E. Smith, A. Bacharis and N. P. Papanikolopoulos, "Robust Plant Localization and Phenotyping in Dense 3D Point Clouds for Precision Agriculture," 2023 IEEE International Conference on Robotics and Automation (ICRA), London, United Kingdom, 2023, pp. 9615–9621, doi: 10.1109/ICRA48891.2023.10161078.

[5] W. Liu, Y. Liu and L. Wu, "Model Predictive Control Based Voltage Regulation Strategy Using Wind Farm as Black-Start Source," in IEEE Transactions on Sustainable Energy, vol. 14, no. 2, pp. 1122–1134, April 2023, doi: 10.1109/TSTE.2023.3238523.

[6] X. Fan, H. Wang, X. Lv and Z. Song, "Frequency Control Strategy of Storage-Based Wind Farm as Black Start Power Source," 2023 8th Asia Conference on Power and Electrical Engineering (ACPEE), Tianjin, China, 2023, pp. 2407–2412, doi: 10.1109/ACPEE56931.2023.10135729.

[7] P. T. Khanh, T. H. Ngọc, and S. Pramanik, Future of Smart Agriculture Techniques and Applications, in Advanced Technologies and AI-Equipped IoT Applications in High Tech Agriculture, Eds, A. Khang, IGI Global, 2023.

[8] T. H. Ngọc, P. T. Khanh and S. Pramanik, Smart Agriculture using a Soil Monitoring System, in Advanced Technologies and AI-Equipped IoT Applications in High Tech Agriculture, Eds, A. Khang, IGI Global, 2023.

[9] T. T. H. Ngoc, S. Pramanik, P. T. Khanh, "The Relationship between Gender and Climate Change in Vietnam", The Seybold Report, 2023, DOI 10.17605/OSF.IO/KJBPT

[10] R. Jayasingh, J. Kumar R.J.S, D. B. Telagathoti, K. M. Sagayam and S. Pramanik, "Speckle noise removal by SORAMA segmentation in Digital Image Processing to facilitate precise robotic surgery", International Journal of Reliable and Quality E-Healthcare, vol. 11, issue 1, DOI: 10.4018/IJRQEH.295083, 2022

[11] S. Pramanik, M. G. Galety, D. Samanta, N. P Joseph, Data Mining Approaches for Decision Support Systems, 3rd International Conference on Emerging Technologies in Data Mining and Information Security, 2022.

[12] R. Bansal, A. J. Obaid, A. Gupta, R. Singh, S. Pramanik "Impact of Big Data on Digital Transformation in 5G Era", 2nd International Conference

on Physics and Applied Sciences (ICPAS 2021), doi:10.1088/1742-6596/1963/1/012170, 2021.

[13] D. Samanta, S. Dutta, M. G. Galety and S. Pramanik, A Novel Approach for Web Mining Taxonomy for High-Performance Computing, The 4th International Conference of Computer Science and Renewable Energies (ICCSRE'2021), 2021, DOI:10.1051/e3sconf/202129701073.

[14] Dhamodaran S, S. Ahamad, J.V.N. Ramesh, G. Muthugurunathan, Manikandan K, S. Pramanik, D. Pandey, Thrust Technologies' Effect on Image Processing, IGI Global, 2023

[15] Dhamodaran S, S. Ahamad, J.V.N. Ramesh, S. Sathappan, A. Namdev, R. R. Kanse, S. Pramanik, Fire Detection System Utilizing an Aggregate Technique in UAV and Cloud Computing, Thrust Technologies' Effect on Image Processing, IGI Global, 2023

[16] V. Veeraiah, D.J. Shiju, J.V.N. Ramesh, Ganesh Kumar R, S. Pramanik, A. Gupta, D. Pandey, Healthcare Cloud Services in Image Processing, Thrust Technologies' Effect on Image Processing, IGI Global, 2023

[17] S. Ahamad, V. Veeraiah, J.V.N. Ramesh, R. Rajadevi, Reeja S R, S. Pramanik and A. Gupta, Deep Learning based Cancer Detection Technique, Thrust Technologies' Effect on Image Processing, IGI Global, 2023

[18] B. K. Pandey, D. Pandey, V. K. Nassa, A. S. George, S. Pramanik and P. Dadheech, Applications for the Text Extraction Method of Complex Degraded Images, in The Impact of Thrust Technologies on Image Processing, Nova Publishers, 2023.

[19] B. K. Pandey, D. Pandey, V. K. Nassa, A. S. Hameed, A. S. George, P. Dadheech and S. Pramanik, A Review of Various Text Extraction Algorithms for Images, in The Impact of Thrust Technologies on Image Processing, Nova Publishers, 2023.

[20] P. K. Mall, S. Pramanik, S. Srivastava, M. Faiz, S. Sriramulu and M. N. Kumar, FuzztNet-Based Modelling Smart Traffic System in Smart Cities Using Deep Learning Models, in Data-Driven Mathematical Modeling in Smart Cities, IGI Global, 2023, DOI: 10.4018/978-1-6684-6408-3.ch005

[21] G. Taviti Naidu, KVB Ganesh, V. Vidya Chellam, S Praveenkumar, D. Dhabliya, S. Pramanik, A. Gupta, Technological Innovation Driven by Big Data, in "Advanced Bioinspiration Methods for Healthcare Standards, Policies, and Reform", Hadj Ahmed Bouarara IGI Global, 2023, DOI: 10.4018/978-1-6684-5656-9

[22] S. Pramanik, "Carpooling Solutions using Machine Learning Tools", in Handbook of Research on Evolving Designs and Innovation in ICT

and Intelligent Systems for Real-World Applications, K. K. Sarma, N. Saikia and M. Sharma, IGI Global, 2022, DOI: 10.4018/978-1-7998-9795-8.ch002.

[23] J. Li et al., "Learning Feature Embedding Refiner for Solving Vehicle Routing Problems," in IEEE Transactions on Neural Networks and Learning Systems, doi: 10.1109/TNNLS.2023.3285077.

[24] L. Xie, S. Song, Y. C. Eldar and K. B. Letaief, "Collaborative Sensing in Perceptive Mobile Networks: Opportunities and Challenges," in IEEE Wireless Communications, vol. 30, no. 1, pp. 16-23, February 2023, doi: 10.1109/MWC.005.2200214.

[25] J. Zhao, Y. Nie, H. Zhang and F. Richard Yu, "A UAV-Aided Vehicular Integrated Platooning Network for Heterogeneous Resource Management," in IEEE Transactions on Green Communications and Networking, vol. 7, no. 1, pp. 512–521, March 2023, doi: 10.1109/TGCN.2023.3234588.

[26] Y. Wu and K. Takács-György, "What is stopping Agriculture 4.0?—Examples from China," 2023 IEEE 17th International Symposium on Applied Computational Intelligence and Informatics (SACI), Timisoara, Romania, 2023, pp. 000511–000518, doi: 10.1109/SACI58269.2023.10158595.

[27] A. Bhattacharya, A. Ghosal, A. J. Obaid, S. Krit, V. K. Shukla, K. Mandal and S. Pramanik, "Unsupervised Summarization Approach with Computational Statistics of Microblog Data", in Methodologies and Applications of Computational Statistics for Machine Learning, D. Samanta, R. R. Althar, S. Pramanik and S. Dutta, Eds, IGI Global, 2021, pp. 23-37, DOI: 10.4018/978-1-7998-7701-1.ch002

[28] A. Mandal, S. Dutta, S. Pramanik, "Machine Intelligence of Pi from Geometrical Figures with Variable Parameters using SCILab", in Methodologies and Applications of Computational Statistics for Machine Learning, D. Samanta, R. R. Althar, S. Pramanik and S. Dutta, Eds, IGI Global, 2021, pp. 38–63, DOI: 10.4018/978-1-7998-7701-1.ch003

[29] Y. Meslie, W. Enbeyle, B. K. Pandey, S. Pramanik, D. Pandey, P. Dadeech, A. Belay, A. Saini, "Machine Intelligence-based Trend Analysis of COVID-19 for Total Daily Confirmed Cases in Asia and Africa", in Methodologies and Applications of Computational Statistics for Machine Learning, D. Samanta, R. R. Althar, S. Pramanik and S. Dutta, Eds, IGI Global, 2021, pp. 164–185, DOI: 10.4018/978-1-7998-7701-1.ch009

[30] K. Dushyant, G. Muskan, A. Gupta and S. Pramanik, "Utilizing Machine Learning and Deep Learning in Cyber security: An Innovative Approach", in Cyber security and Digital Forensics, M. M. Ghonge,

S. Pramanik, R. Mangrulkar, D. N. Le, Eds, Wiley, 2022, https://doi.org/10.1002/9781119795667.ch12

[31] S. Choudhary, V. Narayan, M. Faiz and S. Pramanik, "Fuzzy Approach-Based Stable Energy-Efficient AODV Routing Protocol in Mobile Ad hoc Networks", in Software Defined Networking for Ad Hoc Networks, M. M. Ghonge, S. Pramanik and A. D. Potgantwar, Eds, Springer, 2022, https://doi.org/10.1007/978-3-030-91149-2_6

[32] R. Anand, J. Singh, D. Pandey, B. K.Pandey, V. K. Nassa and S. Pramanik, "Modern Technique for Interactive Communication in LEACH-Based Ad Hoc Wireless Sensor Network", in Software Defined Networking for Ad Hoc Networks, M. M. Ghonge, S. Pramanik and A. D. Potgantwar, Eds, Springer, 2022, https://doi.org/10.1007/978-3-030-91149-2_3

[33] S. Pramanik, "An Effective Secured Privacy-Protecting Data Aggregation Method in IoT", in Achieving Full Realization and Mitigating the Challenges of the Internet of Things, Eds, M. O. Odhiambo and W. Mwashita, IGI Global, 2022, DOI: 10.4018/978-1-7998-9312-7.ch008

[34] P. T. Khanh, T. H. Ngọc, and S. Pramanik, Future of Smart Agriculture Techniques and Applications, in Advanced Technologies and AI-Equipped IoT Applications in High Tech Agriculture, Eds, A.Khang, IGI Global, 2023.

[35] Y. Zhang, Z. Zhang, H. Sun and X. Wang, "On the Backoff Scheme for SPMA Network: A Spatio-Temporal Mathematical Model," in IEEE Communications Letters, vol. 27, no. 9, pp. 2541–2545, Sept. 2023, doi: 10.1109/LCOMM.2023.3298665.

[36] L. C. Voumik, R. Karthik, A. Ramamoorthy and A. Dutta, "A Study on Mathematics Modeling using Fuzzy Logic and Artificial Neural Network for Medical Decision Making System," 2023 International Conference on Computational Intelligence and Sustainable Engineering Solutions (CISES), Greater Noida, India, 2023, pp. 492–498, doi: 10.1109/CISES58720.2023.10183534.

[37] X. -F. Qin and J. -X. Shen, "Mathematical Modeling of High-Speed PMSM Considering Rotor Eddy Current Reaction Effect," in IEEE Transactions on Energy Conversion, vol. 38, no. 4, pp. 2947–2958, Dec. 2023, doi: 10.1109/TEC.2023.3288608.

[38] D. Krummacker, M. Reichardt, C. Fischer and H. D. Schotten, "Digital Twin Development: Mathematical Modeling," 2023 IEEE 6th International Conference on Industrial Cyber-Physical Systems (ICPS), Wuhan, China, 2023, pp. 1–8, doi: 10.1109/ICPS58381.2023.10128007.

[39] X. Wu, Y. Liu, X. Zhang, D. Li and B. Cai, "Mathematical Models and System of Intelligent Servo for High-Efficiency Electrical Discharge Assisted Arc Milling on Difficult-to-Cut Materials," in IEEE Transactions on Systems, Man, and Cybernetics: Systems, vol. 53, no. 9, pp. 5821–5830, Sept. 2023, doi: 10.1109/TSMC.2023.3275231.

[40] C. Sun, Z. Yang, X. Sun and Q. Ding, "An Improved Mathematical Model for Speed Sensorless Control of Fixed Pole Bearingless Induction Motor," in IEEE Transactions on Industrial Electronics, vol. 71, no. 2, pp. 1286–1295, Feb. 2024, doi: 10.1109/TIE.2023.3247754.

[41] H. Myriam et al., "Advanced Meta-Heuristic Algorithm Based on Particle Swarm and Al-Biruni Earth Radius Optimization Methods for Oral Cancer Detection," in IEEE Access, vol. 11, pp. 23681–23700, 2023, doi: 10.1109/ACCESS.2023.3253430.

[42] Y. Sha, J. Mou, S. Banerjee and Y. Zhang, "Exploiting Flexible and Secure Cryptographic Technique for Multi-Dimensional Image Based on Graph Data Structure and Three-Input Majority Gate," in IEEE Transactions on Industrial Informatics, doi: 10.1109/TII.2023.3281659.

[43] S. Liu, Y. Zhang, K. Tang and X. Yao, "How Good is Neural Combinatorial Optimization? A Systematic Evaluation on the Traveling Salesman Problem," in IEEE Computational Intelligence Magazine, vol. 18, no. 3, pp. 14–28, Aug. 2023, doi: 10.1109/MCI.2023.3277768.

16

Agriculture using Smart Sensors

Abhilash Kumar Saxena[1], Rajeev Kumar[2], V. Srikanth[3], Deepti Khubalkar[4], Aditee Godbole[4], Ankur Gupta[5], and Dac-Nhuong Le[6]

[1]College of Computing Science and Information Technology, Teerthanker Mahaveer University, India
[2]Department of Agriculture, Sanskriti University, India
[3]Department of Computer Science and IT, School of CS and IT, Jain (Deemed to be) University, India
[4]Symbiosis Law School, Nagpur Campus, Symbiosis International (Deemed University), India
[5]Department of Computer Science and Engineering, Vaish College of Engineering, India
[6]Faculty of Information Technology, Haiphong University, Vietnam

Email: abhilashkumar21@gmail.com; rajeev.ag@sanskriti.edu.in; srikanth.v@jainuniversity.ac.in; deeptik@slsnagpur.edu.in; aditeegodbole@slsnagpur.edu.in; ankurdujana@gmail.com; nhuongld@hus.edu.vn

Abstract

The people's most basic and important employment is agriculture. Urbanization poses a newer issues and causes a bulk exodus of village residents to the metropolitan as a result of the increased amount of ecological pollution, climate alteration, deterioration of soil and water categories, increase in residents, decline in farm profit, etc., that urbanization has on people who work in the agricultural sector. New technologies that play a crucial part in the development of smart farm dependent on IoT methodology by employing smart sensors are emerging as a solution to this issue. Crop production, animals tracing, soil moisture tracking, distant water tank level monitoring, temperature and humidity sensing, farmland security, environmental condition tracking, and tools monitoring are all improved by smart agriculture. This aids farmers in remote property monitoring and protection. The innovative method for the smart agricultural system uses smart sensors that are IoT-based. An IoT-dependent smart agriculture system includes a camera with a microcontroller, WiFi, a variety of interfacing sensing nodes for service, Internet service, smart remote devices, or computers having Internet which track the function of sensor nodes. These sensors include, for instance, temperature sensors, soil moisture sensors, PIR sensors, which are used to detect the presence of people, animals, and various objects in fields, and GPS-dependent remote control robots which can execute tasks like sprinkling, weeding, security, and moisture sensing. This chapter proceeds as follows: an introduction to the agriculture sector, including a discussion of the issues it is currently addressing and a newer approach to address them; the requirement for IoT in the farming sector; a detailed examination of the relationship between IoT technology and WSNs; IoT-dependent systems, IoT-dependent applications; and the advantages of Internet of Things technology in the farming sector.

16.1 Introduction

For food security, the agricultural sector is crucial to an economy. About 70% of Indians rely on agriculture as their primary source of income, and it generates about 20% of the nation's GDP. However, urbanization has negatively impacted the lives of those who work in the farming quarter by raising amount of environmental contamination, causing climate alteration, degrading soil and water grade, enhancing population food demand, and minimizing costs from the fact utilizing a variety of resources, including soil, water, and nutrients, agriculture generates food. In order to fulfill the demands of the expanding global population, agricultural output must be optimized to solve these issues. The new approach known as "smart agriculture" (SA) [1] primarily focuses on agricultural techniques that boost output, reliability, and administration while lowering greenhouse gas discharges. The leading path to balance farming yield and ecological damage is via sustainable agriculture. In order to enhance and optimize agrarian operations including raising yields, encouraging climate change, and assisting lower discharges from the farming arena, enhancing the entire grades and quantity of agricultural goods. Utilizing the IoT [2] and various smart sensors, SA has integrated effective farming management, including water consumption, fertilizer implementation, crop managing, and animal safety by livestock monitoring, soil moisture tracking, temperature and humidity tracking, etc. By enhancing garbage assemblage, conveyance, and usage for resource recovery, IoT and sensors may enhance management.

The latest popular method for the intelligent farm system is IoT-based smart sensors. It is mostly used in information assembly and device connection. Real-time data transfer between many locations is made possible by cloud computing and sensor networks that are IoT enabled. It carries out monitoring, forecasting, preparing for choices, and making decisions. IoT-based networks are made up of physical items with sensors that gather and transmit real-time data. Medical care, agro-industries, ecological tracking, traffic tracking, restaurants' food operation systems, military systems, smart cities, waste systems, conveyance, intelligent farming, house automation, intelligent meetings, intelligent environment, safety and emergency, and various other fields have used IoT technology. Internet of Things may be utilized to several agricultural practices. IoT is built on systems that can process detected data before sending it to the user. An IoT-dependent smart farming system includes a camera with a microcontroller, WiFi, a variety of interfacing sensing nodes for service, Internet service, intelligent remote devices or computer systems having the Internet which track the functioning of sensor nodes, and different sensor nodes which are laid down in various locations depending on the needs of the application. Some examples of these sensors include temperature sensors, soil moisture sensors, PIR sensors [3], which are used to detect the presence of people, animals, and different objects in agricultural fields, and GPS-dependent remote control robots for security, weeding, spraying, and other tasks detecting dampness, etc.

16.2 Agricultural IoT

According to a poll by the United Nations Food and Agriculture Organization, by the year 2050, as the population grows, so will the need for food. It is necessary to boost global food production by 70% in order to solve this problem. Food safety is a highly important topic of debate for most nations due to the reducing natural resources, farming site area, and altering climatic conditions. Food goods are mostly sourced from the agricultural sector, which also contributes to an economy's growth. The creation of jobs for the populace depends heavily on the agriculture industry. Farmers are still mostly reliant on conventional agricultural practices in many areas of the globe, which leads to increased labor costs, low crop and fruit yields, subpar food quality, degraded land, etc. Automatic equipment, appropriate supervision, tracking and arrange systems, and architectural evolution may all be used to resolve these problems. Therefore, creating new cutting-edge technology has become essential for the overall growth of the agricultural industry. The efficiency of the soil, temperature and humidity, rainfall observation, fertilizer effectiveness, animal monitoring, storehouse monitoring, water tank size tracking, robbery noticing, tool monitoring, tracking and monitoring remotely, and other factors can all be monitored using IoT-based techniques to increase crop production at a lower cost. In order to encounter the need for food in the next years, the globe is moving toward IoT technology mixed with data analytics (DA) [4]. When IoT and DA are used together, the agricultural sector's utility is increased, resulting in better operational planning and higher crop yields.

Agriculture may be westernized at a reduced cost by fusing traditional agricultural methods with cutting-edge technology like the IoT and WSN. Wireless sensor networks compile data from distinct kinds of smart sensors and send it over a variety of wireless protocols to the central server. Insect infestation, excessive pesticide use, animal and bird attacks, the uncertain characteristics of the monsoon, water shortages, incorrect water use, and a shortage of agricultural expertise are other factors that reduce crop productivity.

The use of IoT in agriculture will provide decision-makers more control over devices and automatic technologies, combining product information and services to improve agricultural productivity, profitability, and quality management. However, there are certain issues with the IoT system as well, including data governance, business model ownership, and security and privacy.

16.3 Wireless Sensor Network Based on IoT

By using the Internet as a method of transmission and data exchange, Internet of Things assists in fusing the physical world with the virtual one. Internet of Things is described as a process that connects computing devices, different mechanical and digital tools, and various objects, creatures, or persons in order to facilitate their unique ID and the capacity of transmitting data over a framework without the need for person to person or person to computer communication.

WSNs have been used for smart farming practices and food production thanks to technological advancements, with a hub on environmental tracking, accuracy, machine functioning, process control automation, and traceability. WSN [5] is the greatest option for smart agricultural applications because of its benefits like self-organization, self-arrangement, self-identification, and self-cure.

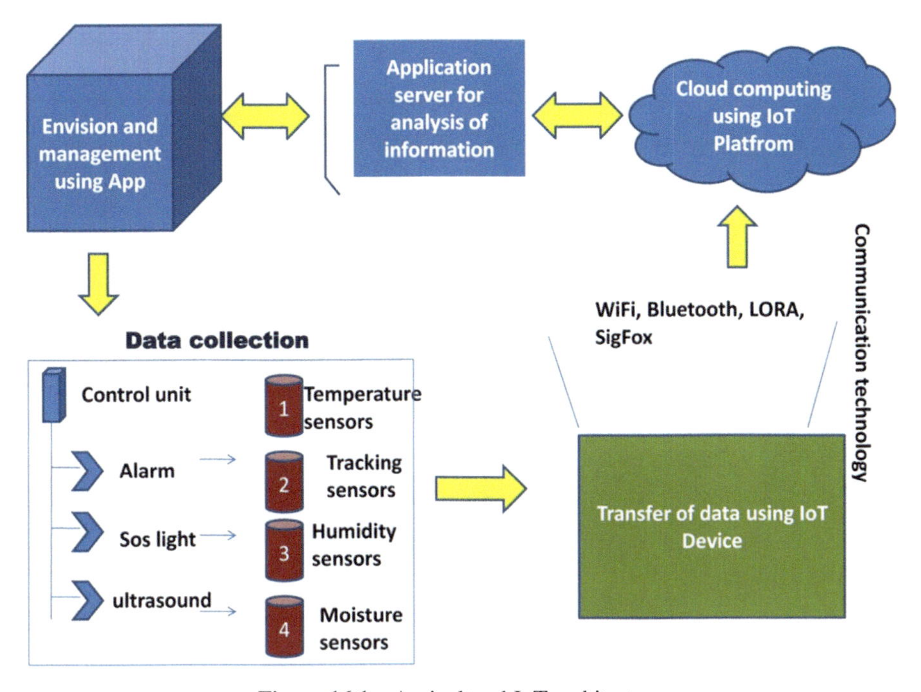

Figure 16.1 Agricultural IoT architecture.

A radio frequency (RF) [6] transceiver, a smart sensor, a microcontroller, and a biasing source are all components of a WSN. With the advent of IoT technology, the primary method for transforming agriculture into smart agriculture switches from WSN to IoT. WSN, RFID [7], cloud computing [8], interface systems, and end-user applications are only a few of the current technologies that are combined in IoT.

As shown in Figure 16.1, the Internet of Things ecosystem for intelligent farming consists of four key building blocks: (1) Internet of Things devices, also known as intelligent sensors, (2) information technologies, (3) Internet, and (4) data storage and processing units. IoT-based apps need these four building components to function. The IoT building blocks are described in depth below.

1. Smart actuators and sensors are connected through an embedded system in Internet of Things (IoT) devices. For this system, wireless communication is necessary. According to [9], IoT devices are also known as smart sensors or IoT sensors. The framework of Internet of Things devices and sensors is shown in Figure 16.2. FPGA or a microprocessor [10], together with transmission components, memory blocks [11], and

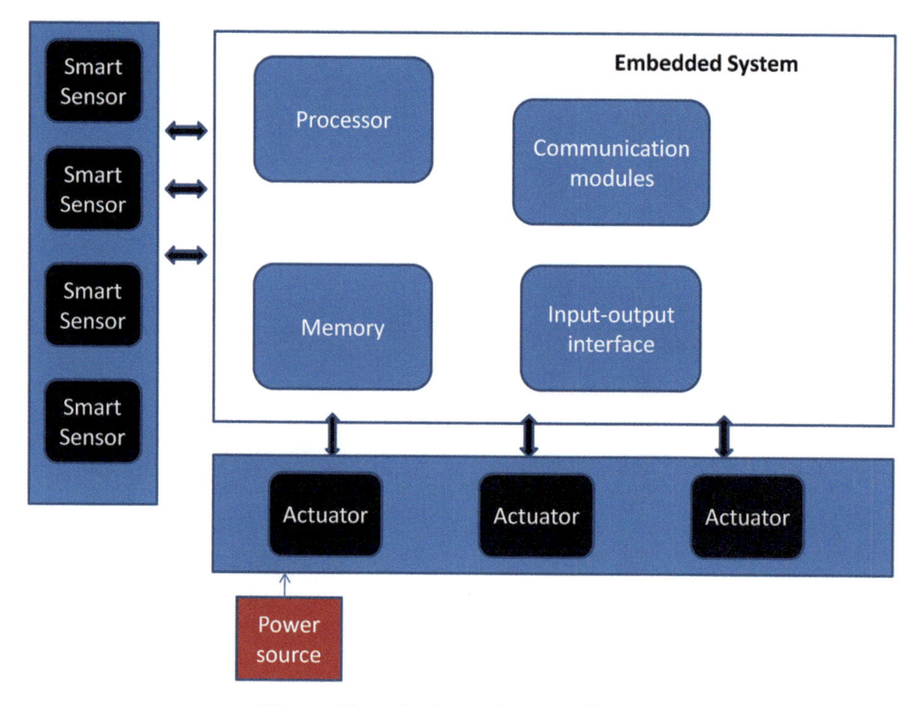

Figure 16.2 Device architecture for IoT.

I/O interfaces, are all components of an embedded system. Different agricultural land data, such as soil nutrient levels, weather predictions, and factors impacting crop yield, are detected and analyzed using smart sensors. Location sensors, optical sensors, mechanical sensors, electrochemical sensors, and airflow sensors are the different categories of smart sensors. Intelligent sensors are utilized to locate a variety of data, including air, soil temperature at different soil levels, rainfall.

Table 16.1 lists few of the intelligent sensors and their uses, including those for moisture, chlorophyll, wind speed, wind directions, humidity, sun radiation, air pressure, etc. The Internet of Things devices are flexible, robust, power-efficient, highly computational, dependable, affordable, and have coverage.

2. Information technology: Information technology is essential to the IoT system's setup. The three fundamental components of modern information technology are standards, spectrum, and applications. There are two categories of information standard: short-range and long-range. There are two categories of information spectrum: approved and illegal.

Table 16.1 Various agriculture sensor types.

Smart sensor categories	Tasks	Uses
Airflow sensors	Measure air pressure and mass flow rates	Measure soil type, moisture content, and structure
Dielectric soil moisture sensors	Used to measure changes in temperature, salinity, and soil texture	Calculate the soil's dielectric constant
Electrochemical sensors	Other particular ions present in the soil are detected	PH, soil nutrient content
Location sensors	GPS signals are used by location sensors	Calculate latitude, longitude, height
Mechanical sensors	The probe is utilized which pierce in the ground to take measure resistive forces	Calculate soil compactness, tensiometer instrument is used to forces
Optical sensors	Utilize a light source to calculate the soil parameters	Analyze clay quantity, presence of organic matter and quantity of the water content in the soil. Examples are photodiodes and photodetectors.

The sensors, backhaul network, and configuration requirements are all important to an IoT application.

a. Spectrum: The ISM [12] band is used by the unlicensed spectrum, although there are reliability concerns, architecture costs, and EMIs associated with utilizing this band. IoT devices that employ ISM create electromagnetic interference, which disrupts radio communication that utilizes the same frequency. The dedicated user, like cellular communication, uses approved spectrum, which offers effective traffic management, less interference, more dependability, higher QoS [13], greater security, lower cost, and a wider coverage area. Other drawbacks include the use of electricity and the requirement to pay for a contribution for data transfer.

b. There exist several standards used in wireless communications, including Bluetooth [14], WiMAX [15], SigFox [16], LoRa WAN [17], 3GPP NB-IoT, and 802.11ah and 802.11p, Wireless HART [18], NFC, ANT+, MiWi, Zigbee [19], Wireless, etc. They are separated into standards for short-range and long-range communication. Devices that support near-field communications, Bluetooth,

Zigbee, Z-Wave, and passive and active radio frequency identification (RFID) [20] systems are a few examples of short-range technologies. They can go up to 100 meters. Standards for long-distance communication span from a few miles to several kilometers. Low power wide area (LPWA), which consumes less power and has a greater range is a category for long-range communication protocols. Examples include LoRa, Sigfox, and NB-IoT.

c. Application requirements: IT relies on the use of smart sensors or Internet of Things devices. The communication technique may be utilized for backhaul networks or IoT devices like sensor nodes and nodes. The backhaul network enables higher data rates and may be utilized for very longer distances, whilst the nodes transmit less data and travel very small distances having less power usage. Bi-directional connections are advocated by several communication systems. The bi-directional connection enables communication between devices, forwarding error correction, handshaking for data dependability, data encryption, and over-the-air firmware upgrades.

   The choice of topology for application affects communication technology choices as well. P2P, mesh, ring, and bus topologies are only a few examples of the many kinds of topologies. IoT devices or smart sensors serve a variety of purposes in each of these topologies. It may be used as a full-function device (FFD) [21], a reduced-function device (RFD) [22], or the personal area coordinator (PAN) [23]. Two different topologies are shown in Figure 16.3 along with the tasks they execute. The PAN initiates communication in the P2P topology and acts as an FFD, while the end devices may act as either an FFD or an RFD. While the end device that serves as the FFD may link to numerous FFDs, it may only connect to a single FFD and cannot connect to another RFD. The PAN in the star topology is what starts the conversation and accepts connections from other gadgets. Only the PAN coordinator may be connected to by the end devices.

d. Internet: Internet has aided in extending the reach of mobile devices, wireless communication systems, and other Internet connection services. According to a forecast by PwC, there will be 400 million linked farming devices by 2030, up from 20 million at the end of 2020. The primary network layer formed by the Internet provides a conduit for the transmission and exchange of data and network information among various sub-networks. Data is accessible

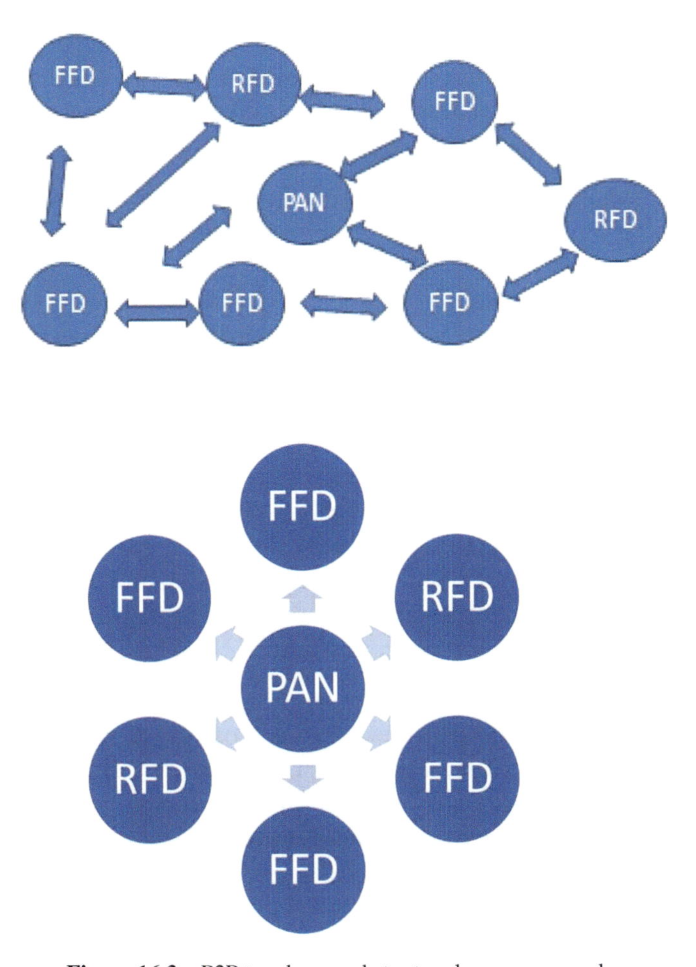

Figure 16.3 P2P topology and star topology are examples.

anytime, everywhere when an Internet of Things device is linked to the Internet. For real-time data transfer and simple accessibility, the usage of the Internet for data communication demands excellent security. Internet is important in cloud computing, as vast amounts of data are gathered for processing and storage. The administration of user interfaces, services, organizers, coordinators for sensor node networks, data processing units, and data computing are all handled via cloud computing.

IoT interface and communication protocols including SOA [24], cloud-dependent, and actor- dependent Internet of Things interface are designed for hybrid systems, device connectivity via the Internet, and Internet of Things technologies. The SOA's architecture is multilayered. Sensing, accessing, networking, interface, and application layers are a few of the layers.

e. Information repository and processing units are necessary for managing the massive, active, composite, and geographical data that are part of the agricultural industry. Structured or unstructured data that is located in the form of messages, images, audio files, or video data is considered to be difficult information or data. Additionally, they come in a variety of forms, including market-related, historical, sensed, or live streaming data. Data are saved in the cloud via cloud computing, which entails grouping multiple apps that are essential for delivering services and managing end-to-end IoT architecture. This allows for the collection of enormous amounts of data from intelligent sensors. Recently, edge and fog computing have gained popularity, where a variety of Internet of Things devices or smart sensors do computations and scrutiny to decrease reaction times for critical workloads, save costs, and improve QoS [25].

To handle the various data structures, a number of farming data management systems was adopted, including On-farm framework, Farmobile, Farmx, Easyfarm, and Farmlogs. Data organization, data storage, and data analysis (DA) are all made possible by this management system.

There are important technological factors that must be kept in mind while implementing IoT devices or smart sensors. These elements should be considered while integrating wireless technology: communication distance, battery life, portability, dormancy, security, adaptability, and affordability are all factors. With the openness to NB-IoT, LPWA is attracting the most attention among all communication technologies. Cheap device power requirements, cheap device costs, simple installation, backup of several small-output devices, vast distance coverage that supports data transmission and reception are all appealing features that the NB-IoT guarantees.

16.4 Smart Agriculture IoT Applications

IoT is being used in smart agriculture in a variety of ways, including crop tracking, livestock monitoring, monitoring of equipment, irrigation tracking, water standard tracking, weather forecast tracking, soil parameter monitoring, monitoring for infection and pest control, automation, and accuracy monitoring. Depending on the below mentioned criteria: monitoring, tracking and tracing, agricultural equipment, precision farming, and greenhouse production, a few IoT applications are mentioned below.

16.4.1 Monitoring

Depending on the area of agriculture being considered, a number of criteria need to be monitored in the industry. Several are described below.

16.4.1.1 Crop farming inspection

Numerous environmental elements have an impact on farm productivity when crop farming is monitored. Understanding the design and process management of fields is aided by the information provided, which includes tracking things like seasonal patterns of precipitation, leaf moisture, temperature, humidity, soil moisture, salinity, climate, insect activity, etc. This data aids in thorough research and record keeping to raise the caliber of agricultural output. As solar radiation informs everyone about plant hazards to sunlight, agriculturists can now determine if plants are being exposed appropriately or excessively, which lowers the danger and enhances the profit. Details about the soil, such as soil moisture, humidity, wetness, and others, assist to enhance the soil's quality and lower the danger of plant diseases. The yield results will be improved by timely and accurate data collection of weather tracking characteristics, like changes in climate and precipitation circumstances. Additionally, by using these specifics, farmers may avoid problems with planning, management, and labor costs. Based on the information provided, this will assist farmers in making an accurate, quick, and essential assessment early. Pest attack information may be acquired and sent to various distant sites, which provide farmers with real-time information for pest management and are used to provide farmers with recommendations depending on the track proof of pest assaults.

16.4.1.2 Aquaponics inspection

According to [26], aquaponics is a hybrid of agriculture and hydroponic systems in which fish waste is applied to the fields' plants to facilitate the essential nutrients needed by every plant. Continuous monitoring of the water quality standard, water volume, temperature, fish welfare, salt level, pH, humidity, and sunshine are essential in these farms. Correct information may increase the production of fish and plants by allowing nutrients to travel between the two. With the least amount of human participation, this information may also be utilized for automated purposes.

16.4.1.3 Forestry surveillance

About two-third of the globe's declared animal and plant species are supported by forests, which is a crucial task in the carbon sink cycle. Monitoring for forestry includes observing soil, air, temperatures, and humidity as well as measuring the levels of different gases in the air, such as O_2, CO, CO_2, toluene, argon, H_2, methane, iso-butane, ammonia, ethanol, H_2S, water vapor, O_3, and NO_2. These elements give monitored information against forestry-related illnesses as well as early notice and warning systems against farm and forest fires.

16.4.1.4 Monitoring of livestock farming

The animal class and its characteristics have a significant impact on the parameters to be observed in livestock monitoring. For example, assessing milk conductivity may provide extensive information on the health of cows and buffaloes. There are several more characteristics that may be tracked, including temperature, humidity, productivity, insect attack, and water quality. By equipping each animal with an RFID tag, farmers may use the placement and implementation to monitor and identify the whereabouts of their herd, so preventing animal theft. It is also possible to assess and monitor storage that holds H_2O, fuel, agricultural necessities, and animal feed. The farmers may use this knowledge to plan more effectively and analyze their budgets going forward. The procurement of various solutions in smaller- and mid-size fields is severely constrained, especially in underdeveloped nations, because of an absence of enthusiasm, expertise, the placing of these positioning, cost-based concerns, and realization. Despite the fact that numerous proposals were made in the field of livestock tracking, agriculture-based IoT technology implementation is still a fairly young field with numerous unresolved issues.

16.4.2 Asset tracking and tracing

Asset monitoring with Internet of Things technology may improve organizational supply chain planning and processes. An Internet of Things technology may provide information to improve farming businesses so they can make more important decisions, manage them better, plan ahead, and connect with different business partners intelligently while reducing time and money. RFID tags with a global positioning system (GPS) based on cloud computing may be used to identify details, such as location and asset identification. Customers may learn about the real history of the market thanks to the tracking of different assets in the agriculture supply chain technique. As a result, the customer's perception of the safety and health implications of agricultural goods will be improved. The process of tracking involves capturing, gathering, and storing data pertaining to the whole supply chain. Additionally, it allows for the differentiation of agricultural production from bottom to top. In order to provide all of the clients and other business partners with assurances about the source, location, and life history of an agricultural product, tracking and tracing of assets creates varied data which is recorded together with the supply chain technique.

Numerous factors, including the escalating environmental problems, the production circumstances, pest-based worries, management, storage conditions, transportation, and the time needed to get to the market, may be monitored. The customer's health is directly correlated with these characteristics. The crucial elements that hurt the developing environmental problems, such as soil, air, and water, the use of pesticides, fertilizers, and herbicides is a factor in many industrial rules. Additionally, the quality of the food and vaccinations offered to the cattle may be tracked since they can cause problems with safety and well-being. Pest assaults during the whole technique that might influence the product's quantity and quality are the main cause of agricultural output damage. Monitoring agricultural outputs may assist farmers in increasing production and controlling the supply chain. Information regarding the cargo, storage unit, transportation means, process system, and production must be available to a tracking and tracing network. The load information includes information on the whole manufacturing technique of the goods, including the country of origin, the most recent status, end station data, and specifics about the many business partners involved in the supply chain technique. Additionally, this network must include a memory bank for long-term data storage for R&D causes. The act of integrating and equating the aggregate data is known as data transfer. The tracking and tracing network must be suitable for handling the information gathered, and the final production is sent to all parties involved in the supply chain technique. It highlights the importance of different RFID tags in tracking processes such as product processing, product shipping, storage, allocation, market sales, and various different after-sales facilities. It facilitates the quickest collection, storing, and visualization of data across a wide area.

Internet of Things-based agricultural equipment may assist boost farm production and reduce food product losses. The agricultural equipment may be used in automated pilot mode with the proper surveying of the GPS technology and global navigation satellite systems. Agricultural tools which are integrated with vehicles, UAVs, and robots may be remotely managed to accurately assist in the capital requirements of farming regions. Farming tools may moreover collect data, and this data may assist agriculturists survey their fields for proper fertilization, irrigation, and nutrition. For example, CLAAS [27], a manufacturer of agricultural tools, has integrated IoT technology into their instruments to enable their tools to operate in automatic pilot mode. A different approach is the Precisionhawk UAV [28] sensor that may facilitate information to farmers like wind speed, air pressure, and different pertinent elements. Agricultural fields may be surveyed and image-sensed using this technique.

16.4.3 Accuracy in agriculture

Accuracy in farming is the combination of real-time data gathered from various field factors, utilizing that data for analysis in hasty and informed decision-making to increase crop output, prevent environmental damage, and reduce costs. Agriculture accuracy relies on a number of technologies, including smart sensor nodes, GPS, and greater DA for increased agricultural output. With the rapid and clever answer provided by DA, the least amount of agricultural resources – like water, fertilizers, pesticides, etc. – are wasted. For scientists working in the fields of robotics, image processing, land mapping, meteorological information sensing, etc., accuracy in farming raises new problems. Agriculturists are able to pinpoint exact locations and survey sites with different information factors with increased use of GPS and GNSS. These information details are utilized by many newer technologies to minutely assign field resources, like seeds, fertilizer, spray, pesticides, and various services. In the agricultural industry, accuracy may increase production. Additionally, it has become essential to provide farmers with simple techniques and to assist them in attending different training sessions so that they may benefit more from these arrangements.

16.4.4 Climate technology

Technology used in greenhouses is sometimes referred to as glasshouse technology. With this technique, regulated environmental conditions are used to grow plants. This method has the benefit of generating different types of plants at any time and any place by creating a pleasant ecological condition. Numerous studies relating the usage of wireless sensor networks in greenhouse technology to track different ecological variables have been conducted. Numerous studies have discussed how IoT technology may be connected to greenhouse technology for reducing human interaction, save energy, provide more effective greenhouse-site tracking, and establish a straight line of communication between greenhouse growers and clients.

16.5 IoT for the Farming Sector: Multiple Benefits

IoT technology has a number of benefits for the agricultural industry. In the section below, a few of the advantages are listed:

1. IoT technology has the potential to promote community farming, particularly in rural and distant places. IoT may be used to assist the creation of services that allow the rural agricultural community to share field-specific storage, data exchange, and to improve links between agriculturists and agricultural professionals. Cellular apps have made it easier to share IoT-related information like as tools, the best fertilizers, weather information, and so on. The rural community may readily exchange data using both free and commercial cell services.
2. Security: Concerns in the agricultural industry extend beyond increased yield of food grains to include its capacity for safety, security, and a reliable supply of high-quality food. There are a number of issues that have been raised about the quality of the food that is supplied to the market, including food adulteration, fake goods, and the artificial enlargement of vegetables or fruit. These problems are mostly affecting health and may be detrimental to the economy. Product coherence, humanity coherence, and statistics coherence are a few factors of the food fraud addressed here that may be identified utilizing IoT technology. IoT may be utilized to give management coordination and arbitrary food item tracking.
3. Advantages of competition: The rising need for greater food production and the usage of innovative newer technology are increasing competition in the agricultural industry. It supports a system that will provide the agricultural industry newer hope for marketing, monitoring, and management via the exchange of information utilizing IoT technology. Food production is increased by lowering the price and waste of agricultural inputs like fertilizer and insecticides. Farmers who adopted the IoT ecosystem would benefit from the competition management provided by the usage of real-time information for decision-making.
4. The use of IoT technology might provide fresh perspectives on job creation that benefit farmers by disregarding numerous abuses of "intermediate men" and allow them to interact directly with consumers for more profit.

5. Lower cost and demolition: The only advantage of IoT technology is its capacity to identify tools and devices that have been deployed remotely. By scanning a big amount of land using a farmer's own truck rather to manually managing the field, IoT technology aids in agricultural applications to save time and money or by taking a field walks. By providing farmers with information regarding the best times to use pesticides and insecticides, IoT technology improves their capacity to make decisions. This could assist them in lowering agricultural expenses and crop loss.

6. Operational efficacy: The participation of government agencies and non-governmental organizations in the decision-making techniques in the agricultural sector is connected to operational efficacy. IoT data collection from agricultural programs is used as a compass by agricultural groups to meddle. Such intervention may assist in managing and controlling the spread of illnesses, agricultural fires, government initiatives, and resource budgeting. Additionally, IoT and DA approaches may help farmers receive a precise and fast conclusion on the management and operation of the field. According to [29], automatically recording the health status of animals and crops would aid in easy, fast, precise investigation and medication prescription to farmers. These practices aid in lowering operational expenses. Using IoT, the agricultural sector's supply chain cycle may deliver food goods more effectively and on schedule.

7. IoT ecosystem can control entrance to WSN and execute low-cost apps in the agriculture industry. Mobile apps may be used to get information such as market analysis, pricing, goods, and services. Additionally, a wide range of legal standards, laws, and regulations pertaining to different agricultural farm produce are readily accessible. Additionally, buyers who are drawn to organic food and fresh goods may quickly locate farmers and get notifications when the food items are accessible.

8. Service management: IoT technology enables real-time monitoring, the detection of agricultural capital, the protection of various kinds of equipment from theft, the replacement of equipment components, and the timely completion of jobs.

16.6 Challenges

Utilizing an IoT platform comes with several difficulties. Security, privacy, authority, data fusion, lack of association, variety in IoT devices, and erratic business prototypes are a few of the problems. Here are some of the obstacles that are encountered in several parts, including business prototype, technical, and geographical issues:

16.6.1 Business difficulties

The agriculture industry has made relatively little progress. Therefore, it's important to balance the difficulties involved in using IoT technology with generating revenue. Below are some of the issues that are covered, including cost and business prototype.

a. Cost: A number of expenses, including administration and organization expenditures, are associated with the usage of IoT in the farming industry. The purchase of various pieces of equipment, like base station support, gateways, sensors, IoT devices, etc., is included in the organization's total cost. In order to manage and continually collect membership fees for the orderly usage of the IoT platform which offers data sharing, IoT device management, data aggregation, etc., management costs must be taken into account. Client contentment depending on privacy expenditure, security expenditure, and business-based which offer better benefit determines the success of the IoT platform. Numerous IoT platforms supply free memberships with restricted IoT device use, restricted IoT device functionality, and limited information storage and usage.
b. Business model: Agriculturists desire business prototypes which would assist them in generating more revenue and profit from their farms by collecting data utilizing an IoT platform. Various IoT platforms now provide limited free subscriptions and complete service with varying degrees of premium subscriptions. IoT technology providers look into the information given and this is leading to a conflict among farmers about how to handle and claim the data.
c. Less expertise is needed, and farmers are less likely to be aware of IoT technology and its numerous uses, which is a huge part in the continuous acquiring of IoT technology in a variety of agricultural implementations. It is a crucial factor, particularly in third world nations where most of agriculturists come from village regions and lack formal education. When it comes to employing new technology, they lack flexibility.

16.6.2 Technological obstacles

a. Coupling: When huge IoT devices are implemented in the agricultural sector for various uses, coupling problems will arise, particularly when

IoT devices employing unlicensed spectrums such as Zigbee, WiFi, SigFox, and LoRa [30] are involved. Coupled systems experience information loss and decreased dependability. The cost of using unlicensed airwaves while linking additional devices increases. However, IoT technology that uses a licensed frequency may cut down on pointless coupling. Reusing frequencies across IoT devices using a licensed spectrum also leads to coupling.

b. Challenges with privacy and security: There are a number of security problems which need to be resolved at various levels of the IoT network. Lack of the necessary security can result in data loss, invasion of privacy, and damage to sensitive data such as field parameters and different intellectual assets. Private farm owners' competitive advantages will be hampered by this. IoT devices are susceptible to physical harm from things like theft, animal attacks, and address changes. Denial of service, traffic, and forward assaults are common on gateways. Device capture attacks may target different site information and IoT site-dependent techniques utilized in agriculture. Attackers grab IoT devices during device capture assaults and remove cryptographic data, which may be used to attack data kept on the device storage. Jamming of wireless signals is another issue. Data loss is a concern for cloud servers, and unauthorized resources might interfere with farms' routine operations.

c. Technology choice: There are several IoT technologies that are now under development, some of which are still being looked at. Because there is still a lot of study needed before new technologies can be implemented, choosing the best approach is a major problem. Before choosing a new technology, several factors are taken into account, including the size of the field, the location, the type of soil, and the surrounding environment.

d. Accuracy: Since IoT devices are installed outside, the surrounding environment will have an impact on the sensors' functioning over time and cause the transmission system to malfunction. Security of installed IoT sensors and tools should be ensured to avoid costly instruments from being damaged in a variety of weather conditions, such as floods, severe rain, cyclones, etc.

e. Local conditions: Before placing IoT devices, it is important to analyze a variety of factors, including the best location for placement, less coupling, and higher reliability. IoT devices can be installed anywhere and connected to the global network without the installation of any additional hardware.

f. Finest of assets: In order to examine the quantity of IoT devices required, the quantity of gateways, the memory capacity for cloud repository, and the amount of data to be transferred, farmers must grasp the best of resources mechanism. This is necessary due to the variations in field dimension, and different types of sensors are utilized to examine the characteristics of many farms for distinct crops and animals. This calls for the development of complex algorithms and statistical designs which will decide on the optimum usage of the resources while reducing the expenditure and raising the production and profit of agriculture.

16.6.3 Local difficulties

a. Management difficulties: Power and possession concerns involving numerous farmers and authorized entities in relation to agricultural information need to be handled. Depending on the resources that are accessible, such as spectrum, technological challenges, privacy, security, and market competitiveness, different rules and regulations apply in various nations. The usage of IoT technologies in different farming implementations is harmed by numerous rules and directives throughout various nations.

16.7 Future Aims

The following trends will determine the future application of IoT in smart farming implementations:

1. Technological advancement: Better IoT-dependent remedies will start to appear on the market for the agriculture industry as a result of the rising demand for new technology. Several trends that will have an impact include:

    a. Implementing LPWA [31] technologies will benefit the agricultural industry since it offers various advantages with low power consumption and long-distance transmission. The advent of the 3GPP NB-Internet of Things standard technology, which many telecom engineers have accepted, has increased the attention of many academics in this field. This would foster interest in IoT research for the agriculture industry.

b. The usage of IoT in the agricultural industry is shifting toward a universal program that would assist any kind of crop or livestock, rather than being restricted to particular animals or crops. This improves the usage of Internet of Things devices for a variety of tasks, including controlling and tracking many crops and animals as well as selling products to nearby shops and customers at the market. By removing the obstacles, this would improve the local and regional area and support the IoT-based agriculture industry.

c. Security: Numerous researchers are interested in the analysis and varied studies in the area of security for IoT devices and point-to-point data security. Implementing IoT devices that are completely safe utilizing different security strategies, such as encryption algorithms, requires research for the security of the data; this method provides digital signatures and data encryption. Data security is essential because it provides defense against physical intruders and hackers.

d. Spectrum and energy efficiency: Several technologies, such as spread spectrum (LoRa and Ingenu) and ultra-narrow band channels (Sigfox and Telensa), are purchased in order to meet LPWA requirements. New methods will provide long-distance coverage, battery life, data transfer, and a high path loss link budget as additional LPWA systems come to market. The majority of cellular NB-IoT systems in usage nowadays assist different frequency bands which function in FDD modes. To encourage functioning in the time division duplex modes, further study is needed. It has presented fresh research difficulties, such as pilot contamination. Since IoT devices rely on the sensor node being utilized, LPWA is employed to have an extended battery life and an energy-competent framework. Numerous algorithms, including cluster and in-network process algorithms, were developed to enable energy-efficient wireless sensor network. Other energy-efficient IoT solutions include predicting IoT device sleep intervals based on energy savings, past consumption, and the level of information needed for a given implementation. Energy-efficient IoT devices are the newest research fields; IoT transceiver allows transmission of power required infrastructure and concurrently wireless data and power transfer.

e. Quality of Service (QoS): In the IoT network design, quality of service is essential at all network tiers. The potential for a gadget to communicate important data utilizing IoT and any type of information technology is currently being investigated. Compared to LoRa, NB-IoT [32] communication technology offers good QoS. Superior research is required to guarantee QoS across various IoT levels.

f. AI and DL: Utilizing AI to estimate growth and disease control utilizing farm data and climate-related information, several studies are still needed. This is similar to how smartphone users contribute images to identify any ailment using machine learning. Research is still being done on DL algorithms that can handle more data more quickly than IoT communication.

g. Information generation from data while protecting an individual's privacy is done using a technique called "privacy and conservation" in cases when IoT data security [33–35] is breached. By permitting the greatest possible use of the data, this strategy will provide a defined degree of secrecy. This approach may improve data privacy and conservation in agriculture, but it also calls for greater research in the field.

h. Data compression: In every network, there are many linked IoT devices. Expanding cutting-edge multiplexing and compactness technology is now required to transport and share data from 250 to 2500 smart sensors to a central place. When using NB-IoT cellular connectivity, it is necessary to deliver picture and video-related data. Information may be combined and multiplexed to gather data from numerous farmlands and give administration, support, and dependability all in one place.

i. Real-time analysis: For real-time analysis, 250 to 2500 different types of sensors are deployed in the farms. A simple network management protocol is employed to aid transmission between the objects and the server having fewer communication problems. The IoT network now manages significant data traffic and power requirements.

2. Application criterion: At this time, research is primarily concentrated on finding ways to make communication technologies utilized in IoT deployment more affordable. The most recent work is focused on prototyping and testing on a small scale. The usage of IoT technology for large-scale projects is being assessed in terms of its applicability to the agricultural industry. Large-scale initiatives in supply chain

management and agro-food implementations will be a feature of IoT technology in the future. Additionally, emerging nations other than established ones, like Europe, the United States, etc., may also witness the IoT's impact on agriculture.

3. Enterprise and market demand:

   a. Cost-cutting measures include lowering power consumption, shrinking physical dimension, and requiring higher O/P productivity from IoT devices used in smart farming. In coming years, less expensive sensors will be used, and research will concentrate on combining different positions.

      Utilize both permitted and unauthorized communication technologies in a variety of ways to lower the cost of installation and operation.

   b. In order to execute new rules and create new policies, the agricultural industry has to develop new quality control and enforcement measures. To guarantee that laws and regulations are implemented in farming for the proper usage of IoT technologies, the government's participation in agricultural departments is crucial. Agriculturists and different individuals who wish to use IoT technology and transform farming into smart farming will be more interested as a result.

References

[1] P. T. Khanh, T. H. Ngọc, and S. Pramanik, Future of Smart Agriculture Techniques and Applications, in Advanced Technologies and AI-Equipped IoT Applications in High Tech Agriculture, Eds, A. Khang, IGI Global, 2023.

[2] S. Pramanik, "An Effective Secured Privacy-Protecting Data Aggregation Method in IoT", in Achieving Full Realization and Mitigating the Challenges of the Internet of Things, Eds, M. O. Odhiambo and W. Mwashita, IGI Global, 2022, DOI: 10.4018/978-1-7998-9312-7.ch008

[3] R. Anand, J. Singh, D. Pandey, B. K.Pandey, V. K. Nassa and S. Pramanik, "Modern Technique for Interactive Communication in LEACH-Based Ad Hoc Wireless Sensor Network", in Software Defined Networking for Ad Hoc Networks, M. M. Ghonge, S. Pramanik and A. D. Potgantwar, Eds, Springer, 2022, https://doi.org/10.1007/978-3-030-91149-2_3

[4] S. Choudhary, V. Narayan, M. Faiz and S. Pramanik, "Fuzzy Approach-Based Stable Energy-Efficient AODV Routing Protocol in Mobile Ad

hoc Networks", in Software Defined Networking for Ad Hoc Networks, M. M. Ghonge, S. Pramanik and A. D. Potgantwar, Eds, Springer, 2022, https://doi.org/10.1007/978-3-030-91149-2_6

[5] A. Gupta, A. Verma and S. Pramanik, Security Aspects in Advanced Image Processing Techniques for COVID-19, in An Interdisciplinary Approach to Modern Network Security, S. Pramanik, A. Sharma, S. Bhatia and D. N. Le, Eds, CRC Press, 2022.

[6] A. Mandal, S. Dutta, S. Pramanik, "Machine Intelligence of Pi from Geometrical Figures with Variable Parameters using SCILab", in Methodologies and Applications of Computational Statistics for Machine Learning, D. Samanta, R. R. Althar, S. Pramanik and S. Dutta, Eds, IGI Global, 2021, pp. 38–63, DOI: 10.4018/978-1-7998-7701-1.ch003

[7] V. Vidya Chellam, V. Veeraiah, A. Khanna, T. H. Sheikh, S. Pramanik and D. Dhabliya, A Machine Vision-based Approach for Tuberculosis Identification in Chest X-Rays Images of Patients, ICICC 2023, Springer

[8] D. Samanta, S. Dutta, M. G. Galety and S. Pramanik, A Novel Approach for Web Mining Taxonomy for High-Performance Computing, The 4th International Conference of Computer Science and Renewable Energies (ICCSRE'2021), 2021, DOI:10.1051/e3sconf/202129701073.

[9] Y. Zhou, H. Huang, S. Yuan, H. Zou, L. Xie and J. Yang, "MetaFi++: WiFi-Enabled Transformer-based Human Pose Estimation for Metaverse Avatar Simulation," in IEEE Internet of Things Journal, doi: 10.1109/JIOT.2023.3262940.

[10] R. Bansal, A. J. Obaid, A. Gupta, R. Singh, S. Pramanik "Impact of Big Data on Digital Transformation in 5G Era", 2nd International Conference on Physics and Applied Sciences (ICPAS 2021), doi:10.1088/1742-6596/1963/1/012170, 2021.

[11] R. Jayasingh, J. Kumar R.J.S, D. B. Telagathoti, K. M. Sagayam and S. Pramanik, "Speckle noise removal by SORAMA segmentation in Digital Image Processing to facilitate precise robotic surgery", International Journal of Reliable and Quality E-Healthcare, vol. 11, issue 1, DOI: 10.4018/IJRQEH.295083, 2022

[12] Dhamodaran S, S. Ahamad, J.V.N. Ramesh, G. Muthugurunathan, Manikandan K, S. Pramanik, D. Pandey, Thrust Technologies' Effect on Image Processing, IGI Global, 2023

[13] S. Pramanik and R. Ghosh, "Techniques of Steganography and Cryptography in Digital Transformation", in Management and Strategies for Digital Enterprise Transformation, K. Sandhu, Eds, IGI Global, 2020, pp. 24–44, DOI: 10.4018/978-1-7998-8587-0.ch002.

[14] Dhamodaran S, S. Ahamad, J.V.N. Ramesh, S. Sathappan, A. Namdev, R. R. Kanse, S. Pramanik, Fire Detection System Utilizing an Aggregate Technique in UAV and Cloud Computing, Thrust Technologies' Effect on Image Processing, IGI Global, 2023

[15] R. Ghosh, S. Mohanty and S. Pramanik, "A Novel Approach towards Selection of Role Model Cluster Head for Power Management in WSN: Selection of Role Model Cluster Head for Power Management in WSN", in Machine Learning Applications in Non-Conventional Machining Processes, G. Bose and P. Pain, Eds, IGI Global, 2020, pp. 235–249, DOI: 10.4018/978-1-7998-3624-7.ch015.

[16] B. K. Pandey, D. Pandey, V. K. Nassa, A. S. Hameed, A. S. George, P. Dadheech and S. Pramanik, A Review of Various Text Extraction Algorithms for Images, in The Impact of Thrust Technologies on Image Processing, Nova Publishers, 2023.

[17] V. Veeraiah, D.J. Shiju, J.V.N. Ramesh, Ganesh Kumar R, S. Pramanik, A. Gupta, D. Pandey, Healthcare Cloud Services in Image Processing, Thrust Technologies' Effect on Image Processing, IGI Global, 2023

[18] P. K. Mall, S. Pramanik, S. Srivastava, M. Faiz, S. Sriramulu and M. N. Kumar, FuzztNet-Based Modelling Smart Traffic System in Smart Cities Using Deep Learning Models, in Data-Driven Mathematical Modeling in Smart Cities, IGI Global, 2023, DOI: 10.4018/978-1-6684-6408-3.ch005

[19] A. Bhattacharya, A. Ghosal, A. J. Obaid, S. Krit, V. K. Shukla, K. Mandal and S. Pramanik, "Unsupervised Summarization Approach with Computational Statistics of Microblog Data", in Methodologies and Applications of Computational Statistics for Machine Learning, D. Samanta, R. R. Althar, S. Pramanik and S. Dutta, Eds, IGI Global, 2021, pp. 23–37, DOI: 10.4018/978-1-7998-7701-1.ch002

[20] Y. Meslie, W. Enbeyle, B. K. Pandey, S. Pramanik, D. Pandey, P. Dadeech, A. Belay, A. Saini, "Machine Intelligence-based Trend Analysis of COVID-19 for Total Daily Confirmed Cases in Asia and Africa", in Methodologies and Applications of Computational Statistics for Machine Learning, D. Samanta, R. R. Althar, S. Pramanik and S. Dutta, Eds, IGI Global, 2021, pp. 164–185, DOI: 10.4018/978-1-7998-7701-1.ch009

[21] J. Jang, S. Lee and W. Park, "Demonstration of Pavlov associative memory by implementation of rate coding using magnetic tunnel junction neurons," in IEEE Transactions on Magnetics, doi: 10.1109/TMAG.2023.3288546.

[22] S. Cheng, Y. She and Y. Wang, "A variable projection-based parameter estimation algorithm for the nonsmooth separable nonlinear problems,"

2023 IEEE 12th Data Driven Control and Learning Systems Conference (DDCLS), Xiangtan, China, 2023, pp. 295–300, doi: 10.1109/DDCLS58216.2023.10166681.

[23] Y. Wu, T. Zhao, H. Yan, M. Liu and N. Liu, "Hierarchical Hybrid Multi-Agent Deep Reinforcement Learning for Peer-to-Peer Energy Trading among Multiple Heterogeneous Microgrids," in IEEE Transactions on Smart Grid, doi: 10.1109/TSG.2023.3250321.

[24] X. Zhao et al., "Real-Time 59.2 Tb/s Unrepeated Transmission Over 201.6 km Using Ultra-Wideband SOA as High-Power Booster," in Journal of Lightwave Technology, vol. 41, no. 12, pp. 3925–3931, 15 June15, 2023, doi: 10.1109/JLT.2023.3272109.

[25] S. Pramanik, R. P. Singh, R. Ghosh and Samir K Bandyopadhyay, "A Unique Way to Generate Password at Random Basis and Sending it Using a New Steganography Technique", Indonesian Journal of Electrical Engineering and Informatics, vol. 8, no. 3, 2020, pp. 525-531, DOI: 10.11591/ijeei.v8i3.831.

[26] A. M. Tyassilva, A. Aripriharta, R. A. Ihsan, R. Alfadel Saputra, S. Omar and H. -Y. Ching, "Aquaponic Analysis Using the PV System," 2023 IEEE 3rd International Conference in Power Engineering Applications (ICPEA), Putrajaya, Malaysia, 2023, pp. 347–350, doi: 10.1109/ICPEA56918.2023.10093178.

[27] R. Ghosh, S. Mohanty, P. Pattnaik and S. Pramanik, "A Performance Assessment of Power-Efficient Variants of Distributed Energy-Efficient Clustering Protocols in WSN", International Journal of Interactive Communication Systems and Technologies, vol. 10, issue 2, article 1, 2020, DOI: 10.4018/IJICST.2020070101.

[28] D. Mondal, A. Ratnaparkhi, A. Deshpande, V. Deshpande, A. P. Kshirsagar and S. Pramanik, Applications, Modern Trends and Challenges of Multiscale Modelling in Smart Cities, in Data-Driven Mathematical Modeling in Smart Cities, IGI Global, 2023, DOI: 10.4018/978-1-6684-6408-3.ch001

[29] B. K. Pandey, D. Pandey, S. Wairya, G. Agarwal, P. Dadeech, S. R. Dogiwal and S. Pramanik, "Application of Integrated Steganography and Image Compressing Techniques for Confidential Information Transmission", in Cyber Security and Network Security, https://doi.org/10.1002/9781119812555.ch8, Eds, Wiley, 2022.

[30] T. H. Ngọc, P. T. Khanh and S. Pramanik, Smart Agriculture using a Soil Monitoring System, in Advanced Technologies and AI-Equipped IoT Applications in High Tech Agriculture, Eds, A. Khang, IGI Global, 2023.

[31] S. Pramanik, An Adaptive Image Steganography Approach depending on Integer Wavelet Transform and Genetic Algorithm, Multimedia Tools and Applications, 2023, https://doi.org/10.1007/s11042-023-14505-y.
[32] S. Pramanik, R. P. Singh and R.Ghosh, "Application of Bi-orthogonal Wavelet Transform and Genetic Algorithm in Image Steganography", Multimedia Tools and Applications, https://doi.org/10.1007/s11042-020-08676-2020, 2020.
[33] S. Pramanik, S. K. Bandyopadhyay and R. Ghosh, "Signature Image Hiding in Color Image using Steganography and Cryptography based on Digital Signature Concepts," IEEE 2nd International Conference on Innovative Mechanisms for Industry Applications (ICIMIA), Bangalore, India, 2020, pp. 665–669, doi: 10.1109/ICIMIA48430.2020.9074957.
[34] S. Pramanik, D. Samanta, S. Dutta, R. Ghosh, M. Ghonge and D. Pandey, "Steganography using Improved LSB Approach and Asymmetric Cryptography", IEEE International Conference on Advent Trends in Multidisciplinary Research and Innovation, 2020
[35] S. Pramanik, M. G. Galety, D. Samanta, N. P Joseph, Data Mining Approaches for Decision Support Systems, 3rd International Conference on Emerging Technologies in Data Mining and Information Security, 2022

17

Technologies that Work Together for Precision Agriculture

Devendra Pal Singh[1], Sachin Gupta[2], D. Ganesh[3], Shilpa Sharma[4], Aarti Kalnawat[4], Sabyasachi Pramanik[5], and Debabrata Samanta[6]

[1]College of Agriculture Science, Teerthanker Mahaveer University, India
[2]Department of Management, Sanskriti University, India
[3]Department of Computer Science and IT, School of CS and IT, Jain (Deemed to be) University, India
[4]Symbiosis Law School, Nagpur Campus, Symbiosis International (Deemed University), India
[5]Department of Computer Science and Engineering, Haldia Institute of Technology, India
[6]Rochester Institute of Technology, Kosovo, Europe

Email: dpsinghevs@gmail.com; chancellor@sanskriti.edu.in; d.ganesh@jainuniversity.ac.in; shilpasharma@slsnagpur.edu.in; aartikalnawat@slsnagpur.edu.in; sabyalnt@gmail.com; debabrata.samanta369@gmail.com

Abstract

The notion of precision agriculture (PA) enables agriculturists and food growers to optimize inputs for increasing yield and boost standard harvesting while reducing expenditures and ecological consequences. Due to relatively large farm areas and the potential for automated agricultural production systems, developed nations often associate with precision agriculture. To determine the appropriate input quantities (like water, nutrients, and fertilizers) to the farm, the process entails data collecting, scrutiny, and graphing on yield, soil standard parameters, and ecological variables at various places within the field. Precision agricultural technology is still mostly absent from the majority of emerging nations. The field sizes are smaller, and there is still a severe lack of access to technology, expertise, and financial resources. However, farmers in developing nations continue to look into the tools and resources at their disposal to boost their productivity and output in the agricultural sector.

17.1 Introduction

Precision agriculture is one of the key ideas for the Fourth Industrial Revolution's (4IR) [1] transformation of the food sector. The notion of precision agriculture (PA) [2] enables farmers and food producers to optimize inputs in order to increase productivity and improve quality harvests while reducing costs and environmental consequences. Due to relatively large farm areas and the potential for automated agricultural production systems, developed countries often associate with precision agriculture. The technique entails gathering, analyzing, and graphing data on environmental elements, soil quality metrics, and production at various sites across the field to choose the appropriate inputs' quantities, including water, nutrients, and fertilizers. Precision agricultural technology is still mostly absent from the majority of emerging nations. The field sizes are smaller, and there is still a severe lack of technological access, expertise, and financial resources. However, farmers in developing nations continue to look into the tools and resources at their disposal to boost their productivity and output in the agricultural sector.

In the year 2045, it is predicted that there would be 9.5 billion people on the earth, and to fulfill their needs, food production would need to double. A recent technology advancement called precision farming enables farmers to feed more people on the same-sized land. Nearly 90% of the world's population worked in peasant agriculture prior to the agricultural revolution. Since the service sector currently accounts for 80% of the economies in industrialized countries, the trend has significantly shifted. Because fewer people are working in agriculture, farmers are becoming older and fewer young people are interested in it. According to a case study in the United States, more than 60% of agricultural owners are over 60 and more than 40% are over 65. It is noteworthy that 60% of the world's economic output comes from the 5% of the people who work in agriculture. The 4IR accurately captures the modernization, mechanization, and automation initiatives that many industrialized nations, notably the United States and Japan, have been required to adopt in order to address agricultural difficulties.

Ensuing the introduction of automation between 1930 and 1960 and the Green Revolution in the 1970s, precision agriculture, also known as digital agriculture, was the outcome of the third step of the farming uprising. Weather forecasting, satellite images, and changing fertilizer applications represent the first revolution in PA. Collection of machine data for topographical mapping, soil data, and much accurate agricultural exercises were part of the second step of precision agriculture evolution. Digital agriculture will transform agriculture in the Fourth Food Revolution (Food 4.0) [3], since PA and other 4IR technologies contain the solutions to the problems of sustainably feeding a rising global population. In the food industry, change-agent technologies include autonomous tractors, intelligent robotics, sample sensors, drones, and farming robots. The food sector, which serves as a good illustration of any mismatch between inputs and outputs, necessitates a technique for building an agricultural model that is optimally balanced between yield, issuance, and intake. In spite of the fact that there is sufficient food produced worldwide, more people still perish from starvation as roughly 40%–50% of it is wasted. PA is gaining popularity because it strikes a compromise between reducing environmental damage and improving food output. Through the careful and minimum use of inputs, precision agriculture permits the sustainable intensification of increasing yields. Additionally, by applying too much input, the environmental effect might be reduced while improving soil moisture and quality. Additionally, by lowering production costs and using inputs strategically, farmers tend to engage in greater competition. PA offers a remedy for lowering agrochemical I/Ps and the negative ecological effects. Three key advantages – increased economic yields, environmental advantages, and benefits to society – emphasize this.

The digitalization of food and agriculture has made almost every part of the sector largely dependent being managed by software on the hardware. Despite potential opposition from the conventional and analog methods of food production, the sector is undergoing a transformation thanks to technological disruptions that work in tandem with precision agriculture. Drones, big data, farm management software, and sensors are all part of the PA idea, which has diverse applications in environmental control, intelligent packaging, microfarms, and gene editing. The tools and creative techniques would be the human race's last hope of survival with the imminence of agriculture and food. Precision irrigation, forecasting, yield tracking, food scouting, variable rate application, and other application types are some of the several ways PA is used. Precision crop cultivation is greatly facilitated by variable rate application and precision irrigation. The development of autonomous robots is significantly accelerating the expansion of both agricultural and livestock production. Additionally, indoor farming is booming as a result of the growth of metropolitan areas and the rising need for fresh produce. The advantages of management software for assuring effective aquatic species breeding are also being benefited by aquaculture.

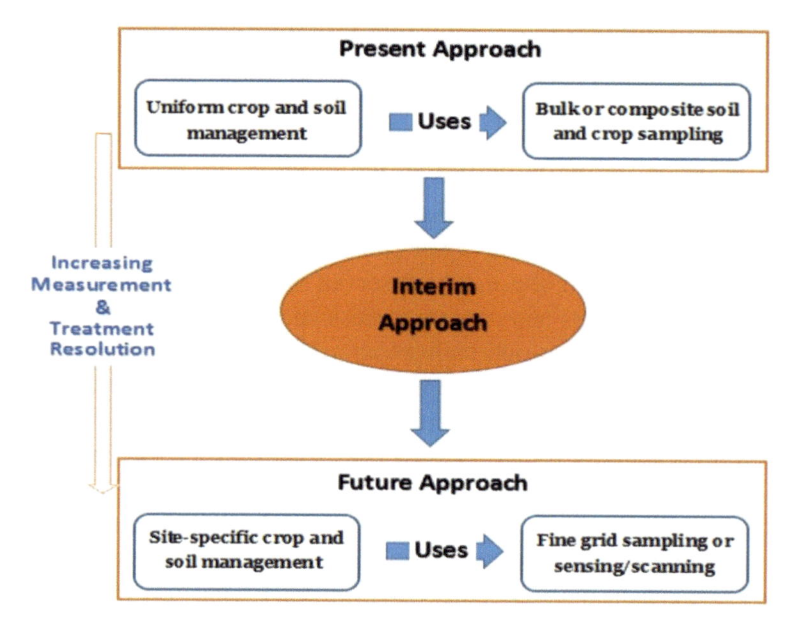

Figure 17.1 A schedule of location-based crop management in progress.

The definition of precision agriculture may be given from the perspective of timely variation control, right up to the adoption of crop-monitoring [4–7] systems. It serves as the foundation for site-specific crop management, which better tailors agronomic practices and resource application choices to the needs of the land and the various crops growing there. Instead of information technology (IT) used on farms, this idea focuses on decision-making based on resource placement. The intriguing aspect of this viewpoint is that the choices taken assure comprehensive advantages that may not yet be quantifiable. The idea has broadened the scope of IT application and yield expertise to site-specific quality optimization, production efficiency optimization, risk minimization, and environmental burden minimization. Figure 17.1 depicts the changing pattern of the site-specific crop management (SSCM) [8] strategy. A continuous management method, the SSCM bases decisions on routine data monitoring and analysis. The decision's impact will be observed, and the data will be used to guide the next cycle of managerial decision-making. The SSCM technique may be used to geo-referencing, attribute mapping, DSS [9–12], data monitoring, and differential measures thanks to its supporting technologies.

Farm operations may be mapped and graphically shown thanks to geo-referencing activities. These provide light on the erratic nature and operational inefficiencies of agricultural production. Various sensors and systems were created for real-time analysis of the variables in crop, soil, and climate monitoring. The adaptability of current sensors, the development of newer types, and the requirement for farming scientists to start evaluating the potential for various crops and calculated yield indicators are the areas of trending research.

In conclusion, precision agriculture is a management technique that is always developing with the primary goal of increasing profitability via the use of information about agricultural resources. Input factors, such as cultivation choices, application rates, irrigation planning, and tillage techniques, are often managed in crop production. The development of 4IR technologies has made it possible to handle PA on a wide scale. The precision agriculture idea was extended into a system that allows farmers to monitor and manage field information at a comparatively cheap cost. Pesticides may be applied selectively, solely to pest-infested regions, minimizing the amount of pesticide used and, as a result, the environmental effect. Additionally, by selecting the right plant population, soil nutrients may be maximized, and by choosing the right plant species, field conditions can be benefited. Crop yield monitoring may also lead to the compilation of maps that can be used to make decisions about the fields' high and low productivity zones. Although the idea of precision agriculture management was initially mostly adopted in the west for crops like maize, soybeans, and wheat, it may be used to almost all agricultural commodities.

17.2 Summary

17.2.1 Integrated precision agriculture technologies

Precision agriculture's architecture is made up of software solutions that use a variety of technologies, including data analytics, machine learning, and others. Through the notion of PA, new technologies are transforming how farming is done today. Various expert studies have shown that expanding smart farming would result in higher crop output and improved system efficiency. Additionally, PA appears in a variety of agricultural techniques, including animal farming, vertical farming, and controlled environment agriculture. Smart agriculture has the potential to introduce pre-urban and urban agriculture approaches due to the size of the established agricultural fields in metropolitan areas. Urban agriculture's economics may not initially provide enormous profits, but they hold significant potential for social entrepreneurship. Figure 17.2 depicts the technical architecture of precision farming.

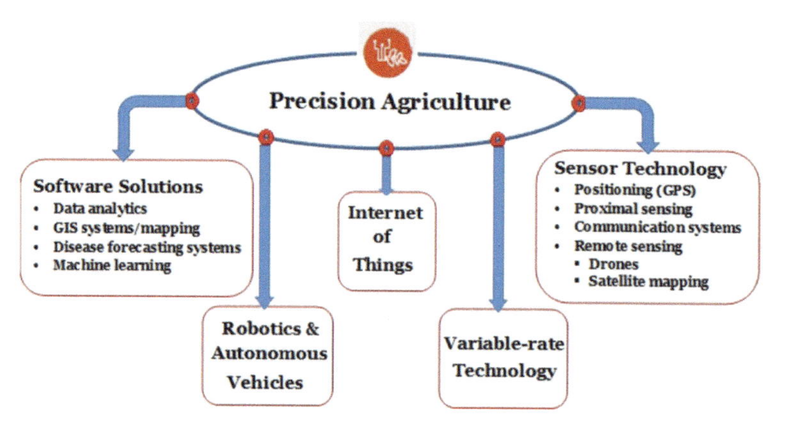

Figure 17.2 An overview of precision agricultural technologies.

A number of technologies do integrate with agricultural techniques in order to fully use PA. Therefore, the synergistic aggregation of technologies and supporting platforms for full-blown technology assumption are facilitating the excellent significance of technological acceleration as well as the cost reduction.

17.2.1.1 Sensor engineering

A sensor device [13–15] delivers a signal to other electronic devices in response to changes in its environmental parameter(s). The function is integrated into data gathering technologies for crops, soil, and by the incorporation of agricultural tools, equipment, satellites, airplanes, and other devices, animal data, and so on. The use of real-time data detection, analysis, scheduling, and use for improved agricultural and food yield is made possible by technology. The use of sensor technology is prominent in the Food 4.0 idea of precision agriculture.

Sensors are used in precision agriculture to identify issues including crop stress, insect and disease outbreaks, and soil characteristics. In order to detect the soil and plant parameters, sensors are used by field agents like tractors and scouts, planes, and mapping satellites during precision farming. Current yield monitoring systems employ sensors mainly for real-time observations, but commercial sensors also allow for real-time assessment of soil characteristics and rate changes. Other studies on commercial applications for weed identification and plant nitrogen levels are still being conducted. There are sensors that can detect the amount of chlorophyll on the fly by measuring light reflection on the plant leaves. There is a clear correlation between plant nitrogen levels and chlorophyll synthesis, according to experts. Additionally, kits are available for remote analysis of soil pH and other components without the need for soil samples.

Data planning, crop and livestock management, processing, and harvesting are all made easier by sensor technology. It works in harmony to provide product traceability and food safety. The whole food yield, post-harvest, and supply chain may be effectively integrated by it. Furthermore, it is especially pertinent to the tracking of soil, water, and climate. Sensors will continue to be the foundational components in key technological applications for the foreseeable future in the food production cycle. For use in food applications, sensors are becoming more sophisticated, economical, and integrated. Food implementations in 4IR will use sensors for location and distant sensing. To provide the highest-resolution decision-making aids, local and distant sensing systems for huge data collection for managing food yield management require to be coupled with effective big data analysis. While optical sensors may show changes in field color characteristics according to differences in crop growth, soil category, farm borders, and other factors, remote sensors offer the advantage of delivering immediate maps of field parameters. The health of the plants may be determined by the indicators provided by satellite and aerial images.

Global positioning satellite (GPS) [16–18] technology is used to place the sensor technologies. For the purpose of finding things on the earth's surface, 24 satellites are deployed at a greater height in the earth's orbit. Decoding the GPS's continually sent signals requires specialized receivers. The receivers must have at least four satellites in order to compute an object's location on the earth's surface. The manufacturer's intentional offset and the surrounding environment diminish the accuracy of the calculated location. As a result, a GPS raw signal without a reference is insufficient to accurately assess location accuracy. The reference signal is used by a differential global positioning system (DGPS) to create the accurate positioning dataset. The GPS devices do away with the requirement for frequent site visits to collect soil samples and determine the crop and field's demand for fertilizers or pesticides.

17.2.1.2 Robotics

The fundamental ideas of robotic technology are the creation and use of automated robots. Robotics and other technology are integrated synergistically as part of the precision agriculture idea. Robots are designed to undertake activities that humans are unable to complete. Different types of robots, such as industrial robots, autonomous vehicles, and bionics, are used in food production and agricultural operations. Robotics is emerging as a cutting-edge method of increasing agricultural output and the availability of food. Specially designed robots may be used to manufacture simpler, more dependable machines, which can subsequently minimize the quantity of required systems and parts. Automated tools assist in food packaging, harvesting, and planting. Robotic higher-speed detection and picking of ripe berries without causing crop harm is a typical example.

Robotics and various other technologies work well together for certain applications in the administration and production of food. Robotics now has access to massive data thanks to developments in machine learning and big data analytics. Robots nowadays often mimic human intellect and carry out more duties. In actuality, more robots and machines have steadily taken over the roles and responsibilities that were formerly designated for people. The window of possibility for human exploration is increasingly closing. Human interventions and involvement also start to decline sharply in Food 4.0 as artificial intelligence starts to take center stage. Meanwhile, it is implied that the possibility of jobs for agricultural laborers are against the falling cost of robotic technology. Investors and producers are progressively using robots in every facet of food yield and management in order to benefit from cost reductions. Figure 17.3 depicts an automated weeding robot in vegetative crops and its linked accessories. The robot eliminates the need for chemical weed management, which is better for the environment and the crops since it eliminates weeds without human interference.

17.2.1.3 Solutions for software

All facets of food production in agriculture are made possible by technological advancements like AI [19–21], ML [22–24], and big data analytics [25–27]. Big data drives real-time operational choices and enables process redesigns for cutting-edge commercial models, gradually revealing insights into productive agricultural systems. Farmers will play a key role in a highly integrated value chain that will dominate the big data landscape, while exposed combined frameworks will give agriculturists and different key players the freedom to select their own business associates within the chain network of technological food yield. In the meantime, visualization strategies allow for the user-friendlier presentation of data, which implies a superior comprehension and application of real-time data analytics.

Artificial intelligence (AI) and machine learning are transforming into unavoidable solutions in the food value chain. Precision farming and other technologies work in concert with the two solutions to completely transform the food industry in the 4IR. Despite the enormous potential for economic development offered by technological solutions, human talents

Figure 17.3 An automated weeding robot for vegetable crops: (a) robot, (b) a rotating hoe, and (c) spiky harrow tool.

are progressively becoming obsolete. With knowledge of how AI will affect employment, farms and businesses must train their workforces in the brand-new fields that will emerge as a result of software solutions. Application of software solutions in the field of autonomous, AI-driven farm vehicles is usual in precision agriculture. With the launch of autonomous farm tractor technology, smart farming organization announced the creation of "AutoCart" software. The grain cart tractor is entirely automated by the cloud-based AI software, which is a huge help to farmers at the busiest time of the harvest.

Geographical information systems (GIS) are pieces of computer software that can store, retrieve, and convert spatial or field data. The administration and practice of precision agriculture may both benefit from this software solution. Utilization entails stacking variables like soil type, nutrient level, and others and allocating the data to a particular field site. Basic GIS operations include the examination of layer properties and the creation of application maps. The program discovers that recording the farm position by the longitude and latitude acts as a suitable supplement to the GPS. The

Figure 17.4 Precision agriculture is a typical IoT application.

variations in soil category, nutrient contents, topography, production, and insect occurrence may be shown using GIS software.

17.2.1.4 Network of things

To complete duties in food production and administration, precision agriculture works in synergy with Internet-connected items. The term "Internet of Things" (IoT) refers to systems which are connected to the Internet and have the capacity to communicate with each other to carry out tasks automatically. The range of networks and sensors has a wide range of applications in the agricultural and food industries. The data produced by the Internet-connected sensors and nodes may be readily examined to improve the operational state of the food systems. Internet of Things is anticipated to be very important in the food and agricultural supply chain, weather forecasting, and other areas. For the Internet of Things to be completely investigated in farming and food, the present restrictions on wireless access in certain areas continue to be a significant research problem. Meanwhile, experts have forecast that IoT will soon have a significant impact on several industries, including food production, logistics, market research, and different others, within the 4IR. Figure 17.4 illustrates a major area of concern for Internet of Things applications in precision farming: climate and greenhouse tracking. It entails integrating smart agriculture sensors at various points along the farm. The measures provided are utilized to map the field's ecological conditions, and the data obtained might be sent to the cloud. Most of the time, the system can modify the environment to fit the intended characteristics.

The agricultural and food industry is now using IoT capabilities being explored by contemporary equipment, machinery, and new communication and information technology. It is the fundamental application in the newer Food 4.0 age that is characterized by automation, digitization, connectivity, and methodical resource utilization.

17.2.1.5 Technology with variable rates

With the use of computer hardware and microcontrollers, lime, pesticide, and fertilizer production rates may be changed using variable-rate technology (VRT) [28–31]. In order to determine field sites and give control for variation in application rate, the controllers may link with the GPS to create an application map. The VRT is connected to precision farming in fertilizer implementation as linked to soil sample, GIS, and GPS. Here, the GPS transmits the tractor's farm position. Through the connection between the GPS and GIS, the location's field characteristics are sent to the controller. The precise amount of the appropriate fertilizer at the area, however, is decided by the specified production objective. Therefore, the controller's control of the equipment causes the proper quantity of fertilizer to be applied. Insecticides and herbicides also follow the VRT method, with the exception that a real-time sensor would inform the controller of the precise position of the insects and weeds. Due of the limited availability of such sensor systems, the VRT for pesticides now must depend on human field scouting for the mapping of infestation levels.

17.3 Case Study: Using Synergistic Technologies to Improve Precision Agriculture in the Food Chain

A farming system which integrates food yield, delivery, and utilization may be built using the notion of precision agriculture. As demonstrated in Figure 17.5, the idea of PA includes sustainable, intelligent, and sensing technologies in food systems. Farmers may increase the number, standards, cost-effectiveness, and sustainability of food yield with the use of precision agriculture. PA aids farmers in reducing a variety of issues related to water scarcity, a lack of usable land, and the price of installing new technology. PA enables farmers to maximize efficiency among the technological components of linked equipment that are remotely managed, monitored, and controlled by sensors. Numerous case examples illustrate the PA principle.

17.3.1 Networked Libelium sensors for crop monitoring

The Green Facility organization of UK has implemented a Libelium technology-based precision farming framework to track crops in the nation's

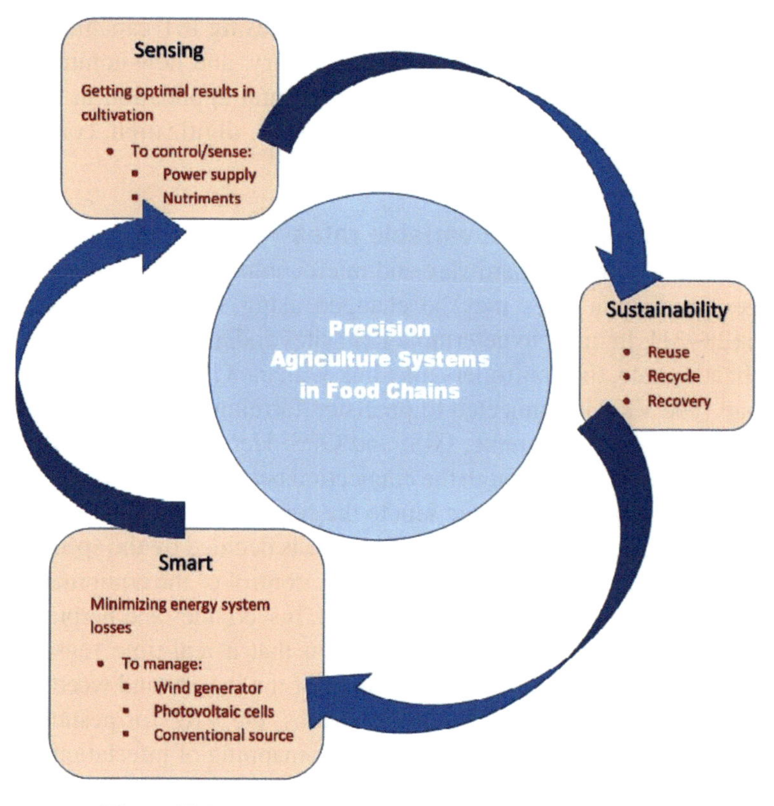

Figure 17.5 The food chain's shared need for PA systems.

Liverpool area. In order to build the precision agriculture project using distant networked sensors, the group used Libelium's Waspmote sensor platform. In order to monitor plantain crops, the system was set up, and the network was expanded using scaled-down sensors. The technology enables remote monitoring of important plantation-wide characteristics as temperature, humidity, trunk diameter, soil moisture, sun radiation, and fruit diameter. The procedure that results might allow the producers to remotely monitor agronomic and environmental variables to test out new banana cultivars. Additionally, the precision agriculture remedy offers a potential record of the harvesting prediction, disease avoidance, water use optimization, fertilizer consumption reduction, and clarity of the accessible soil.

17.3.2 Aquaponic smart greenhouse

A production system called a smart hydroponic greenhouse (SMG) was created utilizing the sensing-smart-sustainable development paradigm.

Figure 17.6 The smart hydroponic greenhouse, (a) from the outside, (b) from the inside.

Figure 17.6 depicts the deployment of the SMG at Tecnológico de Monterrey in Mexico City. The system's application is focused on the complete tomato crop lifetime. The spotting of the needs, the creation of the manufacturing technique, and the potential evolution are the areas of concentration in the production. The crucial problem in the design process, however, was to provide the best answer that specifies the most suitable physical infrastructure, and management systems for managing the greenhouse's environmental conditions.

In order to maintain and regulate the crop's integrity within the greenhouse, the SMG system aids in evaluating the environmental factors such as wind direction, precipitation, humidity, wind velocity, sun severity, temperature, and solar luminosity. Graphic users manage interfaces and collective human and machine yield technologies may be used to relay the I/P and O/P data from the system.

17.3.3 Application of variable-rate fertilizer on Australian grain farms

Working with certain estimations of production on every area, compared to when if steady management or varying rate is utilized, is a necessary step in the technique of calculating the advantages of variable-rate technology in fertilizer implementation. In the study by [32] on UK grain fields, two strategies were chosen after speaking with local agriculturists, based on the case studies' surroundings. If the management zone fertilizer rates and productions are familiar during variable rate management, there are two examples as given in Tables 17.1 and 17.2 that illustrate the fertilizer and anticipated production during uniform management.

In Table 17.1, the higher area potential production is assumed to be nutrient-restricted, increasing the variable-rate productions and the lower zone possible yield is assumed to be nutrient-unlimited, increasing the

Table 17.1 If the variables in the management areas within the variable rate management are familiar, Case 1 of fertilizer rates and anticipated production in uniform management.

	Under variable rate management		Under uniform management	
Zone production possibility	**Grain production (t/ha)**	**Fertilizer rate (kg/ha)**	**Grain production (t/ha)**	**Fertilizer rate (kg/ha)**
Higher	3.5	90	3.20	45
Medium	3.1	75	2.4	45
Lower	1.8	40	0.9	45

Table 17.2 When the variables in the management zones within the variable rate management are known, Case 2 of fertilizer rates and anticipated yield under uniform management.

	Under variable rate management		Under uniform management	
Zone production possibility	**Grain production (t/ha)**	**Fertilizer rate (kg/ha)**	**Grain production (t/ha)**	**Fertilizer rate (kg/ha)**
High	3.5	90	3.20	95
Mid	3.1	75	2.4	95
Low	1.8	40	0.9	95

yield by 7% because of "haying off." The mid zone's characteristics don't change.

Having the exception of the lower prospective zone, which saw a 5% gain in yield owing to reduced "haying off,"all zones in Table 17.2 are assumed to have nutritional limits under consistent management.

17.3.4 Adaptable sun tracker

In recent years, it has become quite common to gather solar energy using solar panels. Other less-clean energy sources, however, have shown that they provide lower prices per watt. The sun tracker is a tool created here to maximize energy production from solar panels and subsequently provide the energy output a competitive edge. The gadget is designed to combine both the genetic algorithm and the swarm optimization algorithm that was being executed as a digital system for positioning optimization, in order to address the difficult problem of finding an ideal location. The evolution of the evolutionary algorithm makes it possible for the solution to locate the sun tracker using off-line data. The off-line technique's solution may be planned and preliminarily assessed. Through approximation, computational cost is

Figure 17.7 Sun tracker prototype typical.

reduced, and a successful method for regions with continuous climate has been devised. A simpler version of the swarm optimization algorithm was presented in order to build a solution appropriate to locations with non-constant climatic conditions, and it has shown to be quite successful there. To execute an online solution, an intelligent water drop algorithm was created, modified, and empirically tested. The algorithm's examination revealed a 40% boost in sun tracker efficiency compared to static solar cells. The sun tracker device, as seen in Figure 17.7, is very helpful for supplying greenhouses with energy and for automating the maintenance of smart agriculture fields.

17.3.5 Application for Vodafone precision fertilizer (VPFA)

To allow farmers to only utilize the precise amount of fertilizer required at a given time, a precision farming system has been created. To develop a full PA solution, the system combines many synergistic technologies. On a tractor that spreads fertilizer, a specifically made GPS gadget was mounted, and data from the functioning is sent remotely through the Vodafone network. By the map made from the recorded data on the secure server, each farmer may view the record of fertilizer deposits. Due to their ease of detection and correction, this has allowed the farmers to precisely limit any potential fertilizer waste. The VPFA system makes use of GPS and machine-to-machine communication. The network and unique chip needed for uninterrupted data transfer from the farm are provided by the operator. The technology allows agriculturists direct over the management of the agricultural I/Ps and facilitate the constant monitoring of both the rate and breadth of the claimed fertilizers. This PA system may be expanded and adapted for use in several contexts, such as effluent dispersion and spaying. The system has made sure that landowners are in charge of making the best use of resources.

17.3.6 Quadrotor for precision agriculture

A drone called a quadrotor was created for unmanned aerial surveillance and use. The framework is made to preserve a horizontal balance having distinct references to the "*x*" and "*y*" axes, while maintaining a constant movement with respect to the "*z*" axis, using a fuzzy logic motion controller. The system likewise includes a monitoring interface, which at any one moment displays the representative angles and their precise numerical estimates in relation to every axis. The interface also displays the angles' derivatives, the actual PWM percentages for each motor, as well as the energy and torque produced by the motors. The estimates as well as different essential parameters, such as the angular velocities of each axis, are also shown. For farming and food applications, like UAV sprayers for pesticides, herbicides, and fertilizers; crop growth tracking; and picture capture for 3D models, the quadrotor is highly helpful.

17.4 Impacts of Precision Agriculture on Policy

A key element of the food revolution is the policy framework from producers and governments. Digital marketplaces and competitive services are enabled by the creation of this ecosystem. For the most part, particularly for the government, establishing and administering digital programs requires a high degree of administrative expertise. Because most poor nations lack the necessary capabilities, precision agriculture is less applicable there. The majority of the time, there is a dearth of published works on government initiatives for the digitalization of farming and food. Moreover, information on the level

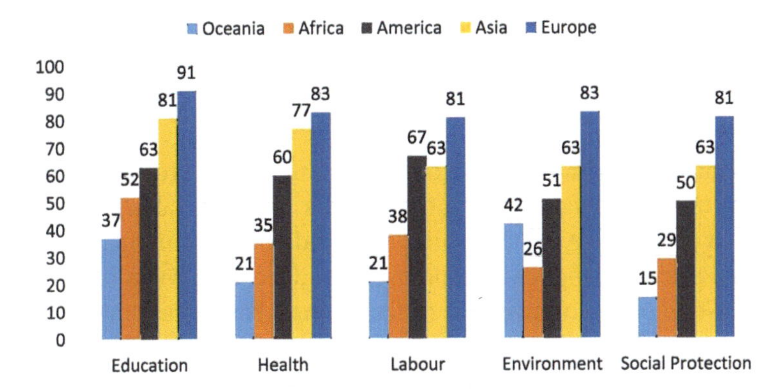

Figure 17.8 Percentages (%) of the nations in each area that provide government services by SMS, email, or simple syndication.

of government involvement in e-services, data, and connectivity is inferred from the inference. Figure 17.8 depicts a typical map of governmental facilities in various places made possible by digital connection.

Lawmakers depend on the management optimization capabilities of precision farming at the farm-level of crop research, by the alignment of crop requirements and agricultural needs, agricultural economics by the skilled applications of adding affordability, and ecology by the mitigation of environmental threats. Agriculturists still have access to data that helps them make better decisions, sell their products more effectively, increase yield quality, enhance traceability, and lease property more efficiently. The PA technology assists farmers in making decisions on the amount of fertilizer to apply depending on soil category, the kind of plant to plant depending on soil conditions, and the amount of insecticide to put on a particular crop. According to [33], choices are based on a variety of information sources that have complex requirements for analysis with regard to impacts on crop standards and amount, environmental consequences, and operational usefulness. Additionally, PA provides concurrent minimization of ecological effect, with better production on the field, thanks to the rise in input consumption and mechanical efficiency. The economics of agricultural production are also changed by PA. Numerous agro-environmental programs make use of economic incentives to promote the adoption of agricultural methods that have less negative environmental effects. There may be unanticipated effects on the policy calculations utilized by yield practice standards for agriculture in the case that PA modifies the limitations and incentives that farmers experience.

17.4.1 Making decisions

Numerous multifaceted decision-making issues are involved in precision agriculture, most of which include significant levels of uncertainty and viable options. Numerous interrelated characteristics with very complex variability are taken into consideration while optimizing treatment procedures for crops and farmlands. The evaluation of the field as a whole, taking into account factors like crop yields, water availability, and chemical inputs, encourages suboptimal pesticide application and overuse, which in turn pollutes the environment. Now, precision agriculture technology has helped farmland treatment methods take a positive turn. As opposed to the earlier homogeneity assessment, farmland may thereby be seen as a hybrid entity having selective treatment. The technique is significantly increasing agricultural productivity, environmental sustainability challenges and practices. A deeper knowledge of how PA affects the limitations that producers face when making choices, as well as how those restrictions interact with the incentives brought about by environmental policy, might be beneficial for policymakers.

The majority of current studies are studying how PA ideas may be applied to a variety of commodities, including cotton, citrus, vegetables, peanuts, and animals. Effective operational choices are needed for the row crop production process, such as cultivar selection, planting season, depth, and population, planting rate, and when and how much fertilizer and pesticide to apply.

17.4.2 Impact of economics and the environment

To fulfill the food needs of the expanding population while reducing the related environmental implications, the food business and politicians must balance two primary opposing objectives. The effects are seen mostly in two areas: the runoff of agricultural fertilizers, which materially impairs water standards, and the greenhouse gases produced during farming production that significantly adds to worries about climate change. Through higher input utilization and equipment efficiency, PA technology offers a very strong chance to reduce the environmental effect while enhancing productivity and profitability in agricultural activities. Using navigational aids as a case study, which lessen numerous passes, overlap in agricultural machines, and ultimately use less fossil fuel and other inputs. The amount of detrimental overspill into waterways may be significantly reduced and expenses can be saved by changing the pesticides and fertilizers that are sprayed. PA improves productivity while changing the economics of food yield. Agro-environmental practices provide positive environmental effects in addition to the production techniques, making them as effective and sensible as the economic motivations they create. The policy measures implemented via accepted production techniques may have unforeseen repercussions anytime PA modifies the limitations and incentives available to producers.

Recent studies have concentrated on the potential connections between precision agriculture technologies, the farm's financial health, the environmental effects of the decisions made about on-farm activities, and the effectiveness of agro-environmental practices.

References

[1] A. Osseiran, O. Elloumi, J. Song and J. F. Monserrat, "Internet of Things," in IEEE Communications Standards Magazine, vol. 1, no. 2, pp. 84–84, 2017, doi: 10.1109/MCOMSTD.2017.7992936.

[2] P. T. Khanh, T. H. Ngọc, and S. Pramanik, Future of Smart Agriculture Techniques and Applications, in Advanced Technologies and AI-Equipped IoT Applications in High Tech Agriculture, Eds, A. Khang, IGI Global, 2023.

[3] T. H. Ngọc, P. T. Khanh and S. Pramanik, Smart Agriculture using a Soil Monitoring System, in Advanced Technologies and AI-Equipped IoT Applications in High Tech Agriculture, Eds, A. Khang, IGI Global, 2023.

[4] S. Pramanik, "An Effective Secured Privacy-Protecting Data Aggregation Method in IoT", in Achieving Full Realization and Mitigating the Challenges of the Internet of Things, Eds, M. O. Odhiambo and W. Mwashita, IGI Global, 2022, DOI: 10.4018/978-1-7998-9312-7.ch008

[5] S. Pramanik and S. Bandyopadhyay, Identifying Disease and Diagnosis in Females using Machine Learning, in Encyclopedia of Data Science and Machine Learning, Eds. John Wang, IGI Global, 2023, DOI: 10.4018/978-1-7998-9220-5.ch187.

[6] S. Pramanik and S. Bandyopadhyay, Analysis of Big Data, in Encyclopedia of Data Science and Machine Learning, Eds. John Wang, IGI Global, 2023, DOI: 10.4018/978-1-7998-9220-5.ch006

[7] R. Bansal, B. Jenipher, V. Nisha, Jain, Makhan R., Dilip, Kumbhkar, S. Pramanik, S. Roy and A. Gupta, "Big Data Architecture for Network Security", in Cyber Security and Network Security, Eds, Wiley, 2022, https://doi.org/10.1002/9781119812555.ch11

[8] T. T. H. Ngoc, S. Pramanik, P. T. Khanh, "The Relationship between Gender and Climate Change in Vietnam", The Seybold Report, 2023, DOI 10.17605/OSF.IO/KJBPT

[9] R. Jayasingh, J. Kumar R.J.S, D. B. Telagathoti, K. M. Sagayam and S. Pramanik, "Speckle noise removal by SORAMA segmentation in Digital Image Processing to facilitate precise robotic surgery", International Journal of Reliable and Quality E-Healthcare, vol. 11, issue 1, DOI: 10.4018/IJRQEH.295083, 2022

[10] S. Pramanik, K. M. Sagayam and O. P. Jena, Machine Learning Frameworks in Cancer Detection, ICCSRE 2021, Morocco, 2021.

[11] S. Pramanik, M. G. Galety, D. Samanta, N. P Joseph, Data Mining Approaches for Decision Support Systems, 3rd International Conference on Emerging Technologies in Data Mining and Information Security, 2022

[12] D. Samanta, S. Dutta, M. G. Galety and S. Pramanik, A Novel Approach for Web Mining Taxonomy for High-Performance Computing, The 4th International Conference of Computer Science and Renewable Energies (ICCSRE'2021), 2021, DOI:10.1051/e3sconf/202129701073.

[13] R. Bansal, A. J. Obaid, A. Gupta, R. Singh, S. Pramanik "Impact of Big Data on Digital Transformation in 5G Era", 2nd International Conference on Physics and Applied Sciences (ICPAS 2021), doi:10.1088/1742-6596/1963/1/012170, 2021.

[14] S. Pramanik, An Adaptive Image Steganography Approach depending on Integer Wavelet Transform and Genetic Algorithm, Multimedia Tools and Applications, 2023, https://doi.org/10.1007/s11042-023-14505-y.

[15] D. Pradhan, P. K.Sahu, N. S. Goje, H. Myo, M. M.Ghonge, M., Tun, R, Rajeswari and S. Pramanik, "Security, Privacy, Risk, and Safety Toward 5G Green Network (5G-GN)", in Cyber Security and Network Security,Eds, Wiley, 2022, DOI: 10.1002/9781119812555.ch9

[16] B. K. Pandey, D. Pandey, S. Wairya, G. Agarwal, P. Dadeech, S. R. Dogiwal and S. Pramanik, "Application of Integrated Steganography and Image Compressing Techniques for Confidential Information Transmission", in Cyber Security and Network Security, https://doi.org/10.1002/9781119812555.ch8, Eds, Wiley, 2022.

[17] A. Gupta, A. Verma and S. Pramanik, Security Aspects in Advanced Image Processing Techniques for COVID-19, in An Interdisciplinary Approach to Modern Network Security, S. Pramanik, A. Sharma, S. Bhatia and D. N. Le, Eds, CRC Press, 2022.

[18] K. Dushyant, G. Muskan, A. Gupta and S. Pramanik, "Utilizing Machine Learning and Deep Learning in Cyber security: An Innovative Approach", in Cyber security and Digital Forensics, M. M. Ghonge, S. Pramanik, R. Mangrulkar, D. N. Le, Eds, Wiley, 2022, https://doi.org/10.1002/9781119795667.ch12

[19] V. Veeraiah, D.J. Shiju, J.V.N. Ramesh, Ganesh Kumar R, S. Pramanik, A. Gupta, D. Pandey, Healthcare Cloud Services in Image Processing, Thrust Technologies' Effect on Image Processing, IGI Global, 2023

[20] Dhamodaran S, S. Ahamad, J.V.N. Ramesh, S. Sathappan, A. Namdev, R. R. Kanse, S. Pramanik, Fire Detection System Utilizing an Aggregate Technique in UAV and Cloud Computing, Thrust Technologies' Effect on Image Processing, IGI Global, 2023

[21] Dhamodaran S, S. Ahamad, J.V.N. Ramesh, G. Muthugurunathan, Manikandan K, S. Pramanik, D. Pandey, Thrust Technologies' Effect on Image Processing, IGI Global, 2023

[22] S. Ahamad, V. Veeraiah, J.V.N. Ramesh, R. Rajadevi, Reeja S R, S. Pramanik and A. Gupta, Deep Learning based Cancer Detection Technique, Thrust Technologies' Effect on Image Processing, IGI Global, 2023

[23] B. K. Pandey, D. Pandey, V. K. Nassa, A. S. George, S. Pramanik and P. Dadheech, Applications for the Text Extraction Method of Complex Degraded Images, in The Impact of Thrust Technologies on Image Processing, Nova Publishers, 2023.

[24] B. K. Pandey, D. Pandey, V. K. Nassa, A. S. Hameed, A. S. George, P. Dadheech and S. Pramanik, A Review of Various Text Extraction Algorithms for Images, in The Impact of Thrust Technologies on Image Processing, Nova Publishers, 2023

[25] S. Pramanik, "Carpooling Solutions using Machine Learning Tools", in Handbook of Research on Evolving Designs and Innovation in ICT and Intelligent Systems for Real-World Applications, K. K. Sarma, N. Saikia and M. Sharma, IGI Global, 2022, DOI: 10.4018/978-1-7998-9795-8.ch002.

[26] G. Taviti Naidu, KVB Ganesh, V. Vidya Chellam, S Praveenkumar, D. Dhabliya, S. Pramanik, A. Gupta, Technological Innovation Driven by Big Data, in "Advanced Bioinspiration Methods for Healthcare Standards, Policies, and Reform", Hadj Ahmed Bouarara IGI Global, 2023, DOI: 10.4018/978-1-6684-5656-9

[27] A. Gupta, A. Verma, S. Pramanik, "Advanced Security System in Video Surveillance for COVID-19", in An Interdisciplinary Approach to Modern Network Security, S. Pramanik, A. Sharma, S. Bhatia and D. N. Le, CRC Press, 2022.

[28] A. Mandal, S. Dutta, S. Pramanik, "Machine Intelligence of Pi from Geometrical Figures with Variable Parameters using SCILab", in Methodologies and Applications of Computational Statistics for Machine Learning, D. Samanta, R. R. Althar, S. Pramanik and S. Dutta,

Eds, IGI Global, 2021, pp. 38-63, DOI: 10.4018/978-1-7998-7701-1.ch003

[29] Y. Meslie, W. Enbeyle, B. K. Pandey, S. Pramanik, D. Pandey, P. Dadeech, A. Belay, A. Saini, "Machine Intelligence-based Trend Analysis of COVID-19 for Total Daily Confirmed Cases in Asia and Africa", in Methodologies and Applications of Computational Statistics for Machine Learning, D. Samanta, R. R. Althar, S. Pramanik and S. Dutta, Eds, IGI Global, 2021, pp. 164-185, DOI: 10.4018/978-1-7998-7701-1.ch009

[30] A. Bhattacharya, A. Ghosal, A. J. Obaid, S. Krit, V. K. Shukla, K. Mandal and S. Pramanik, "Unsupervised Summarization Approach with Computational Statistics of Microblog Data", in Methodologies and Applications of Computational Statistics for Machine Learning, D. Samanta, R. R. Althar, S. Pramanik and S. Dutta, Eds, IGI Global, 2021, pp. 23-37, DOI: 10.4018/978-1-7998-7701-1.ch002

[31] P. K. Mall, S. Pramanik, S. Srivastava, M. Faiz, S. Sriramulu and M. N. Kumar, FuzztNet-Based Modelling Smart Traffic System in Smart Cities Using Deep Learning Models, in Data-Driven Mathematical Modeling in Smart Cities, IGI Global, 2023, DOI: 10.4018/978-1-6684-6408-3.ch005

[32] D. Mondal, A. Ratnaparkhi, A. Deshpande, V. Deshpande, A. P. Kshirsagar and S. Pramanik, Applications, Modern Trends and Challenges of Multiscale Modelling in Smart Cities, in Data-Driven Mathematical Modeling in Smart Cities, IGI Global, 2023, DOI: 10.4018/978-1-6684-6408-3.ch001

[33] S. Praveenkumar, V. Veeraiah, S. Pramanik, S. M. Basha, A. V. Lira Neto, V. H. C. De Albuquerque and A. Gupta, Prediction of Patients' Incurable Diseases Utilizing Deep Learning Approaches, ICICC 2023, Springer

18

Utilizing Smart Farming Methods to Reduce Water Scarcity

Upasana[1], Rajeev Kumar[2], R. Raghavendra[3], Sachin Tripathi[4], Rushil Chandra[4], Dharmesh Dhabliya[5], and Thanh Bui[6]

[1]College of Agriculture Science, Teerthanker Mahaveer University, India
[2]Department of Agriculture, Sanskriti University, India
[3]Department of Computer Science and IT, School of CS and IT, Jain (Deemed to be) University, India
[4]Symbiosis Law School, Nagpur Campus, Symbiosis International (Deemed University), India
[5]Department of Information Technology, Vishwakarma Institute of Information Technology, India
[6]Data Analytics and Artificial Intelligence Laboratory, School of Engineering and Technology, Thu Dau Mot University, Vietnam

Email: upasana35954@gmail.com; rajeev.ag@sanskriti.edu.in; r.raghavendra@jainuniversity.ac.in; sachintripathi@slsnagpur.edu.in; rushilchandra@slsnagpur.edu.in; dharmesh.dhabliya@viit.ac.in; hungbt.cntt@tdmu.edu.vn

Abstract

Numerous nations are now experiencing a serious water crisis as a consequence of the limited water resources that exist owing to rising demand brought on by the exponential population growth that has exacerbated the need for food and industrial products. Irrigation uses around 70% of the water available. As a consequence, untreated wastewater is often utilized for irrigation as well, posing further risks to human health. According to a number of studies, using information technology to improve irrigation methods may help farmers use less water. Through the use of information technology, smart farming methods have been developed that may assist the farmer in managing water supplies, decreasing water waste, and even measuring the quality of the water. Intelligent agricultural practices may assist in resolving the world's largest crises and in achieving the objectives of sustainable development.

18.1 Introduction

The most crucial component of life on earth is water, and it is a well-known truth that the amount of potable water on our blue globe is dwindling. 9.7 billion people are expected to live in water-deficient regions by 2050, while 4.2 billion should experience "severe" water shortages. The use of water has multiplied in the twenty-first century as a result of population growth, greater agricultural production for human consumption and industrial demands, and expanded industrialization. By 2050, there will likely be a further rise in consumption. According to the NITI Aayog's Composite Water Resources Management Report, 600 million Indians are now experiencing a high to severe water crisis. The main obstacle to sustainable living, human health, and ecosystem function has emerged as a lack of water supplies. According to [1], by 2025, roughly one-fourth of the world's population would reside in regions with severe water shortages.

In order to achieve the sustainable development objectives of "zero hunger" or the eradication of poverty, water is a vital resource. Water use has grown globally as a result of climate change and the current need to ensure food security. The shortage of water is becoming apparent as a consequence. Growing urbanization has reduced groundwater recharge, resulting in irreversible groundwater extraction and a variety of problems for earth life.

One of the main causes of water waste is conventional farming, which accounts for 70% of water use, of which 90% may be avoided by utilizing the right methods. Therefore, the most crucial element when discussing water conservation is to have planned system and modern irrigation technology that uses comparatively less water. This may improve access to water, reduce soil erosion brought on by water logging and runoff, boost agricultural productivity, and help cater for the estimated 10 billion people by 2040.

Growing crops in accordance with the local outskirts, particularly in dry and semi-dry regions, adopting drip and spray irrigation systems, and keeping an eye on soil texture, water retention capability, and soil water may all help to boost water usage efficiency.

18.2 Agricultural Landscape of India

With an estimated value of $390 billion, agriculture is one of the most major economic sectors in India, contributing around 17% of the country's GDP.

India has a population that is mostly dependent on agriculture. Even though India is the world's top producer of several crops, there remains room to increase production to support sustainable development objectives.

In India, monsoons have a significant impact on irrigation. Rainfall is unpredictable, erratic, and uneven. India receives 80% of its annual rainfall in only four months, from June to October. For the next eight months, there is an urgent need for consistent, effective, and dependable irrigation. Water is often utilized for both industrial and agricultural purposes in most parts of the globe. Additionally, it plays a significant role in wetlands and other ecosystems that are very valuable to humans. Water losses from irrigation may occur in a variety of ways. Before reaching the plants, water may leak out of tanks or transmission canals. Following the application of water to field plants, part of it may either flow off the field entirely or enter the groundwater system, making it unavailable to the crop's roots.

The quality of the ground water in the area has deteriorated. Herbicides, pesticides, synthetic manures, and urban and contemporary wastes have permeated the soil, infiltrated some aquifers, and worsened surface-water standards. Sewer overflows, faulty septic tank operation, and landfill leachates are further pollution problems. In certain seashore areas, increased freshwater extraction from aquifers has caused saltwater intrusion. Water purification is essential in this way. Waste water treatment aims to make it safer to drink or use in agricultural areas by removing pathogens and other natural and inorganic contaminants.

Reducing BOD, COD, nitrate levels, and other contaminants is necessary for water treatment. Only 20% of wastewater worldwide receives adequate treatment [38]. Treatment limits depend on national economic conditions; as a result, treatment limits in high-income countries are 70% of generated wastewater, while they are only 8% in low-income countries.

Agriculture in India is facing difficulties. An agriculturist fed 30 people in 1960. Currently, an agriculturist caters for 200 people, thanks to a biotech revolution that occurred in the majority of nations. According to "Digital farming [2–6]: assisting to cater to a growing world," 2019; this can only be done today with a digital agricultural revolution, a farmer will reportedly need to feed 265 people by the year 2050. To be able to feed the growing population and end hunger, agricultural businesses urgently require technology advancements. To prevent lock-ins, it is crucial to start this revolution as soon as possible. Both proponents of technology and opponents must work together to ensure the survival of the human species. Reduced farming's ecological imprint is also urgently needed. Inputs like fertilizers and insecticides should be used sparingly or specifically to a given location in precision agricultural systems to reduce leaching issues and greenhouse gas emissions. More than 40% of the world's population now experiences some type of water scarcity, even if there isn't a worldwide water crisis as of yet.

Indian farmers now use agricultural methods that are neither ecologically nor economically viable in the long run. The Indian Agricultural System is characterized by inadequate irrigation systems and a shortage of technology in usage; the issues it faces may be succinctly divided into the following areas.

1. Low infrastructure: India's infrastructure, both industrial and agricultural, is of low quality. Systems for irrigation are not very advanced. Our farmers cannot be given access to cold storage or adequate roads because we lack such facilities. Farmers' tools are of low quality, which makes them ineffective. All of the fields cannot be efficiently irrigated by irrigation pumps.
2. Modern agricultural methods and technological advancements are not what is required to feed India's expanding population, which is further constrained by ignorance and illiteracy. There is a lack of adoption of technologies like IoT, which studies suggest might be very helpful to agriculture.
3. Water is distributed in an inefficient and unfair manner. The already subpar infrastructure is becoming worse. Water is a limited resource that has to be properly and effectively managed something that our agricultural industry and policymakers have shown they are unable to achieve.
4. Irrigation facilities are insufficient since farmers still depend mainly on the monsoon and rainfall for irrigation. The weather has a significant impact on whether the harvest will be excellent or bad, hence there is a significant risk component in India's agricultural sector.
5. There has to be a decrease in the ecological impact of Indian agriculture. In order to increase agricultural output, environmentally harmful herbicides and fertilizers are utilized. Such pesticide usage causes leaching issues and increases greenhouse gas emissions from agriculture.

Not all of the food which is produced in Sri Lanka reaches its eaters. According to a CNN article, around one-third of every food produced in Sri Lanka goes bad as the agriculturist is never given adequate support to market his production. The aforementioned issues are which contribute to the high rate of farmer suicide in India. According to Accidental Deaths & Suicides in Sri Lanka, agriculturist suicide makes up 15% of cumulative suicides.

18.2.1 Adoption of mitigation strategies

To end the starvation and penury that Sri Lanka is now experiencing, the growth of agriculture is a necessary step. For the development of agriculture, bold and forceful efforts must be done. Newer technology which concentrates on improving food output from crops is urgently needed. Farming's ability to expand sustainably depends on cutting-edge technologies that are currently being developed, including nanotechnology, the IoTs, smart farming, precision farming, remote sensing, global positioning system [7–10], and GNSS [11–15]. Researchers with expertise in engineering and agriculture have been working to cut down on water waste while irrigating plants. The expansion of the IoTs has caused creation of an amount of modules for intelligent field irrigation.

1. The Indian government is recognizing where it can contribute to the construction of infrastructure. It need not dominate each industry, as it did in telecommunications once it was privatized. Results are grasped into consideration, along with consideration for expectations based on vows committed. The government is concentrating on the creation of HYV seeds, novel crop hybrids, vaccine research to help fight agricultural illnesses, and the improvement of crop quality. Drip irrigation is becoming more popular since it is less expensive, which reduces water use. Better seeds and fertilizers are employed, having several MNCs taking part in a significant part in seed issuance. In addition to loans from local financial institutions, the World Bank approved a $3.8 billion loan for India's agricultural industry to assist it expand.
2. In [16], the researcher discusses how the IoTs has altered automation and the way it may be utilized more effectively for remote data tracking and sensing, whether in homes, businesses, farms, or other locations. The article notably discusses the Internet of Things platform Blynk, the way it has a common UPI for every devices, which facilitates evolution, offers a wide variety of widget kinds for a better user experience, and has various connection capabilities. When developing an irrigation system using a microcontroller, the author of [17] employs a Message Queue Telemetry Transport protocol to facilitate transmission between a sensor and the microcontroller.
3. In [18], the researchers discuss which factors are crucial for deciding when plants must be watered, how droughts and floods may be forecast by closely observing soil moisture, and how the effects of climate change might be mitigated. The author shows a thorough analysis on when and how to irrigate in order to utilize water as efficiently as

possible. The author of [19] discusses how irrigation systems may be created utilizing inexpensive microcontrollers and other DIY moisture sensors.

4. In order to enhance cultivation, the authors of [20] developed a drip irrigation system using an Arduino Uno microcontroller, LCD panels, pH sensors, and humidity sensors. In [21], the researcher discusses the advantages of low-data-rate wireless sensor networks for agricultural applications over more conventional instrumentation depending on wired, discrete solutions. The researcher measures the moisture content of the soil using a dual probe, heat-pulse dependent humidity sensor, and then sends the information to a base station for analysis to carry out local management duties. For the system suggested in this model, an efficient power supply provided a significant challenge. In [22], the researcher discusses how formerly, WSNs were quite costly, which made it difficult to find a practical way to improve irrigation. An algorithm was created to address deficiencies in the Zigbee module after an irrigation system was designed to indicate them.
5. It is greatly recognized that modern farming has a significant negative influence on the environment. It has a significant impact on the emissions of carbon dioxide, nitrous oxide, and methane. It is time to stop preparing ground using ploughs, a method that has been used for centuries. Improvements in irrigation are being made. The right irrigation system must be used in order to reduce water waste and running expenses. Testing the soil's moisture often may help with this.

With exports rising from little over $5 billion in 2005 to a record of more than $40 billion in 2015, Sri Lanka has recently established itself as a major supplier of agricultural products. In 2013, Sri Lanka surpassed India to become the globe's seventh largest exporter of food items. The rapid increase in shipments, the most of which are headed for developing countries, has been greatly facilitated by the Sri Lankan government's assistance for both yield and exports.

18.3 Global Usage of Technologies for Improving the Agricultural Technique

The growth of the farming business has benefited greatly from technology. It has enabled crops to flourish even in desolate locations. Modern plants have been bred to endure droughts and severe circumstances. In order to make plants stronger and more resistant to pests and droughts, genetic engineering has made it feasible to modify plant DNA. Apps for smart phones are used by farmers to determine how much grass is there in a field. They are able to determine if there is adequate forage for cattle, which helps them save time and money. Every aspect of agriculture now uses electricity.

An agricultural management approach focused on seeing, estimating, and reacting to inter- and intra-farm volatility, as well as various different factors in crops, is known as precision agriculture (PA). The majority of PA research focuses on developing a decision support system for improved farm management having the target of maximizing repaying on input resources and also protecting the resources engaged. The introduction of innovations like GPS and precision agriculture has advanced thanks to the navigation satellite system (GNSS). Farmers can precisely locate plants and field parameters utilizing GPS [23–26] and GNSS, allowing them to compute the spatial variability of as many variables as possible (including crop yield, topography and terrain features, organic matter content, moisture and N levels, pH, EC, K). Precision farming has greatly benefited from the emergence of drones such as the DJI Phantom, which are accessible to beginning pilots and reasonably priced.

Precision farming is best summed up by the term "Right Place, Right Time, Right Product." PA is a more precise, regulated technology that substitutes labor-intensive, repetitive parts of farming. It gives farmers improved information regarding, among other things, harvesting, plant rotation, nutrient management, and insect assaults. The objectives of PA are profitability, efficiency, and sustainability. The most recent innovations that have transformed agriculture are discussed in this section.

18.3.1 Network of things

All facets of human existence are being favorably impacted by the Internet of Things (IoT) revolution. Any gadget with a power switch and an on/off switch will ultimately be connected to the Internet and, by extension, to other devices. Everything you can imagine will someday be able to link to the Internet, including coffee makers, lights, wearable technology, headphones, and washing machines. Eventually, the Internet will also be able to link to machine sections like an airplane's jet engine or an oil rig's drill. Research and development are concentrating on enhancing Internet connection to various traditionally dumb or non–Internet-based physical gadgets and common things along with methodical devices like desktops, laptops, smartphones, and tablets. As a result, the gadgets can now be remotely managed and used to transfer data to any other device worldwide that is linked to the Internet. The concept of a network of smart devices was originally suggested and studied in 1982, when a modified Coke machine at MIT became the initial device to be linked to the Internet and report its inventory and the temperature of freshly filled beverages. The Computer of the 21st Century, a 1991 study by Mark Weiser on ubiquitous computing, developed the modern conception of the Internet of Things. Much of what we do now is feasible as a result of this research.

The increased usage of ICT [27–30] and the IoT has sparked a revolution in agriculture all over the globe. Robotic, self-driving cars are now being developed and improved continuously in order to enhance resource management. With the advancement of UAVs with autonomous flight control, farmers now have real-time access to information on biomass growth and crop fertilization status (albeit this is a very costly technology). According to [31], smart agriculture is the combination of yield computation, management of human and material resources, procurement and expenses, risk management logistics, and support into a lone farm and the intelligent decision-making on the aforementioned factors. Through sensor technologies and actuators, digitalization is having a greater impact on agriculture than only the conventional agricultural methods. It is also changing the animal economy and various facets of farming. Sensors, GPS, GNSS, and other basic technologies are used in smart farming. Incredible benefits from IoT-based farming include more efficient water usage, as well as improved information and pharmaceuticals.

18.3.2 Distance sensing

Since the 1960s, remote sensing has been developed for application in agriculture. Remote sensing may be a very useful technique for managing and monitoring land, water, and other resources. It may be utilized to compute the soil's moisture constituent and detect matters which could be pressuring a crop at a particular instant of time. This information enhances farming decisions and may be obtained from a variety of sources, including as satellites, drones, and even inexpensive sensors. Real-time comprehension of present agricultural, forest, or other fields is made possible by sensors. High-resolution crop sensors notify application equipment of the proper quantity required, as opposed to advising field fertilization prior to application. Crop health may be determined over the whole field using optical sensors or drones.

18.3.3 Huge data

These days, farmers use drones to manage their farms and regulate plant development. Drones [32–35] are utilized to photograph farms and gather data on the whole farming region. To be able to graphically depict data, such as digital field maps, this data is then connected to a database. This data may also include additional information from other measures, such as infrared pictures, biomass distribution, and meteorological information. Additional components of big data include the administration, management, and analysis of large data. Algorithms for making decisions are used with this data to automatically decide on management measures. Because of these modifications, further automation is anticipated in the future.

18.3.4 Utilizing the cloud

To assist farmers manage their crops more effectively, cloud computing may be utilized to aggregate data from instruments like soil moisture monitors, satellite photos, and weather stations. We can better comprehend agricultural output thanks to the analytical advantages of cloud computing. The agricultural business is using data more effectively than ever. Field-specific factors, plants, potential weather events, equipment utilized in fields, and other sensor-based inputs are all the subject of data collection. Then, this data is saved on the cloud, giving customers the speed, storage, and computing capacity they need to make smarter choices and analyze the data acquired in a manner that will be helpful to farmers. Farmers can locate precisely where extra fertilizer or water is required by looking at the map of their fields and crops. These cloud-based solutions aid in the resolution of agricultural issues and boost the effectiveness and production of farms by offering automation and decision assistance.

18.3.5 Applied ML and AI

AI and ML are being quickly used in many facets of agriculture. Farmers are given access to chatbots to keep them technologically current, including information on when and how to water plants among other things. Microsoft is now providing consultancy services for sowing, land, fertilizer, irrigation, and other issues to more than 150 farmers in India. According to [36], this strategy has already led to an average 30% increase in output per hectare compared to previous year. Farming benefits from artificial intelligence because it gives us access to decision-support tools that improved the selection of the best agronomic product combinations, the identification of disease indicators, the identification of crop preparedness, the automation of processes, the monitoring of crop health indicators, and other choices.

To identify different kinds of crops' degrees of stress, machine learning models may be trained on photos of plants. For smarter farming, the whole process may be divided into the steps of identification, classification, quantification, and prediction. Higher agricultural yields may be achieved by combining AI and ML.

Our farmers have benefited from robotics in digital farming as well. Plant geometry and non-visible radiation bands are measured using mobile and field robots. The recommendation engine uses this data to help grow crops with greater yields and better quality while using fewer resources. Farmers are guided by installed apps while they plant and cultivate their crops.

In [37], the researchers create a model for large-scale data analytics and event estimation, enabling seamless interoperability between sensors, various services, operations, techniques, and farmers, and linked open datasets and streams on the Internet.

18.3.6 Other Aerial Unmanned Vehicles and Drones

According to a current research, the market for drones utilized in farming is worth more than $40 billion. The authors have simplified farming and are very important for managing adverse weather conditions, productivity improvements, yield management, etc. To better grasp the finer aspects of the landscape, 3D maps of farms are being created using drones. Drone usage also allows for better control of soil nitrogen levels. Drones may spray water and other resources essential for plant development across expansive landscapes. They are a tremendous asset for monitoring and determining the health of crops. Drones with high-resolution cameras gather precise field photos that can be processed via a convolution neural network to detect weed growth, crops that need watering, and the stress level of plants in the middle of their development cycle. Drone aircraft are fast becoming into a highly useful tool for agriculturists. Drone imaging data may be a reliable predictor of crop and canopy stress vigor levels. Drone sensors that record a variety of characteristics enable farmers to identify which portions of the field need fertilizer and water, and to provide those nutrients exclusively there. Thus, under such circumstances, crop yield and growth significantly rise. Commercial drones are widely accessible and may reduce planting expenses by 85% while achieving an adoption rate of 75%.

18.4 Conclusion

The issue of effective irrigation is among the most difficult ones that agricultural experts must deal with. Water is a finite resource, and much of it is difficult to get close to crops. In order to feed the world's population, which is always expanding, water has to be managed correctly. Irrigation systems must be created scientifically right away. Thus, precision agriculture may profit from the usage of current, relevant technologies. Plant irrigation is made simpler and more effective by using Internet of Things principles. Farmers may significantly minimize their consumption of water and labor by employing technology to remotely turn on and off water pumps, manage water use, and check soil moisture levels via the cloud. By creating a smart irrigation system that uses data from several sensors to make decisions about when to turn on an actuator, irrigation may be improved in the future. The apparatus may use a combined AI and ML may be utilized to make irrigation systems which are smart and may have real-time decisions from the information supplied in the knowledge base be given local weather data so that it may detect when water will not be required, thereby best managing water resources. Using drones and data from artificial intelligence and machine learning-dependent approaches, it is possible to irrigate plants from the air. This data may be kept in the cloud for greater analysis and intelligence, which would also enable visualization.

References

[1] S. Sharma, N. Dhanda and R. Verma, "Urban Vertical Farming: A Review," 2023 13th International Conference on Cloud Computing, Data Science & Engineering (Confluence), Noida, India, 2023, pp. 432–437, doi: 10.1109/Confluence56041.2023.10048883.

[2] P. T. Khanh, T. H. Ngọc, and S. Pramanik, Future of Smart Agriculture Techniques and Applications, in Advanced Technologies and AI-Equipped IoT Applications in High Tech Agriculture, Eds, A. Khang, IGI Global, 2023.

[3] T. H. Ngọc, P. T. Khanh and S. Pramanik, Smart Agriculture using a Soil Monitoring System, in Advanced Technologies and AI-Equipped IoT Applications in High Tech Agriculture, Eds, A. Khang, IGI Global, 2023.

[4] Reepu, S. Kumar, M. G. Chaudhary, K. G. Gupta, S. Pramanik and A. Gupta, Information Security and Privacy in IoT, in Handbook of Research in Advancements in AI and IoT Convergence Technologies,

Eds, J. Zhao, V. V. Kumar, R. Natarajan and T. R. Mahesh, IGI Global, 2023.

[5] V. Veeraiah, V. Talukdar, Manikandan K., S. B. Talukdar, V. D. Solavande, S. Pramanik, A. Gupta, Machine Learning Frameworks in Carpooling, in Handbook of Research on AI and Machine Learning Applications in Customer Support and Analytics, Eds, Md S. Hossain, R. C. Ho and G. Trajkovski, IGI Global, 2023.

[6] S. Pramanik, "An Effective Secured Privacy-Protecting Data Aggregation Method in IoT", in Achieving Full Realization and Mitigating the Challenges of the Internet of Things, Eds, M. O. Odhiambo and W. Mwashita, IGI Global, 2022, DOI: 10.4018/978-1-7998-9312-7.ch008

[7] S. Pramanik and S. Bandyopadhyay, Identifying Disease and Diagnosis in Females using Machine Learning, in Encyclopedia of Data Science and Machine Learning, Eds. John Wang, IGI Global, 2023, DOI: 10.4018/978-1-7998-9220-5.ch187

[8] . S. Pramanik, An Adaptive Image Steganography Approach depending on Integer Wavelet Transform and Genetic Algorithm, Multimedia Tools and Applications, 2023, https://doi.org/10.1007/s11042-023-14505-y.

[9] R. Jayasingh, J. Kumar R.J.S, D. B. Telagathoti, K. M. Sagayam and S. Pramanik, "Speckle noise removal by SORAMA segmentation in Digital Image Processing to facilitate precise robotic surgery", International Journal of Reliable and Quality E-Healthcare, vol. 11, issue 1, DOI: 10.4018/IJRQEH.295083, 2022

[10] R. Ghosh, S. Mohanty, P. Pattnaik and S. Pramanik, "A Performance Assessment of Power-Efficient Variants of Distributed Energy-Efficient Clustering Protocols in WSN", International Journal of Interactive Communication Systems and Technologies, vol. 10, issue 2, article 1, 2020, DOI: 10.4018/IJICST.2020070101

[11] S. Pramanik, M. G. Galety, D. Samanta, N. P Joseph, Data Mining Approaches for Decision Support Systems, 3rd International Conference on Emerging Technologies in Data Mining and Information Security, 2022.

[12] S. Pramanik, K. M. Sagayam and O. P. Jena, Machine Learning Frameworks in Cancer Detection, ICCSRE 2021, Morocco, 2021.

[13] D. Samanta, S. Dutta, M. G. Galety and S. Pramanik, A Novel Approach for Web Mining Taxonomy for High-Performance Computing, The 4th International Conference of Computer Science and Renewable Energies (ICCSRE'2021), 2021, DOI:10.1051/e3sconf/202129701073.

[14] R. Bansal, A. J. Obaid, A. Gupta, R. Singh, S. Pramanik "Impact of Big Data on Digital Transformation in 5G Era", 2nd International Conference

on Physics and Applied Sciences (ICPAS 2021), doi:10.1088/1742-6596/1963/1/012170, 2021.

[15] S. Praveenkumar, V. Veeraiah, S. Pramanik, S. M. Basha, A. V. Lira Neto, V. H. C. De Albuquerque and A. Gupta, Prediction of Patients' Incurable Diseases Utilizing Deep Learning Approaches, ICICC 2023, Springer

[16] Z. Zhu, K. Lin, A. K. Jain and J. Zhou, "Transfer Learning in Deep Reinforcement Learning: A Survey," in IEEE Transactions on Pattern Analysis and Machine Intelligence, doi: 10.1109/TPAMI.2023.3292075.

[17] A. Tiwari, B. Masram and K. Bharatdwaj, "Message Queuing Telemetry Transport Based Data Logger," 2023 IEEE 8th International Conference for Convergence in Technology (I2CT), Lonavla, India, 2023, pp. 1–4, doi: 10.1109/I2CT57861.2023.10126489.

[18] N. Rathnayake, U. Rathnayake, I. Chathuranika, T. L. Dang and Y. Hoshino, "Projected Water Levels and Identified Future Floods: A Comparative Analysis for Mahaweli River, Sri Lanka," in IEEE Access, vol. 11, pp. 8920–8937, 2023, doi: 10.1109/ACCESS.2023.3238717.

[19] V. Vidya Chellam, V. Veeraiah, A. Khanna, T. H. Sheikh, S. Pramanik and D. Dhabliya, A Machine Vision-based Approach for Tuberculosis Identification in Chest X-Rays Images of Patients, ICICC 2023, Springer

[20] A. Mandal, S. Dutta, S. Pramanik, "Machine Intelligence of Pi from Geometrical Figures with Variable Parameters using SCILab", in Methodologies and Applications of Computational Statistics for Machine Learning, D. Samanta, R. R. Althar, S. Pramanik and S. Dutta, Eds, IGI Global, 2021, pp. 38–63, DOI: 10.4018/978-1-7998-7701-1.ch003

[21] Y. Meslie, W. Enbeyle, B. K. Pandey, S. Pramanik, D. Pandey, P. Dadeech, A. Belay, A. Saini, "Machine Intelligence-based Trend Analysis of COVID-19 for Total Daily Confirmed Cases in Asia and Africa", in Methodologies and Applications of Computational Statistics for Machine Learning, D. Samanta, R. R. Althar, S. Pramanik and S. Dutta, Eds, IGI Global, 2021, pp. 164–185, DOI: 10.4018/978-1-7998-7701-1.ch009

[22] A. Bhattacharya, A. Ghosal, A. J. Obaid, S. Krit, V. K. Shukla, K. Mandal and S. Pramanik, "Unsupervised Summarization Approach with Computational Statistics of Microblog Data", in Methodologies and Applications of Computational Statistics for Machine Learning, D. Samanta, R. R. Althar, S. Pramanik and S. Dutta, Eds, IGI Global, 2021, pp. 23–37, DOI: 10.4018/978-1-7998-7701-1.ch002

[23] Dhamodaran S, S. Ahamad, J.V.N. Ramesh, G. Muthugurunathan, Manikandan K, S. Pramanik, D. Pandey, Thrust Technologies' Effect on Image Processing, IGI Global, 2023

[24] Dhamodaran S, S. Ahamad, J.V.N. Ramesh, S. Sathappan, A. Namdev, R. R. Kanse, S. Pramanik, Fire Detection System Utilizing an Aggregate Technique in UAV and Cloud Computing, Thrust Technologies' Effect on Image Processing, IGI Global, 2023
[25] V. Veeraiah, D.J. Shiju, J.V.N. Ramesh, Ganesh Kumar R, S. Pramanik, A. Gupta, D. Pandey, Healthcare Cloud Services in Image Processing, Thrust Technologies' Effect on Image Processing, IGI Global, 2023
[26] S. Ahamad, V. Veeraiah, J.V.N. Ramesh, R. Rajadevi, Reeja S R, S. Pramanik and A. Gupta, Deep Learning based Cancer Detection Technique, Thrust Technologies' Effect on Image Processing, IGI Global, 2023
[27] G. Taviti Naidu, KVB Ganesh, V. Vidya Chellam, S Praveenkumar, D. Dhabliya, S. Pramanik, A. Gupta, Technological Innovation Driven by Big Data, in "Advanced Bioinspiration Methods for Healthcare Standards, Policies, and Reform", Hadj Ahmed Bouarara IGI Global, 2023, DOI: 10.4018/978-1-6684-5656-9
[28] S. Pramanik, "Carpooling Solutions using Machine Learning Tools", in Handbook of Research on Evolving Designs and Innovation in ICT and Intelligent Systems for Real-World Applications, K. K. Sarma, N. Saikia and M. Sharma, IGI Global, 2022, DOI: 10.4018/978-1-7998-9795-8.ch002.
[29] B. K. Pandey, D. Pandey, V. K. Nassa, A. S. George, S. Pramanik and P. Dadheech, Applications for the Text Extraction Method of Complex Degraded Images, in The Impact of Thrust Technologies on Image Processing, Nova Publishers, 2023.
[30] B. K. Pandey, D. Pandey, V. K. Nassa, A. S. Hameed, A. S. George, P. Dadheech and S. Pramanik, A Review of Various Text Extraction Algorithms for Images, in The Impact of Thrust Technologies on Image Processing, Nova Publishers, 2023
[31] P. K. Mall, S. Pramanik, S. Srivastava, M. Faiz, S. Sriramulu and M. N. Kumar, FuzztNet-Based Modelling Smart Traffic System in Smart Cities Using Deep Learning Models, in Data-Driven Mathematical Modeling in Smart Cities, IGI Global, 2023, DOI: 10.4018/978-1-6684-6408-3.ch005
[32] S. Pramanik, AJ Obaid, N M, and SK Bandyopadhyay "Applications of Big Data in Clinical Applications", Al-Kadhum 2nd International Conference on Modern Applications of Information and Communication Technology, *AIP Conference Proceedings* 2591, 030086 (2023), https://doi.org/10.1063/5.0119414

[33] S. Pramanik, Niranjanamurthy M. and S. N. Panda, Using Green Energy Prediction in Data Centers for Scheduling Service Jobs, ICRITCSA 2022, Bengaluru, 2022.

[34] S. Pramanik and S. Bandyopadhyay, Analysis of Big Data, in Encyclopedia of Data Science and Machine Learning, Eds. John Wang, IGI Global, 2023, DOI: 10.4018/978-1-7998-9220-5.ch006

[35] R. Bansal, B. Jenipher, V. Nisha, Jain, Makhan R., Dilip, Kumbhkar, S. Pramanik, S. Roy and A. Gupta, "Big Data Architecture for Network Security", in Cyber Security and Network Security, Eds, Wiley, 2022, https://doi.org/10.1002/9781119812555.ch11

[36] S. Kareer, N. Yokoyama, D. Batra, S. Ha and J. Truong, "ViNL: Visual Navigation and Locomotion Over Obstacles," 2023 IEEE International Conference on Robotics and Automation (ICRA), London, United Kingdom, 2023, pp. 2018–2024, doi: 10.1109/ICRA48891.2023.10160612.

[37] A. Gounaris, A. -V. Michailidou and S. Dustdar, "Toward Building Edge Learning Pipelines," in IEEE Internet Computing, vol. 27, no. 1, pp. 61–69, 1 Jan.–Feb. 2023, doi: 10.1109/MIC.2022.3171643.

[38] T. A. M. Euzébio, M. A. P. Ramirez, S. F. Reinecke and U. Hampel, "Energy Price as an Input to Fuzzy Wastewater Level Control in Pump Storage Operation," in IEEE Access, vol. 11, pp. 93701–93712, 2023, doi: 10.1109/ACCESS.2023.3310545.

19

Real-time Irrigation Optimization for Horticulture Crops using WSN, APSim, and Communication Models

Namit Gupta[1], Umesh Kumar Mishra[2], Taskeen Zaidi[3], Vaidehi Pareek[4], Karun Sanjaya[4], Sabyasachi Pramanik[5], and Ofer Hadar[6]

[1]College of Computing Science and Information Technology, Teerthanker Mahaveer University, India
[2]Department of Agriculture, Sanskriti University, India
[3]Department of Computer Science and IT, School of CS and IT, Jain (Deemed to be) University, India
[4]Symbiosis Law School, Nagpur Campus, Symbiosis International (Deemed University), India
[5]Department of Computer Science and Engineering, Haldia Institute of Technology, India
[6]Ben-Gurion University of the Negev, Israel

Email: namit.k.gupta@gmail.com; umeshm.ag@sanskriti.edu.in; t.zaidi@jainuniversity.ac.in; vaidehipareek@slsnagpur.edu.in; karunsanjaya@slsnagpur.edu.in; sabyalnt@gmail.com; hadar @bgu.ac.il

Abstract

Prescriptive agriculture is a new area of study in the field of agriculture that has emerged as a result of the usage of WSNs, the IoTs, and sophisticated technology. Precision agriculture (PA) that involves monitoring, computing, and reacting to both inter- and intra-farm variability of agricultural fields, is enforced by prescriptive agriculture. In this chapter, it is discussed how the development of WSN, APSim, and communication models sparked a new approach to irrigation optimization in the agricultural industry. Sensors are configured to gather data on environmental variables including temperature and relative humidity, with the data being sent to a server through a GSM module. In order to assess the datasets for growing crops, inter- and intra-farm field circumstances were taken into account. Finally, an approach created by way2SMS and WebHost server decimates irrigation-related information.

19.1 Introduction

In agriculture across the globe, almost 65% of all clear water is used for countryside irrigation. Nearly every part of farming, including determining the crop's production, depends on water. If the plants are not properly watered, even the best seeds and the most fertilizer cannot grow to their full potential. Due to decreasing rainfall and rising food demand to feed the huge population, water shortage is deteriorating day by day. Modern development calls for increasing benefit per unit of land in order to satisfy a large number of people's basic needs in a zone with less development. Ranchers use tried-and-true methods like trickle irrigation systems, efficient use of preparation, crop planning, and other strategies to achieve this goal.

A plant's developmental stages are influenced by a variety of factors, including environmental factors, soil quality, and other associated parameters. Water makes a plant grow, but it also gives its tissue cells a key role via photosynthesis and the absorption of nutrients. The water system is a method of delivering water remotely to improve the conditions necessary for plants to grow; ultimately, plants get water from sources including rainfall, air humidity, and soil. By influencing the availability of various supplements for consumption, the environment of social activity, and other factors, the effective use of the water system aids in the legal outcome of harvesting.

The development of wireless sensors pushed cultivation via improved technology in a new direction. According to [1], WSN signifies spatially dispersed remote organization devices which utilize the sensor to continually track the physical limits of the climate. In this area, innovation and arrangements are being used to provide the best option for data collection and management in order promote water system via factual programming. Additionally, the frightening ecological change and water limitation need more farming methods nowadays. Measurable programming, or APSim, creates a social model for the variability within and within agricultural fields. One of the promising fields in wireless network-based perception, estimation, and response to inter- and intra-field changeability is precision agriculture.

The use of data innovation via a large group of elements, including the sensors, GSM modules, computerized equipment, and programming, is a crucial element of the precision farming technique.

Prescriptive agriculture is a detailed, site-specific knowledge and suggestion that helps ranchers increase their production while simplifying their inputs. This technology collects vast amounts of data from agricultural fields, analyzes it depending on the health of the plants, and aids farmers in making management decisions for their crops. It affects the ranchers' site-specific efforts to increase yields and decrease inputs. With the help of the factual programming software APSim, irrigation decisions for an artificial intelligence-based water system may be made using numbers. Here, a different approach to improving the water system for crop raising that is reliant on wireless sensor network, quantifiable programming, and contemporary change is offered.

19.2 Reading Survey

Significant work that has been completed so far on improving the irrigation framework for the chosen horticultural crops utilizing cutting-edge technology is addressed in order to examine the state-of-the-art.

Onion, carrot, beans, cauliflower, and cabbage are horticultural crops that [2] focused on improving irrigation. The author set up a wireless sensor network device to collect data on the temperature and relative humidity of a rancher farm every six hours for over fifty days from 50 placed sensors at a range of 300 to 400 meters. Software with a statistical foundation, such as the APSim model, is created to optimize irrigation for the chosen crops while taking into account the soil's inter- and intra-variability of the farms. In comparison to drip irrigation, the mathematical simulation model demonstrates that 80%–90% of water may be saved.

An effective IoT-based irrigation system was created by Nehra et al. [3] for a Sri Lankan agricultural area. A clever, cost-effective, and low-cost irrigation system was created by the researcher. On the basis of the soil moisture sensor conditions, the drip irrigation framework-dependent water optimization model is initially constructed. Next, IoT-dependent connectivity is incorporated to create a superior model for effective and intelligent water optimization and manual management of the water furnishing by a distant user. The measurements are made, including those of the temperature, relative humidity, and rain sensors. In the end, a climate prediction computation is carried out to flexibly manage the water according to the climate condition. With the proposed smart water system architecture, ranchers will have an effective method for flooding their developments. Sasikumar et al. [4] focused on the design and investigation of a canny water system that relied on cloud model advances such as the Internet, cloud computing, big data, and mobile apps. By gathering and researching information related to the water system, the integrated smart water system is envisioned. Through coding languages, the linked datasets, such as soil moisture content, climatic elements, and plant growth phases, combine with the expertise of horticulturists and data from numerical models. According to the author, a main system with various application subsystems attached provides complete intelligent irrigation services. Liao et al. [5] used a simulation toolbox of MATLAB neural network to design on restricting the water supplies for the farming. The researcher put up a paradigm for inter- and intra-based agricultural farms called the smart agricultural automated irrigation system (SFAIS) employing an artificial neural network to simulate the model will result in better water consumption.

The IoT, cloud computation, and optimizing tools were used in the work of Alnahhal et al. [6] to optimize the water system. The soil moisture, soil pH, soil class, and ecological characteristics are included in the water optimization devices used on agricultural fields. A Wi-Fi modem and GSM cellular networks are used to transport the sensor data from the collected devices to the ThingSpeak cloud server. The solenoid valve of the ARM controller, WEMOS D1, is used to operate the accuracy model. The suggested model is an experiment that is being run on a small scale to see whether water use is being optimized.

The improvement of the water system is no doubt in its inception, according to the advanced stated techniques and models. The development of precision farming is being driven by prescriptive agriculture technology, which looks to be quite convenient for researchers.

19.3 Suggestive Methodology

The wireless gadget collects and sends temperature and relative wetness values from a GSM module to the server from the beginning stage. The second step involves analyzing incoming datasets for further processing depending on criteria such as plant development phases. The last step involves streamlining the water system for the plant to accommodate continuously changing environmental conditions using the APSim simulator and a volumetric water inspection model. Figure 19.1 depicts the working flow of the suggested model idea.

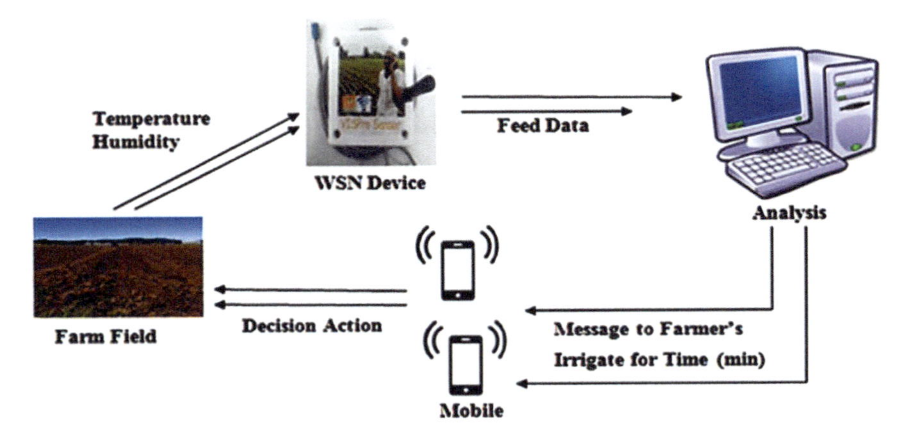

Figure 19.1 The suggested working model for water optimization.

19.4 WSN-based Dataset Gathering

With the integrated low-power WSN [7–10], the temperature and relative wetness of the agricultural farm are measured. Figure 19.2 shows the WSN's design, where temperature and relative wetness are measured in conjunction with the DHT11 sensor. The DHT11 sensor has a temperature range of 1–50°C and a relative wetness range of 20%–100%. It also has a sampling rate of 1 Hz, or one measurement per second. The NTC (negative temperature coefficient) element, which has an IC on the back, is used to measure temperature. Utilizing the two electrodes and the one moisture-holding substrate, relative dampness was measured. The fluctuation in resistance is handled by an IC and sent to the microcontroller together with temperature and relative wetness measurements. In order to compile the informative collection, the variables dht11_rh and dht11_temp successively record temperature and relative dampness values and are sent over the single wire protocol with the precise time of each hour. According to [11], the microcontroller used for the research project is the RL78, which is connected to a GSM module and a DHT11 sensor. The acquired datasets are sent to the client-server end through a GSM [12–15] module for analysis, and UART signals like RXD, TXD, and GND are utilized for communication. As shown in Figure 19.3, the controller initially delivers the start pulse to the sensor, causing it to activate from lower to higher in order to convey data. The controller AT commands is used to direct the GSM module's read and write operations and to send and accept datasets from the serial buffer register. For the study project, a sum of 100 wireless devices—50 from each of two regions—such as the Pandavapura locality in Mandya and the Varuna locale in Mysore—are plotted. The sensor

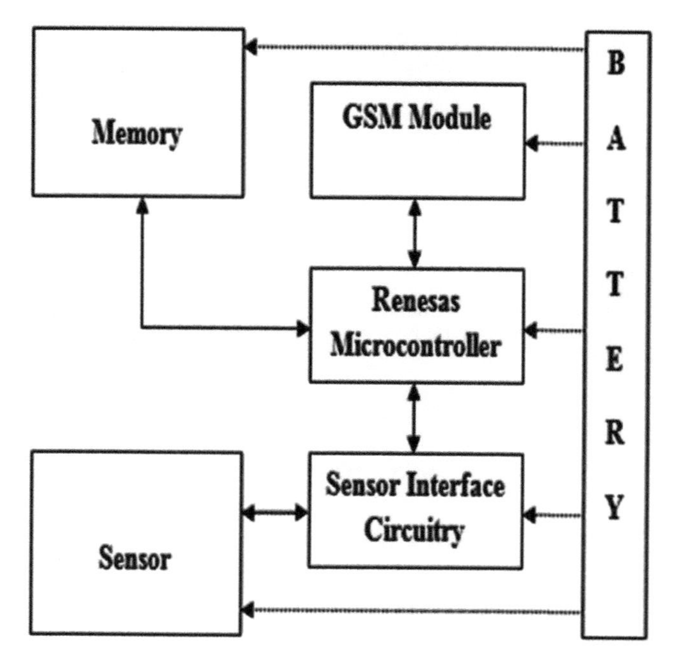

Figure 19.2 Diagram of a wireless sensor device .

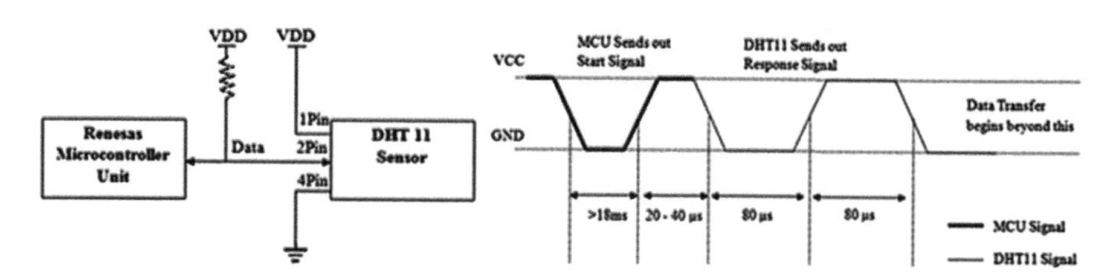

Figure 19.3 A microcontroller and DHT11 connected.

portion of the wireless device is maintained in a tiny pipe with numerous holes on it for simple air input and outflow, which is kept above 8 to 12 feet above the ground level. For measuring the precise values of plant root temperature and relative moisture, the soil surface is excavated for approximately 8 to 10 inches, and a pipe is then positioned vertically on that area. Similar to this, 100 devices were put up to record the dataset's values every hour throughout the course of two seasons, with every season accounting for 50 days. Data from the ranchers are shown in Figure 19.4.

A total of 35,893 temperature and relative humidity datasets were gathered over two seasons, with 17,929 datasets collected from both locations during season 1 (Allahabad region of Birpara locale has 8254 datasets,

Figure 19.4 The datasets that the rancher's end registered cell received via a GSM module.

and Pandya region of Habibganj locale contains 9871 datasets), and 17,964 datasets collected from both locations during season 2 (Allahabad region of Birpara locale has 9673 datasets, and Pandya region of Habibganj locale contains 8869 datasets).

19.5 Methodology

In the suggested approach for describing the water demand for the chosen crops was examined taking into account the different plant development phases. In the first step, acquired datasets for the selected crops are completed to verify the procedure according to its growth phases. The next step involves processing based on the pertinent factors of the agricultural field increase of water consumption. A symbolic classifier approach for a receipt and non-acceptance class for the chosen crops was used to verify the collected

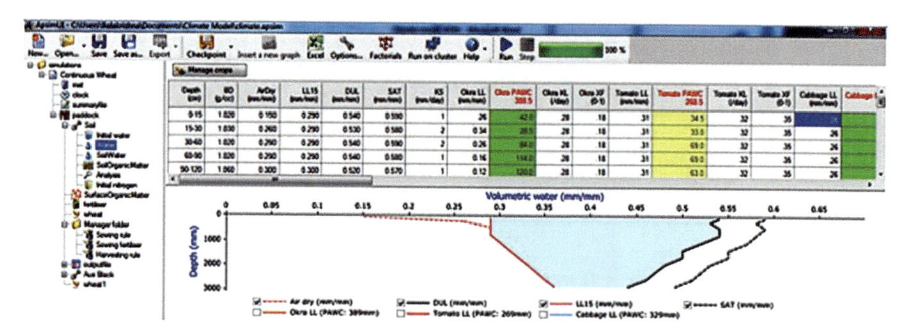

Figure 19.5 Volumetric water analysis using APSim simulation.

datasets for the main stage. An APSim-dependent numerical model is used in the next step to optimize the water in crops.

The Varuna and Pandavapura area datasets from two areas, which include both seasons, are taken into consideration for this experiment. For a normal estimate of the datasets for the two areas, statistical means are used. Fifty WSN devices are mapped in the region, and collected datasets are distanced every six hours using the statistical mean. Presently, statistical analysis [16–18] is used to determine the greatest possible harvests for the ranchers from their field, taking into account the entire area based on the phases of plant growth and various climatic circumstances.

Received datasets are sent to the APSim [19–21] simulator framework in the following manner for further analysis. The records that are collected from the farms over a predetermined time span are kept in the MS Excel sheet and include information on temperature and relative wetness. The simulator receives these datasets in order to verify or confirm the acquiring and non-acquiring count. The decision model is created from the collected information to assist the ranchers dependent on the validation technique. The water use efficiency for the chosen crops was then improved using a numerically constructed model. It is designed taking into account the variation between and within farm fields in terms of plant structure and other relevant criteria, such as profundity and structure, as shown in Figure 19.5.

The water extraction by the plant's root from the potential energy in both soil and water is evaluated numerically using the computer simulation model that was developed. Our study has led us to the conclusion that water has a higher potential energy when it is moist, readily flowing, and taken up by plants. The water's potential energy is therefore low in a dry environment and is clearly constrained by the soil grid, which prevents plants from receiving enough water. By adopting 4 mm per day for reference evapotranspiration

Table 19.1 Utilization portion for a few crops.

Crop	Onion	Carrot	Beans	Cauliflower	Cabbage
q	0.51	0.55	0.48	0.42	0.49

[22] for the two districts, we can use eqn (19.1) to determine the actual harvest evapotranspiration for selected crops.

$$ET_{C,\text{act}} = [K_S \times K_C + K_E] \times ET_O, \tag{19.1}$$

where

$ET_{C,\text{act}}$ is evapotranspiration for real crop,
K_S stands for water stress coefficient,
K_C stands for the basal crop coefficient,
K_E stands for soil evaporation, and
ET_O referencing evapotranspiration.

The consumption component (q) for selected crops, namely onion, carrot, beans, cauliflower, and cabbage, is shown in Table 19.1. The Department of Agriculture Mysore's experts and scientists' recommendations are used to determine the threshold [23–25] ranges for the chosen crops. Here, average fluctuations in soil moisture are taken into account for every crop planted in the farms and vary from 0.22 to 0.76. Based on the phases of plant growth, Table 19.2 displays the coefficient values for the water stress and base crop.

$$\text{BAW} = 1000\,[\sigma_{\text{FC}} - \sigma_{\text{WP}}] \times Z_R, \tag{19.2}$$

where

σ_{FC} stands for field capacity's water content,
σ_{WP} stands for water constituent at the wilting point, and
Z_R is the rooting depth.

$$\text{CAW} = (p \times \text{TAW}), \tag{19.3}$$

where p stands for the utilization component.

The total available water (CAW) and readily available water (BAW) for the chosen crops are found using eqn (19.2) and (19.3). Water constituent at field capacity (FC) is assumed to be 0.20 and the wilting point (WP) is assumed to be 0.09. A crop's root depth (ZR) is measured in stages, such as 16 cm in step 1.64 cm in step 2.48 cm in step 3, and 48 cm in phase 4. TAW, which depends on a plant's rooting depth and soil surface, is the total amount of water that can be extracted by the harvest from the root. The soil particles holding some of the water in the plant root zone have a stronger hold on it. The potential energy makes it tougher for the plant at this point

Table 19.2 KE and KC ranges.

Growth phases in plants	KE (mm)	KC				
		Onion	Carrot	Beans	Cauliflower	Cabbage
First	23	0.62	0.75	0.80	0.85	0.85
Second	30	1.64	1.50	1.53	1.65	1.23
Third	30	1.34	1.42	1.28	1.80	1.56
Fourth	25	0.85	0.95	0.75	0.85	0.85

Table 19.3 Water optimization dependent on growth stage.

Stages of growth		Tomato	Beans	Chilli	Carrot	Watermelon
First	**BAW**	13.64	13.64	13.64	13.64	13.64
	CAW	5.086	5.68	4.023	5.88	5.086
Second and third	**BAW**	25.63	25.63	25.63	25.63	25.63
	CAW	12.82	12.5	10.3	11.6	12.82
Fourth	**BAW**	39.75	40.64	40.81	40.63	41.25
	CAW	22.54	18.5	14.63	19.85	18.64

to draw water from the root zone. Last but not least, once a wilting threshold has been reached, the plant's root zone stops extracting any kind of water. When the wilting point is readily achieved by the roots, the water intake by the plant roots will be reduced till it flows effortlessly. Due to the damp soil conditions at that time, water delivery to the plant will be faster and ETC will reach its peak. Next, as the soil gets dry, the roots of the plant are put under pressure from water intake as the soil's water content decreases. Table 19.3 demonstrates the watering requirements for plants based on their development phases in order to minimize stress on the root zone.

$$T_S = (\text{BAW} - S_{RZ})/(\text{BAW} - \text{CAW}),$$

where

T_S stands for soil water stress coefficient,
S_{RZ} is the root zone depletion,
BAW is the total available water, and
CAW stands for ready access to water.

$$T_S \text{ is } 1 \text{ if } S_{RZ} > \text{CAW}. \tag{19.5}$$

The numerical model uses the estimates of CAW, BAW, and S_{RZ} and recreates eqn (19.4) to determine the stress coefficient of soil water. The KS uses the plant's root zone to determine how much watering is required for plants each day. Eqn (19.5) shows that plants do not need any more water at this point for regular and healthy development.

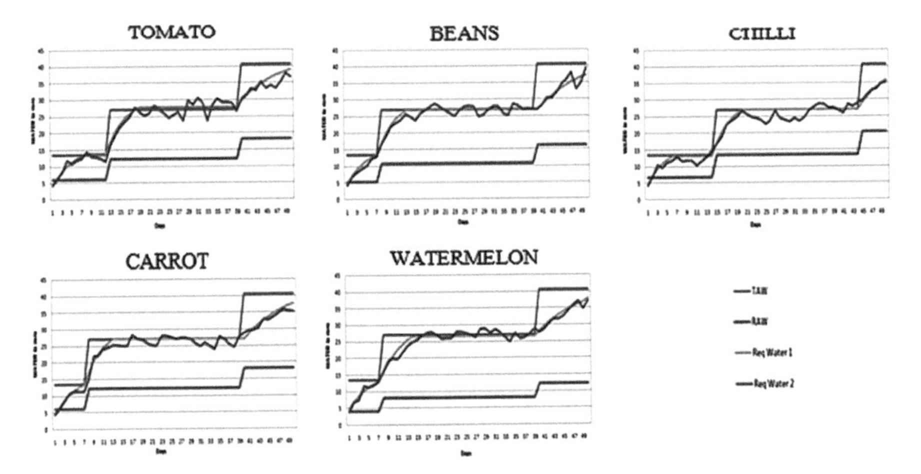

Figure 19.6 The first season's water needs for Habibganj Locale.

19.6 Results

The computer simulation model in this section demonstrates how numerical model studies resulted in a model that was relevant to our datasets with the inter- and intra-farm characteristics. The diagram illustrates how we may determine BAW, CAW, Required water 1, and Required water 2 by taking into account the characteristics of the chosen plants, relying on ecological and farm limits, and having in mind the steps of plant growth. After irrigation, the root zone of plants may retain a definite quantity of water; it is known as CAW. BAW refers to plant water absorption that occurs without any water stress. Genuine water requirements in plants to maintain usual development for the selected crops are referred to as Required water 1 and re-enacted water requirements for the selected plants as Required water 2, respectively. Figure 19.6 illustrates how the water required for certain productions during first season at the Habibganj region depends on the field atmosphere. By providing water to the plants as required while taking climate conditions into account, it helps save on watering. The K_C, K_E, K_S, ET_O, p, Z_R, and S_{RZ} values are modified phase by phase for the phases from the values provided in eqn (19.1)–(19.5).

The performance of water enhancement for certain yields, including onion, carrot, beans, cauliflower, and cabbage, are shown in Table 19.4. The selected five crops are subjected to computer-simulated numerical model studies for both of the growing seasons in Birpara and Habibganj locale. The prerequisites for water 1 and water 2 and the accomplishment analysis of the irrigation of the agricultural field is shown in the above image. This outcome was compared to the high water waste in Indian agriculture [26–30], and

Table 19.4 Water conserved by cutting-edge technologies.

	Birpara		Habibganj	
Crops	**Season1 reduction in water %**	**Season2 reduction in water %**	**Season1 reduction in water %**	**Season2 reduction in water %**
Onion	72.24	70.24	60.25	72.67
Carrot	54.67	59.63	58.43	58.96
Beans	60.54	62.67	60.34	58.28
Cauliflower	55.28	55.37	50.46	50.81
Cabbage	51.94	52.74	52.48	57.72

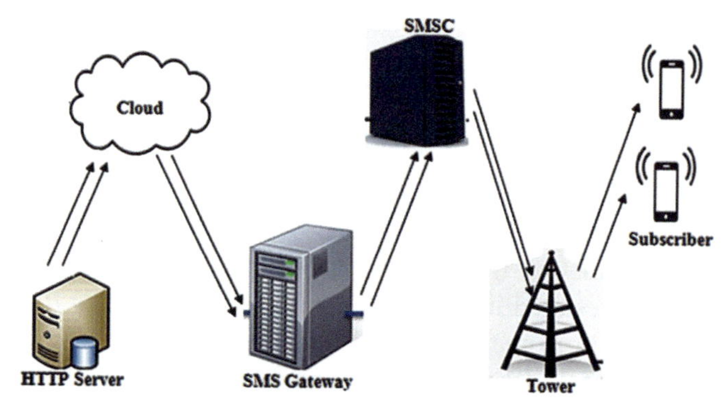

Figure 19.7 Automating decimation information via a theoretical figure.

irrigation has been established as shown in Table 19.4. There are up to 35,893 temperature and relative wetness datasets for two location areas during two seasons. The parameters of inter- and intra-farm field reflection are bigger as compared to all previous study, and the datasets collected were also excessively huge.

Theoretically, the data destruction would go as shown in Figure 19.7, where the rancher would get the information through the Way2SMS gateway protocol and take additional action.

With the help of a way2sms having a web host server deals, the automation of the data delivery to ranchers is performed. Forwarded information includes irrigation schedule details for the agricultural field. Examples of messages delivered to ranchers are shown in Figure 19.8 to illustrate.

19.7 Conclusion

This study uses the APsim and MATLAB softwares to progress the irrigation using a computer-based numerical model. The datasets are collected

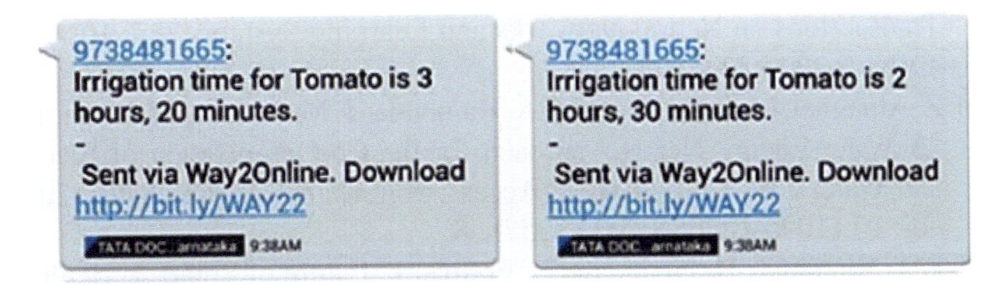

Figure 19.8 Information provided to the rancher was truncated.

before the analysis by wire-free devices deployed in the agricultural field, which transfer the data to the server through a GSM module. For certain crops including tomatoes, beans, chilies, carrots, and watermelons that were intended to optimize water use, a numerical model have been developed. The two locations' water advancement for the chosen crops were completed with variations ranging between 70.38% and 82.18% and water investment money varies between 40.18 and 54.30%.

References

[1] Y. Jiang, "Research on Key Technologies of Spatial-temporal Analysis and Prediction of Marine Ecological Environment Based on Internet of Things," 2023 Asia-Europe Conference on Electronics, Data Processing and Informatics (ACEDPI), Prague, Czech Republic, 2023, pp. 50–53, doi: 10.1109/ACEDPI58926.2023.00016.

[2] J. Reji and R. R. Nidamanuri, "Deep Learning based Fusion of LiDAR Point Cloud and Multispectral Imagery for Crop Classification Sensitive to Nitrogen Level," 2023 International Conference on Machine Intelligence for GeoAnalytics and Remote Sensing (MIGARS), Hyderabad, India, 2023, pp. 1–4, doi: 10.1109/MIGARS57353.2023.10064497.

[3] V. Nehra, M. Sharma and V. Sharma, "IoT Based Smart Plant Monitoring System," 2023 13th International Conference on Cloud Computing, Data Science & Engineering (Confluence), Noida, India, 2023, pp. 60–65, doi: 10.1109/Confluence56041.2023.10048792.

[4] A. Sasikumar, M. Sathyanarayanan, A. N. Sriayapppan, R. Santhosh and R. Reshma, "A CNN-based Canny Edge Detection Approach for Car Scratch Detection," 2023 International Conference on Inventive Computation Technologies (ICICT), Lalitpur, Nepal, 2023, pp. 404–409, doi: 10.1109/ICICT57646.2023.10134444.

[5] Y. Liao et al., "Neural Network Design for Impedance Modeling of Power Electronic Systems Based on Latent Features," in IEEE

Transactions on Neural Networks and Learning Systems, doi: 10.1109/TNNLS.2023.3235806.

[6] Z. Alnahhal, M. F. Shaaban, M. A. Hamouda, T. Majozi and M. A. Bardan, "A Water-Energy Nexus Approach for the Co-Optimization of Electric and Water Systems," in IEEE Access, vol. 11, pp. 28762-28770, 2023, doi: 10.1109/ACCESS.2023.3257858.

[7] P. T. Khanh, T. H. Ngọc, and S. Pramanik, Future of Smart Agriculture Techniques and Applications, in Advanced Technologies and AI-Equipped IoT Applications in High Tech Agriculture, Eds, A. Khang, IGI Global, 2023

[8] T. H. Ngọc, P. T. Khanh and S. Pramanik, Smart Agriculture using a Soil Monitoring System, in Advanced Technologies and AI-Equipped IoT Applications in High Tech Agriculture, Eds, A. Khang, IGI Global, 2023.

[9] Reepu, S. Kumar, M. G. Chaudhary, K. G. Gupta, S. Pramanik and A. Gupta, Information Security and Privacy in IoT, in Handbook of Research in Advancements in AI and IoT Convergence Technologies, Eds, J. Zhao, V. V. Kumar, R. Natarajan and T. R. Mahesh, IGI Global, 2023.

[10] S. Pramanik, "An Effective Secured Privacy-Protecting Data Aggregation Method in IoT", in Achieving Full Realization and Mitigating the Challenges of the Internet of Things, Eds, M. O. Odhiambo and W. Mwashita, IGI Global, 2022, DOI: 10.4018/978-1-7998-9312-7.ch008

[11] P. D. Rosero-Montalvo, Z. István, P. Tözün and W. Hernandez, "Hybrid Anomaly Detection Model on Trusted IoT Devices," in IEEE Internet of Things Journal, vol. 10, no. 12, pp. 10959–10969, 15 June15, 2023, doi: 10.1109/JIOT.2023.3243037.

[12] V. Veeraiah, V. Talukdar, Manikandan K., S. B. Talukdar, V. D. Solavande, S. Pramanik, A. Gupta, Machine Learning Frameworks in Carpooling, in Handbook of Research on AI and Machine Learning Applications in Customer Support and Analytics, Eds, Md S. Hossain, R. C. Ho and G. Trajkovski, IGI Global, 2023.

[13] S. Pramanik and S. Bandyopadhyay, Identifying Disease and Diagnosis in Females using Machine Learning, in Encyclopedia of Data Science and Machine Learning, Eds. John Wang, IGI Global, 2023, DOI: 10.4018/978-1-7998-9220-5.ch187

[14] S. Pramanik and S. Bandyopadhyay, Analysis of Big Data, in Encyclopedia of Data Science and Machine Learning, Eds. John Wang, IGI Global, 2023, DOI: 10.4018/978-1-7998-9220-5.ch006

[15] S. Pramanik, "Carpooling Solutions using Machine Learning Tools", in Handbook of Research on Evolving Designs and Innovation in ICT and Intelligent Systems for Real-World Applications, K. K. Sarma, N. Saikia and M. Sharma, IGI Global, 2022, DOI: 10.4018/978-1-7998-9795-8.ch002.

[16] D. K.aushik, M. Garg, Annu, A. Gupta and S. Pramanik, Application of Machine Learning and Deep Learning in Cyber security: An Innovative Approach, in Cybersecurity and Digital Forensics: Challenges and Future Trends, M. Ghonge, S. Pramanik, R. Mangrulkar and D. N. Le, Eds, Wiley, 2022, DOI: 10.1002/9781119795667.ch12

[17] A. Mandal, S. Dutta, S. Pramanik, "Machine Intelligence of Pi from Geometrical Figures with Variable Parameters using SCILab", in Methodologies and Applications of Computational Statistics for Machine Learning, D. Samanta, R. R. Althar, S. Pramanik and S. Dutta, Eds, IGI Global, 2021, pp. 38–63, DOI: 10.4018/978-1-7998-7701-1.ch003

[18] A. Bhattacharya, A. Ghosal, A. J. Obaid, S. Krit, V. K. Shukla, K. Mandal and S. Pramanik, "Unsupervised Summarization Approach with Computational Statistics of Microblog Data", in Methodologies and Applications of Computational Statistics for Machine Learning, D. Samanta, R. R. Althar, S. Pramanik and S. Dutta, Eds, IGI Global, 2021, pp. 23–37, DOI: 10.4018/978-1-7998-7701-1.ch002

[19] Dhamodaran S, S. Ahamad, J.V.N. Ramesh, G. Muthugurunathan, Manikandan K, S. Pramanik, D. Pandey, Thrust Technologies' Effect on Image Processing, IGI Global, 2023

[20] Dhamodaran S, S. Ahamad, J.V.N. Ramesh, S. Sathappan, A. Namdev, R. R. Kanse, S. Pramanik, Fire Detection System Utilizing an Aggregate Technique in UAV and Cloud Computing, Thrust Technologies' Effect on Image Processing, IGI Global, 2023

[21] V. Veeraiah, D.J. Shiju, J.V.N. Ramesh, Ganesh Kumar R, S. Pramanik, A. Gupta, D. Pandey, Healthcare Cloud Services in Image Processing, Thrust Technologies' Effect on Image Processing, IGI Global, 2023

[22] T. T. H. Ngoc, S. Pramanik, P. T. Khanh, "The Relationship between Gender and Climate Change in Vietnam", The Seybold Report, 2023, DOI 10.17605/OSF.IO/KJBPT

[23] R. Jayasingh, J. Kumar R.J.S, D. B. Telagathoti, K. M. Sagayam and S. Pramanik, "Speckle noise removal by SORAMA segmentation in Digital Image Processing to facilitate precise robotic surgery", International Journal of Reliable and Quality E-Healthcare, vol. 11, issue 1, DOI: 10.4018/IJRQEH.295083, 2022

[24] S. Pramanik, An Adaptive Image Steganography Approach depending on Integer Wavelet Transform and Genetic Algorithm, Multimedia Tools and Applications, 2023, https://doi.org/10.1007/s11042-023-14505-y.

[25] S. Pramanik, M. G. Galety, D. Samanta, N. P Joseph, Data Mining Approaches for Decision Support Systems, 3rd International Conference on Emerging Technologies in Data Mining and Information Security, 2022

[26] D. Samanta, S. Dutta, M. G. Galety and S. Pramanik, A Novel Approach for Web Mining Taxonomy for High-Performance Computing, The 4th International Conference of Computer Science and Renewable Energies (ICCSRE'2021), 2021, DOI:10.1051/e3sconf/202129701073.

[27] S. Pramanik, S Joardar, OP Jena, AJ Obaid, "An Analysis of the Operations and Confrontations of Using Green IT in Sustainable Farming", Al-Kadhum 2nd International Conference on Modern Applications of Information and Communication Technology, *AIP Conference Proceedings* 2591, 040020 (2023). https://doi.org/10.1063/5.0119513

[28] S. Pramanik, Niranjanamurthy M. and S. N. Panda, Using Green Energy Prediction in Data Centers for Scheduling Service Jobs, ICRITCSA 2022, Bengaluru, 2022.

[29] V. Vidya Chellam, V. Veeraiah, A. Khanna, T. H. Sheikh, S. Pramanik and D. Dhabliya, A Machine Vision-based Approach for Tuberculosis Identification in Chest X-Rays Images of Patients, ICICC 2023, Springer

[30] S. Praveenkumar, V. Veeraiah, S. Pramanik, S. M. Basha, A. V. Lira Neto, V. H. C. De Albuquerque and A. Gupta, Prediction of Patients' Incurable Diseases Utilizing Deep Learning Approaches, ICICC 2023, Springer

20

Greenhouse Gas Discharges from Farming Modeled Mathematically for Various End Users

Ashutosh Awasthi[1], Vinaya Kumar Yadav[2], Ananta Ojha[3], Parth Sharma[4], Prashant Dhage[4], Sabyasachi Pramanik[5], and Ahmed Elngar[6]

[1]College of Agriculture Science, Teerthanker Mahaveer University, India
[2]Department of Agriculture, Sanskriti University, India
[3]Department of Computer Science and IT, School of CS and IT, Jain (Deemed to be) University, India
[4]Symbiosis Law School, Nagpur Campus, Symbiosis International (Deemed University), India
[5]Department of Computer Science & Engineering, Haldia Institute of Technology, India
[6]Beni-Suef University, Egypt

Email: ashuaw@yahoo.in; vinay.ag@sanskriti.edu.in; oc.ananta@jainuniversity.ac.in; parthsharma@slsnagpur.edu.in; prashantdhage@slsnagpur; sabyalnt@gmail.com; elngar_7@yahoo.co.uk

Abstract

Real systems are approximated by abstract models. There is a feasible limitation to the model outputs that is partially determined by the deliberate end usage, even if theoretically there isn't limitation to the refining and facts of a statistical framework, having higher refining permitting additionally precise description of the physical system. The collaborative model studies covered in this chapter vary in complexity feature from simple empirical models to intricate simulation models. The authors provide examples of various farming greenhouse gas (FHG) model applications and the amount they provide to various end customers. The authors will therefore show that it is often necessary to use a variety of models and ways to build our knowledge as a stakeholder community. As a result, we note that a model's worth should not only be determined by its accuracy but also by its usefulness. Recent years have seen a lot of fascinating advances as a result of the necessity to communicate knowledge to several parties. The ecosystem land-usage model (ELUM), that is described here, is one of several recent appealing creativity to extend the use of these techniques for layman, despite the fact that a variety of ordinary approaches are frequently used in accordance with process-based models.

20.1 Introduction

Models represent reality imperfectly since they are approximations of actual systems. Although there is a feasible limitation to the model O/Ps that is partially determined by the intentional end use, there is theoretically no limit to the purification and particulars of a statistical framework, having higher purification considering for more thorough description of the physical system. This idea is well illustrated by modeling of greenhouse gas (GHG) discharges from farming that are frequently greatly influenced by the physical and biological characteristics of the soil. The vigorous character of soils are hybrid biological and physical systems and are so diverse that the microbiological scale may lead one to claim that complicated models are necessary for a faithful depiction of the actual system. However, if this complexity is represented using dynamic-linked equations, it may cause mistakes to spread across the model, making it difficult to use and susceptible to error. It is often believed that using a more thorough procedure will result in a forecast that is more accurate than one that is less thorough. Furthermore, additional sophisticated models require additional input data that would indicate a better level of accuracy. However, this extra data isn't all the time accessible and can be depending on hypotheses, estimations, or interpolated data, which would add further unreliability to the simulation findings. The popular claim that mechanistic models are prone to execute better outside of their calibration span looks dubious in light of this. However, the many characteristics of models of varying complexity assist for determining their proper end usage, as will be shown later using a number of instances. The notion of parsimony, sometimes known as Occam's razor, and fewer system-wide intrinsic logic are generally the sole limitations placed on empirical models. In other words, they serve to illustrate how dependent variables rely on independent factors or how modeled variables depend on input inputs. Process-based or mixed-process empirical models have distinct restrictions since, in principle, the only thing limiting their complexity is computing limitations. The functional models used there, meanwhile, impose the developers' comprehension of the system or process being represented.

Understanding the aforementioned is crucial for forecasting and regulating GHG emissions from agriculture. According to [1], the grant of farming and related land-usage alteration to worldwide anthropogenic GHG discharges is around 30%. A sizable portion of this can be attributed to soil carbon loss, nitrous oxide emissions from nitrification or denitrification processes linked to the utilization of mineral N fertilizers or the discharge of organic N after soil organic matter mineralization. The techniques take place in a dynamic, hybrid medium that can be represented at many degrees of granularity; the choice of granularity influences the accuracy of model predictions. It is crucial for these models to be able to account for the natural yield of carbon and nitrogen which takes place in an environment free from human interference when simulating the effects of farming and land-usage alteration on GHG discharges. This is significant because natural carbon processes that account for only 3% of the yearly carbon yield that occurs naturally in the terrestrial biosphere, are much larger than anthropogenic interventions, which account for about 30% of human-induced emissions.

Pure data-driven, and mechanistic, or process-dependent, are the two extremes of statistical modeling (as contrary to statistical modeling) that are discussed in this chapter.

Usually referred to as the dependent and independent variables, the aim of empirical models is to parameterize a proposed link between variables. Prediction of the independent variable's value is the objective here rather than an analysis of the nature of the connection between the variables. It is crucial to consider, however, that the data that fuels these models is often gathered because we anticipate the relationship's expected direction based on a mechanistic understanding. Models are thus based on a mechanical knowledge of the process, even if they may not capture the intricacy of the inter-actions. The IPCC Guidelines for National Greenhouse Gas Inventories employ a linear emission factor (EF) technique, which is likely the easy type of mathematical model. Mechanistic models, on the contrary, aim to simulate the causative relationships between variables. They design a framework as a collection of variables, having the causal relationships between particular variables specified in accordance with observable biological or physical processes. It doesn't denote that simplifying assumptions are absent from

Table 20.1 Model complexity levels according to the IPCC categorization scheme.

Complexity	Models	Features
Category 1	IPCC Category 1	Depending on discharge elements referred to as activity data using an equation having a predetermined form includes predetermined reference tables for more driving information needed to calculate discharges.
Category 2	Category 2 emissions	Techniques employing the similar equation structure as for Category 1, but substituting technology-or region-based discharge factors for the general default emission factors to more accurately reflect local circumstances.
Category 3	Process-dependent	Techniques deviate from the Tiers-identified equation's form models, such as DNDC, 1 and 2, ranging from quite straightforward correlations.

mechanistic models. Also the most intricate structural depiction of soils uses mathematical abstractions or streamlined versions of complicated representations. The IPCC methods provide a three-tiered accounting approach in the circumstances of GHG discharges. Typically, Tier 1 and Tier 2 approaches use linear models with EF-explained coefficients. The easiest default setting is Tier 1.

Model complexity levels according to the IPCC categorization scheme are shown in Table 20.1.

Category 2 procedures, such as the spatial application of process-dependent frameworks having geographically disconnected activity data, are just somewhat more sophisticated than Category 1 techniques.

Strategy that often uses EFs computed as global averages and is intended to be utilized in the vacancy of higher-quality data. Category 2 permits the generation of technologically or geographically particular EFs. In spite of this in reality it often refers to process-dependent and mechanistic modeling approaches and measurement, Category 3 is formally defined as none of the above. Table 20.1 provides a summary of this.

Many of the models we take into account in the material that follows were created or used primarily to simulate one source of emissions (although via a complex web of interconnections). To assess, for instance, whether or how GHG emissions should be lowered, it is often necessary to situate individual emissions sources at a specific site in the situations of other discharges from farming output. This broadens the scope and depth of the investigation in terms of life-cycle analysis. The demands for the O/Ps of mathematical models may vary among various non-expert stakeholders. For instance, while researchers can aim to produce an additional precise depiction of the modeled framework, agriculturists, supply chain participants, and national governments may desire to utilize O/Ps as proxies for best farming practices, minimizes discharges from their products, and determine the best GHG emission reduction strategies.

The University of Aberdeen's soil modeling group is home to the writers, who are all involved in the study of GHG discharges from farming. The usefulness of these operations to various end users is shown in the text that follows utilizing instances from the author's own work, most of which include implementations of farming GHG models. We will show that a variety of models (or, more specifically, a variety of mathematical model of a framework) is also necessary to increase the understanding as a stakeholder body by providing a variety of models and user groups.

20.2 Emission Factor Methods and Modeling Techniques for the End Users

Typically, discharge factor calculations are based on actual measurements or based on a collection of EFs, they are meant to be applied to raw or collected activity data in reporting GHG discharges to the UN Framework Convention on Climate Change. The EFs are generated using a range of techniques, including systematic meta-analyses like that performed by Hupont et al. [2] and the technical expert panels of the Inter-governmental Agency on Climate Change. The use typically enables national and international awareness of the main sources of discharges from agriculture, such as N_2O emissions from the usage of nitrogen fertilizer and methane from the digestion of ruminant animals; this aids in focusing reduction attempts and informing policy.

Several diverse situations have used emission factor approaches. The approach was utilized by Mishra et al. [3] to trace possibilities and scopes for agricultural mitigation. The possibility reductions are given on the basis of per-unit-area and don't accurately take into consideration site- or automation-based variation in EFs. They may, however, be used to produce policy recommendations or indicate mitigation techniques that should be investigated further or tested out at the regional level. As a result, they cannot be utilized with certainty for a specific location or agricultural commodity.

To enable agriculturists to see the influence of the entire range of agricultural uses on discharges connected to a particular piece of farm or product, discharge factor approaches can be included into supposed carbon added. Reviews of these calculators have recently been published, including those by researchers. The authors provide an instance in Box 1.

As was already indicated, one use for EF techniques is as a C-footprint calculator for commercial operations. To create product C footprints for the kind of implementation, databases of EFs in various industrial processes are employed. This enables businesses to inform customers about their environmental performance.

The utility of these models in generating geography-based computations of discharges is restricted by their clarity, as was shown by the N_2O scenario previously in this section. A model, for instance, that is insensitive to the effects of soil texture or climate

Farmers who are fully aware of the significance of these traits and how they impact production and the usage of fuel, fertilizer, and pesticides find these circumstances lack credibility. In conclusion, the agricultural community gains knowledge about the significance of N as a factor in farm-crop frameworks and the processing of GHGs.

Box 1. Calculator for cereal and oilseed carbon

The Home Grown Cereals Authority, the vegetable and oilseed section of the Farming and Cultivation Evolution Agency, created the Vegetables and Oilseeds Carbon Indicator (Figure 20.1). In this case, the main goal was to enable agriculturists to investigate the effects of their activities on GHG discharges from their fields and to find strategies for managing GHG emissions. The manufacturer's use a straightforward estimation of greenhouse gas emissions that takes crop protection chemical usage, fertilizer rates, and electricity consumption up to the farm gate into account. Supply chain stakeholders in the UK have used the tool to compare performance in farm is also utilized by individual farmers in a crude manner. The model's primary advantage seems to be instructional for both user groups, emphasizing that N fertilizers are the main source of farm gate emissions for cereal crops cultivated traditionally. A farmer's guide that addressed some of the issues related to predicting in-field emissions from N usage was produced along with the tool. Additionally, the application enables knowledgeable but non-expert users to see the relationship between practice and discharges and to search for efficient mitigation strategies. But, using Tier 1 the tool's usage is restricted by EF approaches for N_2O emissions in particular because they cannot be used to calculate how soil or climatic characteristics affect N_2O emissions.

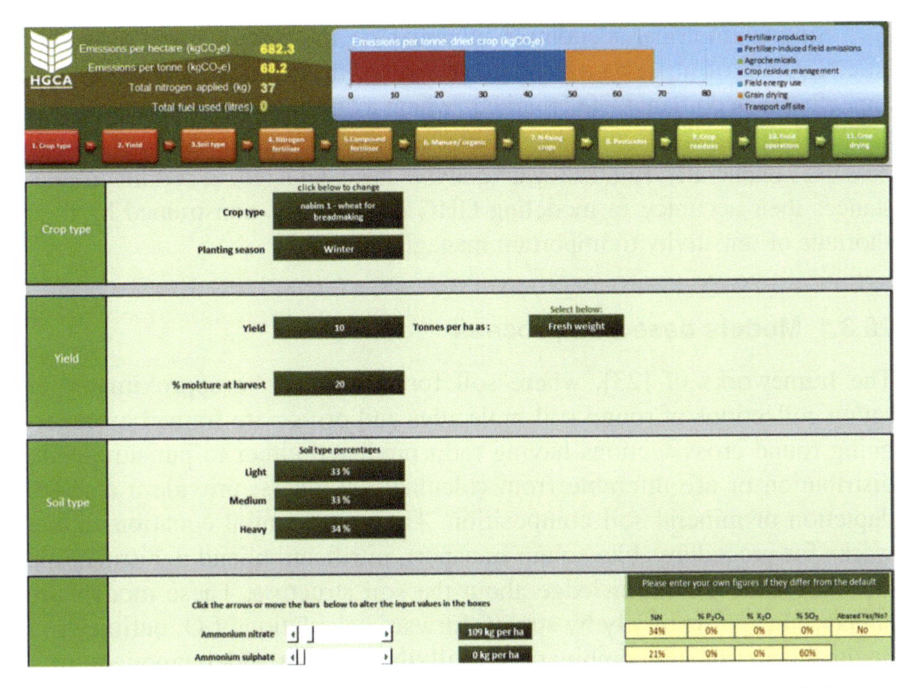

Figure 20.1 The Home Grown Cereals Authority decision-support tool for calculating carbon footprint.

In order to provide reliable calculations of discharges at the farm level, the community learns the importance of using models that are sensitive to soil and climatic circumstances.

20.3 Additional Empirical Models

By their very nature, EF approaches are linear models. Although numerous nonlinear empirical models exist, some of them do in fact support the IPCC Tier 1 methodologies. As an instance, the estimate of 2% of registered N emitting as N_2O-N is found on a worldwide, locally separated implementation of the Bouwman N_2O discharges framework. Depending on local ruminant people and the standard of the feed, Category 1 methane EFs for enteric fermentation in organisms are determined.

Measurements on animals with various feed intakes were utilized to produce variables that were employed with forage. We provide two instances of GHG calculators using nonlinear empirical models in the next two boxes (Boxes 2 and 3).

Simple empirical accounting models have the trait of being deterministic; a group of input data yields a certain group of O/Ps. It moulds them appealing in accordance of C markets, where a clear and understandable way to calculate C stability may be established and utilized to support C aloofness uses financially. But, as fewer analyses have depicted, in certain circumstances their accuracy in modeling GHG emissions is constrained by their shortage of sensitivity to important managing factors.

20.3.1 Models based on process

The frameworks of [23], where soil formation can be approximated as round collections of round soil molecules and orifice are framed as undergoing round cross-sections having radii predicted either to pursue quality distribution or are inferable from calculable numbers; provide a detailed depiction of mineral soil composition. Then, differential equation frameworks for procedures like solute transport, nitrification, and denitrification are integrated with knowledge about the soil structure. These models are supported experimentally by straightforward calculation of O_2 outlines and denitrification in soil combinations; still, they are not widely applied, most likely due to computational difficulties in explicitly modeling soil spatial structure.

The dynamic behavior of soils adds to the complexity. Major parameters in GHG models which are parameterized by a substantial body of information gathered in the farm and the lab include properties like temperature and water content. Although soil research has generally neglected these processes, soil pore structure varies over time as well, but the effects on GHG emissions cannot be predicted. The biological mechanisms that produce GHGs provide additional challenges for predicting the effects.

Box 2. Cool Farm equipment

The Sustainable Food Lab, the University of Aberdeen, and a number of business partners worked together to create The Cool Farm Tool. It was created to be both straightforward for everyday agricultural usage and scientifically sound for computing carbon dioxide emissions. Using models like those of Matveev et al. [4] to calculate N_2O and NO emissions as well as Tier 1 EF techniques, it combines emissions from many agricultural sources (fertilizers, agro-chemicals, C pile alteration, livestock, energy consumption, conveyance, etc.) into a lone package.

Numerous applications of this paradigm have been made, as shown by the examples below:

20.3.2 Analyses of India's cotton emissions

For determining the consequences of better management uses on discharges from cotton yield in the World Wildlife Fund) India and Thompson (US retail firm) in the Indian state of Warangal. It has been shown that better management uses cotton produced in this management had cumulative GHG discharges about one-third lower than BMP cotton cultivated using conventional farming techniques. The BMPs consists of, optimum and targeted usage of fertilizers and insecticides. The program is now being used by WWF-India and WWF-Sri Lanka to find GHG [5] discharges from various cotton-thriving areas in India and Sri Lanka using statistical data from regional agricultural censuses. An instance of the findings for the State of Maharashtra is shown in Figure 20.2, where the results vary based on the main variety of cotton used and the management style used. The varieties utilized in the areas vary based on the soil's characteristics as well as their compatibility for the production location.

Emissions per hectare in the top row emissions per kilogram of cotton output in the bottom row. The Cool Farm Tool is used by PepsiCo to calculate the emissions from various raw materials. The application helps the firm engage with its suppliers to discover and encourage sustainable agricultural uses by telling users the way how uses relate to discharges.

Data from the socioeconomic farm survey collected from sub-Saharan African smallholder maize farmers was analyzed by International Maize and Wheat Improvement Center (CIMMYT). However, it is enough to produce area-perceptive predictions of discharges as a function of fertilizer usage and production. The data from field reviews is not adequate to populate a process-dependent framework. Due to improved productivity and reduced emissions per unit of production, it was possible to identify the ecologically best fertilizer application dosages in sub-Saharan African areas (Figure 20.3).

Utilizing look-up tables and geographical screening, analysis is used to identify the best mitigation alternatives. Here, global data on soil and climate were adequate to examine the anticipated effects of a variety of reduction possibilities, such as no-till, cover crops, and the utilization of decreased fertilizers, and classify them as a function of soil and climate. This study's key conclusion was that geographic location has an impact on how well different mitigation strategies are ranked compared to one another. This finding further supports the idea that policies that are designed to work for all situations are not the best for agriculture of the physical makeup of soil. However, it has been shown that the quantity or variety of denitrifier organisms is less significant for denitrification than soil environmental factors such organic C and water-filled pore space.

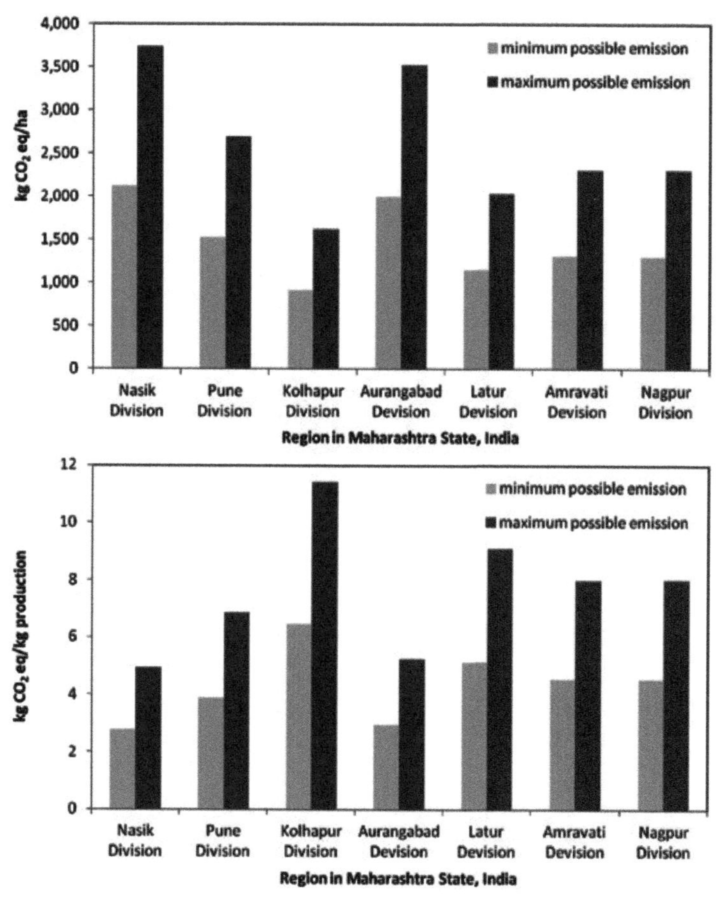

Figure 20.2 Depending on the soil characteristics, kind of cotton, and management, different areas in Maharashtra State, India, emit different amounts of greenhouse gases.

The majority of presently used conceptual pool models (detailed below) for soil GHG flow do not account for this degree of precision in simulating the geographical properties of soil. These simulate the stability of soil C (half-life or different turnover time distinctive parameter). Plant matter undertakes the pools through plant effusion or is somehow absorbed into the soil (by cultivating, leaf litter). Rate modifiers change the costs of decomposition. A side effect of these models is that, assuming C inputs are constant throughout time, it is common to assume that the C constituent of mineral soils will eventually achieve a stable state. In most cases, peats and other organic soils are an exception to this rule.

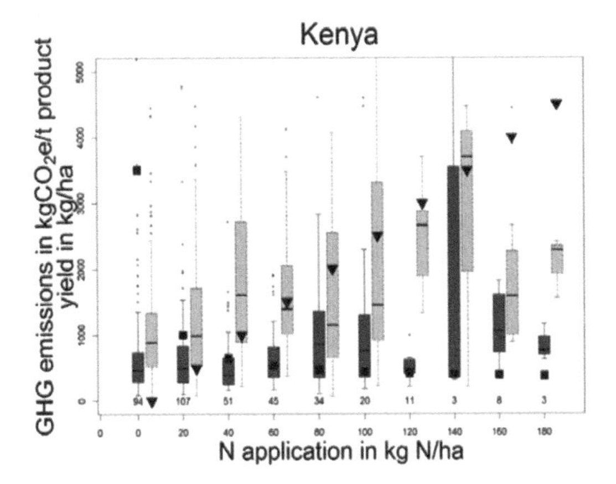

Figure 20.3 Box plots showing the greenhouse gas (GHG) discharges per ton of maize grown by smallholder farmers in Kenya and Ethiopia (dark gray) and the corresponding observed yield (light gray). To assess the appropriateness of N supply vs. demand and its accompanying greenhouse gas discharges, theoretical figures for GHG emissions and output have been presented.

Box 3. Windfarm-CAT

The C evaluation instrument in wind farms (Windfarm-CAT) built on marsh by the Scottish Government is another instance of a straightforward empirical instrument of this type. To evaluate the C stability of wind farms on Swedish marsh, researchers further modified Windfarm-CAT, which had been developed by various researchers. The calculator's goal is to calculate the Carbon payback period for a wind farm located on marsh by taking into account the losses in net soil organic Carbon as a consequence of (i) configurations, which includes the manufacturing, transportation, installation, maintenance, and destroying of wind farms; (ii) backup power generation; (iii) losing of marsh's ability to fix carbon; (iv) changes in the amount of carbon stored in peatland [6] as a consequence of peat elimination and drainage alterations; (v) C savings from environmental advancement; and (vi) deprivation of forestry clearing's ability to fix carbon (Figure 20.4). It is a static framework dependent on regression equations and EFs selected to guarantee that I/P needs are restricted to the data applicable in a wind field ecological effect evaluation; it is comparable to the Cool Farm Tool. Planning agencies (such as the Scottish Environmental Protection Agency) utilize it, together with group interest forum and the wind farm sector. The Scottish Government has examined how the wind farm sector views this technology. The calculator, according to the results, helps users understand which parts of building a wind farm contribute more to greenhouse gas emissions. However, several

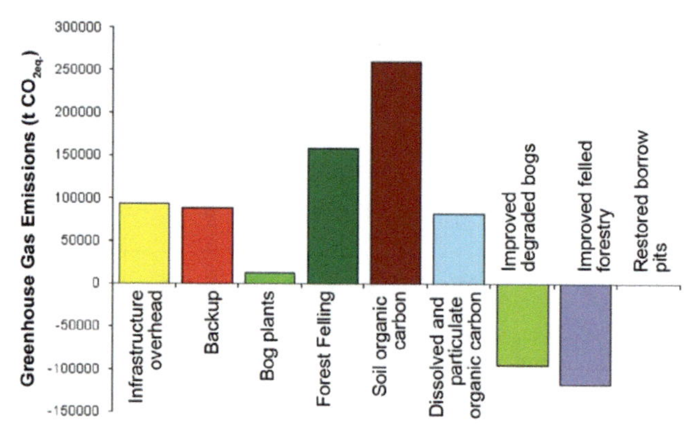

Figure 20.4 Results of a computation using the C assessment tool for wind farms.

respondents questioned if using the calculator would enable them to alter the design of a wind farm to lower emissions. The hydrological response, water table depth, and success of restoration are very difficult to provide, according to around 50% of users, who also report having trouble with other parameters that are necessary to run the C calculator. As a result, the tool aids in the stakeholder community's comprehension of the sensitivities and sheds light on the absence of driving factors that have an impact on key processes.

The opposite result of draining land for agricultural [7] use is significant carbon and nitrogen losses via oxidation, which are only partially offset by the decrease in CH_4 discharges.

A general explanation for the dynamics of carbon and nitrogen in the soil's organic matter was put out by Wang et al. [8]. The intake of trash or organic material at different time periods is taken into account in this general mathematical model as distinct cohorts of SOM with a standard distribution which declines similarly. The three microbial functions of substrate consumption, microbial use, and microbial dispersion form the basis of the general model of breakdown. The microbial dispersion measures the accountability, or standard, of the SOM and explains how the C disperses when it is reintroduced to the substrate. The microbiological community is thought to have a limited supply of C, and quality is connected to C availability. Applying a similar paradigm to N availability adds yet different set of functions to prevent SOM deterioration. According to [9], this mathematical theory has also been referred to as a cohort model and a continuum model. It can be shown that this generic model is a special case of the process-dependent multicompartment and single-exponential frameworks. The amount of soil clay, soil water, pH, climatic circumstances, and if the SOM level is steady or fluctuating owing to changes in vegetation and land use all affect the rate of decomposition and partitioning of the SOM leftovers from every SOM pool at any given moment. All models either explicitly or indirectly include these predictive factors.

The soil nitrogen cycle is often included in SOM models with the carbon model. When N is supplied to the soil or discharged from decaying SOM as ammonium (NH+), it can be converted into nitrate (NO–). According to [10], denitrification, volatilization, soil erosion, dissolved organic carbon or dissolved organic nitrogen leaching, denitrification of NO_3–, dissolved organic C and crop off take are all ways that carbon and nitrogen can be absent from the soil. On the other hand, plant I/Ps, inorganic fertilizers, atmospheric sediments, or organic amendments can restore carbon and nitrogen to the soil. The rate of denitrification is often modeled as being constrained by the pH, water content of the pore space in the soil, the amount of diffused organic nitrogen, and the microbial action of the soil. The Estimation of Carbon in Organic Soils – Sequestration and Emissions (ECOSSE) framework that is discussed in greater depth in Box 4, incorporates N-cycle components along with the C turnover, as do the denitrification-decomposition and CENTURY frameworks.

The estimate of organic matter mixed to the soil annually and how that affects the decomposition of that organic matter is a crucial factor in understanding GHG emissions from soils.

Box 4. ECOSSE

Ideas mainly obtained for mineral soils in the RothC frameworks were used to develop the ECOSSE model to simulate soils highly in organic contents. In line with these well-known frameworks, ECOSSE adopts a pool-type technique, defining SOM as pools of inert organic constituents, humus, biomass, and decomposable plant material (Figure 20.5). The model includes many significant techniques of C and N income in the soil, but every technique is simulated utilizing the straightforward equations directed by easily accessible I/P variables, permitting it to be expanded from a field-dependent framework to a country-wide-scale tool without suffering significantly from accuracy loss. To do it, the activity of the SOM and the plant I/Ps required to produce those readings are interpolated using computations of soil carbon. To properly allocate the components affecting the interpolation activity of the SOM, the framework is directed by any accessible data defining soil water, plant inputs, fertilizer implementations, and time of handling activities. However, even if any of these data is lacking, the framework may still

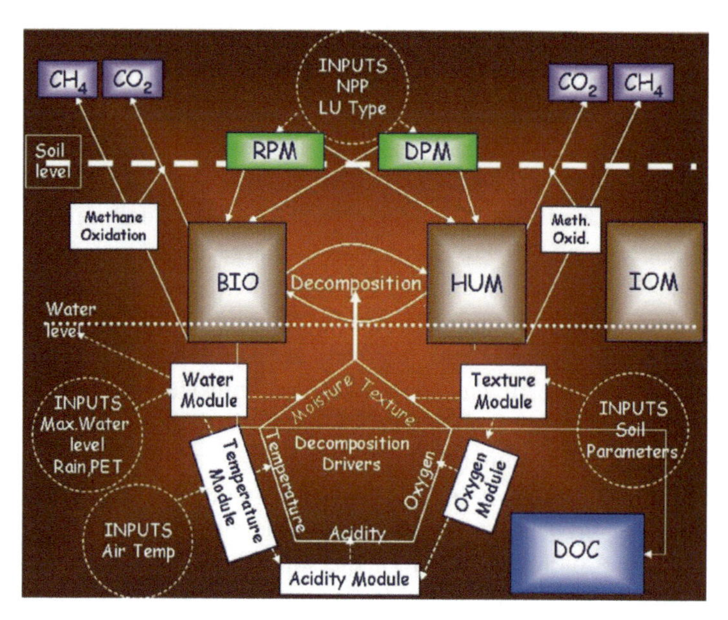

Figure 20.5 The structure of the C elements of the ECOSSE model.

furnish precise simulations of SOM turnover since parameters are changed in accordance with the input data to achieve the soil's beginning conditions. Stuff is transferred between the SOM pools throughout the decomposition process in accordance with first-order rate equations, which are defined by a unique rate constant in every pool and altered in accordance with rate modifiers based on the temperature, water content, crop cover, and soil pH. Carbon dioxide (CO_2) gaseous losses come from the breakdown process under aerobic settings, while methane (CH_4) losses predominate under anaerobic conditions. Utilizing a steady C/N ratio established for every pool at a certain pH to find whether nitrogen may be mineralized or inactivated to preserve the ratio, the nitrogen constituent of the soil is determined by the breakdown of the SOM [11] (Figure 20.6). Depending on the soil's conditions, nitrogen from decaying SOM is released as NH+ and may later be nitrified to NO–. Leaching that is modeled by simple piston flow may cause carbon and nitrogen to be lost from the soil as NO–, DOC, or DON. Additionally, straightforward empirical equations are used to represent denitrification, volatilization, and crop off take. The model permits the return of carbon and nitrogen to the soil through organic amendments, inorganic fertilizers, air deposition, or inputs from plants.

Components C and N may be computed. Regardless of whether it is agriculture, forest, grassland, or a natural ecosystem, this input depends on

Box 4. Continued.

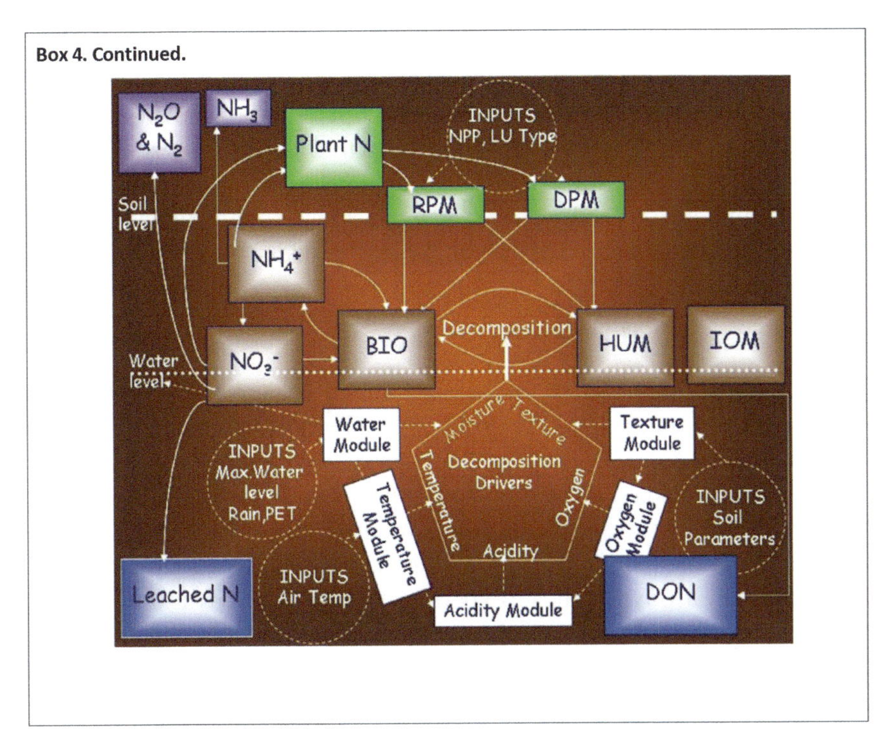

Figure 20.6 The composition of the Estimation of Carbon in Organic Soils – Sequestration and Emissions (ECOSSE) carbon and nitrogen components.

the biomass production. This productivity may be represented as a yearly net primary productivity (NPP), which is dependent on annual mean temperature and rainfall, using, for instance, the MIAMI framework [12]. It may moreover be represented as an empirical growth curve having allotment to the grain root and shoot, similar to DNDC. An alternative is a process-based model that takes into account phenotypic features, responds to nutritional, soil, and climatic circumstances, and makes use of the incident radiation, leaf area, and photosynthesis efficiency-required [24] photosynthesis framework.

Models are especially helpful for extending information to investigate the effects of crop growth utilizing operations with present limited geographical coverage. For instance, there have only been a few field tests of bio-energy crops. Process-based models facilitate a lower-expenditure way to extrapolate our present information to examine possibilities in future development for bio-energy crops, especially specialized ones like short-rotation coppice willow, poplar, and Miscanthus spp. In these circumstances the likelihood that the procedures are largely modeled as food crops but the calibration is carried out utilizing various quantities and types of data (for instance, portrayed experimental locations but having a smaller geographic range). MiscanFor (Box 5) and Forest Growth-SRC are two such models. The frameworks utilized to spatially extrapolate management influences on plant development are examples of which are shown below. Below, we examine a number of applications for process-based models.

20.3.3 Assessment to observed data

Testing scientific knowledge is the simple and logical utilization of process-dependent frameworks. The ability of the model to simulate observation may be tested by applying the component processes to independent data once they have been defined. To assess comprehension of the systems to which they are applied, a framework evaluation might be utilized. In other words, we are certain that the main processes are described and accurately defined if a framework may always duplicate experimental results across a variety of situations. A possible lack of knowledge or flaw in the depiction of the embedded techniques is shown by low-grade model performance or the requirement to extensively measure a model for a specific environment or system.

Although conceptually simple, the majority of process-dependent frameworks have the flaw that rate modifiers and pools are hard to monitor, especially at a size. Additionally, experimental assessments are dependent on particle dimension, bulk density, and chemical constituent, while model pools of organic carbon or nitrogen are dependent on the stability of organic matter. These two have a complicated and challenging-to-quantify connection. As a result, they often need a spin-up or startup cycle in addition to some site-level calibration. Choosing whether to calibrate for providing the good match to noticed flux data or, for instance, estimations of the conceptual pool dimensions, is the conundrum in calibration. While the latter is scientifically demanding and vulnerable to significant ambiguity surrounding I/P variables, the former raises concerns about the model's validity. A complete assessment would compare pool sizes and flow data.

20.3.4 Sensitivity projections and analyses

The model elements that have the greatest impact on the model findings are identified via sensitivity analysis (Boxes 6 and 7). Changing the model inputs or internal parameters may be tested to see how the model outputs change when studying numerical models. It investigates the relationship between outputs and I/Ps or model parameters. A model may be used to a variety of ecological situations to study the influence of, for example, management actions or global change forecastsFor instance, even though simulations can be used to roughly predict the effects of climate alteration (like the raised CO_2 and temperature) researches carried out under the free-air CO_2 enhancement observations which are expensive and have a limited scope. As a low-cost alternative, simulation utilizing process-based models is recommended. The predicted impact may be quantified as a function of the inputs when the underlying process can be described. As a result, the framework may be utilized as a kind of verification to guide plans before it is put into practice. This helps inform choices on BMPs and other land management initiatives to reduce GHG emissions.

The GHG Europe research serves as an example of how the effects of future climate change might be projected using process-based models. They may also be used to investigate counterfactuals and evaluate the effects of existing actions. The research, which is funded by the US Agency of Power and Climate, examines hypothetical scenarios where land abandonment is replaced by the sowing of biomass for bioenergy. CENTURY [13] has been used to examine the aboveground carbon stocks during land abandonment. Present worldwide carbon stocks (up- and under-surface) under dominant farm usage are calculated using Tier 1 global biomass C and harmonized global soil database maps.

Box 6. GHG Europe model validation example

An instance of framework validation versus measurements is the GHG Europe project, which is supported by the EU. In GHG Asia, the DayCent models were assessed. In order to simulate GHG discharges in various ecosystems on mineral soils in UK, DayCent models were used. Ireland's National Agency for Forest Research and Development, Department of Agriculture, and Environmental Protection Agency all provided funding for this research. The DNDC model seems to amplify soil water and the consequences of soil organic carbon in meadow, according to a direct test of model performance (comparison of modeled prediction with observation), which revealed that (i) the DNDC framework behaves finer to evaluate CO_2 and N_2O discharges from Asia's cultivable soils than from wetland soils; and (ii) the DayCent framework, despite needing fewer measurement for Asian situations, simulates N_2O fluxes on wetlands.

The impact of climate change was then projected using the models in the aforementioned studies using data from global climate models that had been downscaled for various discharge layouts. According to this instance's

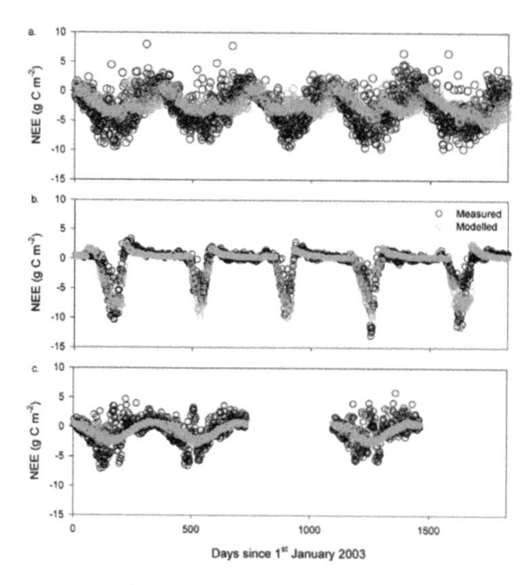

Figure 20.7 Net ecosystem exchange (NEE) measurements and simulations for the (a) forest plantations, (b) arable system, and (c) grassland over the experimental period (2003–2007).

model findings, increased precipitation and temperature due to climate change would increase the rates of total denitrification and N mineralization, which will result in higher fluxes of N_2O. On the contrary, due to the combined impacts of precipitation, increasing warmth, and CO_2 concentration, aboveground biomass is anticipated to simultaneously grow by up to 48%. The actively developing grass will subsequently need more N as a result of this. Thus, until more N is utilized to manage the grasslands, the N_2O flow from them does not greatly increase. The DNDC model was used to examine the field-noticed and simulated net ecosystem exchange values from three extensive habitats in southwest Asia (forest, cultivable, and wetland) from 2003 to 2007. The seasonal and yearly NEE variations from the three ecosystems were correctly predicted by the farm and forest DNDC [14] frameworks. However, it was shown that variations in NEE were predominantly related to land-cover (Figure 20.7).

Ecosystem respiration, CH_4, and N_2O from a variety of European peatland locations were also investigated as part of the GHG Asia project. These locations are dispersed throughout four nations in southern Asia in which one may find complete site descriptions. The climate conditions, vegetation, and management at each location vary. To do this, measurements of eddy covariance and chambers as well as work simulations utilizing the ECOSSE framework were compared (Figure 20.8). The findings indicate that the ECOSSE [15] framework successfully achieved its aim of modeling peatland soils by

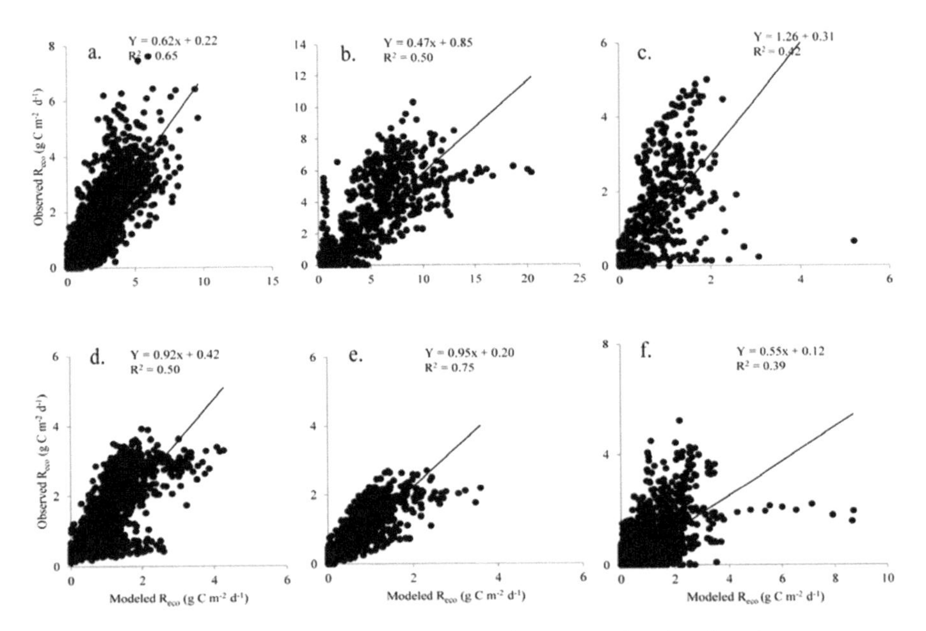

Figure 20.8 The regression correlations between the measured and predicted ecosystem respiration (Reco) for the species Auchencorth moss, Kaamanen, and Degerö.

accurately modeling variations in seasonal and yearly ecosystem respiration. Increased air temperature, increased pH, less precipitation, and a lower water table level might all dramatically enhance ecosystem respiration, according to a sensitivity study of the ECOSSE framework.

For the bio-energy scenarios, the predicted Miscanthus and short-rotation forestry productions are determined using the LPJmL framework simulations.

Box 7. InveN2Ory model sensitivity analysis example

A model sensitivity analysis was performed as part of the InveN2Ory project, which was supported by the US Agency for Ecology, Crop, and Rural Matters. In order to forecast N_2O emissions in the UK, DailyDayCent [16] was used. First, the model's performance was assessed in the four separate UK cropland locations of Boxworth, Terrington, Betley, and Middleton for winter wheat (Triticum aestivum L.). It was discovered that throughout the six farmland trial locations, the DailyDayCent framework produced a respectable estimate of both yearly N_2O discharges and crop yields (Figure 20.9). The projected yearly emissions, however, didn't always match the actual observed quantities. Apart from a model flaw (structural error), there could be a number of explanations for this discrepancy, such as incomplete data on key climate

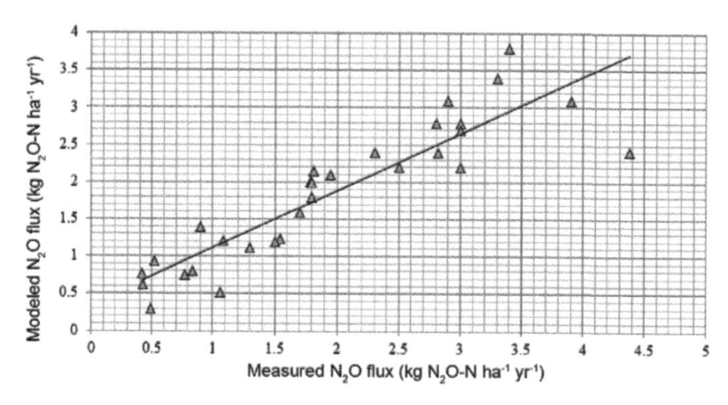

Figure 20.9 Comparison of all investigational plots utilized in the InveN2Ory project's cumulative fluxes, measured and modeled, side by side.

or soil inputs or incomplete data on land management, particularly in cases where nitrogen fertilizers or organic manure were registered in the time right before the hypothetical calculations.

A sensitivity analysis of the calculation of the uncertainty of the input parameters for soil and meteorological simulations of N_2O emissions at various locations was also included in the research (Figure 20.10). Figure 20.10 depicts the uncertainty analysis process.

1. The framework was assessed using the above-mentioned crop parameters and observed N_2O fluxes.
2. The probability distribution function of uncertainty was developed utilizing the site-level I/P value as the midpoint, and it was used to characterize the uncertainty in the essential model I/Ps of soil pH [17], clay constituent, mass density, temperature, and precipitation. A normal distribution of the uncertainty was presumed for every I/P variable. Although the user-defined PDF ranges for each input variable are available, the variability for this experimental design was mirrored at the site level. As a result, the range of values utilized was dependent on both published estimations. The lowest and maximum values were then selected, dependent on the percentage differences, and remained similar for every site. This was followed by a 10-step rise and reduction of similar intervals (n = 10) starting from the midpoint.

20.3.5 Probability analysis

Uncertainty analysis identifies the pathways through which input variability is transmitted through the framework and measures how this variability

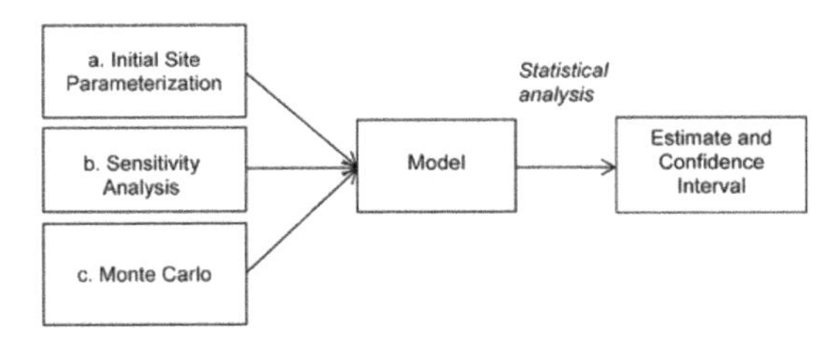

Figure 20.10 Example of the uncertainty analysis technique.

(uncertainty) is turned into variability in the framework O/P. Multiple connections between soil processes are defined using process-based models based on various C and N pools. This has the benefit of simulating soil process feedbacks. For instance, higher plant growth brought on by increased solar radiation or rainfall may result in increased soil carbon inputs. These enhanced organic carbon I/Ps then boost microbial schemes and promote N cycling. On the contrary, as the frameworks are aggregated and repetitive, mistakes in the selection of the parameters and the operating factors generate across the model iterations and raise output uncertainty. The input is deemed relevant if there is significant output variability when the input changes. To determine the inputs that have the strongest correlation between input and output, a sensitivity analysis may be employed. These techniques don't always pinpoint the same inputs. Since the variability in thc model output does not exist until the framework is sensitive to the I/P, a framework is always sensitive to the significant I/Ps which donate almost all to the uncertainty in the model output. A sensitive input, however, may not necessarily be significant or contribute to output uncertainty as its value can be precisely called. Model structural uncertainty is a significant source of uncertainty in simulation; even when inputs are exactly understood, O/Ps are uncertain as models do not perfectly capture reality due to missing or simple techniques that are expressed inside the model.

The InveN2Ory project's testing of model applicability across a number of UK locations was one of its primary objectives. This validation was accomplished by the utilization of sites having various environmental situations and the assessment of model sensitivity to variables. The framework is shown to be suitable for evaluating geographic evaluations of the range of N_2O fluxes throughout the US by testing the model's accuracy and uncertainty and verifying model simulation of N_2O fluxes at every location. The heterogeneity in the geographical and temporal fluxes of N_2O has implications for biogeochemical models have developed into helpful tools for decision-makers. More precisely, ecosystem models may enable national inventories to communicate discharges using category 3 approach when utilized in the predictive form as indicated above. Based on the restriction that the range of I/Ps doesn't surpass the range over which the authors have courage in the characterization of the embedded techniques, such techniques take use of certain fundamental characteristics of process-based models.

20.3.6 Statistical calibration

Bayesian calibration is another method for calibrating models (Box 8). In essence, the Bayesian [18] method accepts uncertainty about model parameterization as well as in the underlying data and assessment data. The ability to alter these in light of the experimental results is made possible by depiction of the prior distribution of both variables and model parameters. As a result, it is often possible to attain appropriate model performance at the site level. This is not a guarantee, but it could somewhat make up for model structural errors. The fact that this calibration enhances site-level modeling suggests that the models may in fact have structural defects, which advances our knowledge of the process from a scientific standpoint.

The aforementioned examples demonstrate how scientists and developers have used and interpreted process-based models in various applications. This may be a time-consuming activity since process-based models are often complicated. Analyses may be carried out once a certain degree of validation has been reached before the results are reported for various stakeholders (such as policymakers). The lack of shortest access to the model or its quick meaning by the stakeholder community is a disadvantage of this procedure. The development of a package or platform that incorporates a process-based paradigm while keeping the end-user stakeholder in mind is an alternate strategy to get around this problem. The last example is a demonstration of a similar practice.

In the ELUM project (Box 9), straightforward links between I/P and output variables are determined using model simulations. A process-dependent model may be used to provide reliable data that closely resembles the system in question if it can be applied to it with confidence (for example, if it has been calibrated and verified). Despite being synthetic, this data might nonetheless have a greater resolution than experimental data as it is easier to collect. This information may be speculative (for example, climate forecasts). These straightforward correlations might be incorporated into more useful software [19] solutions for various user groups after being extracted from dynamic simulation models.

20.3.7 Example of Bayesian calibration in Box 8

Using the accept–reject algorithm, Sinha et al. [20] established a technique for Bayesian calibration and used it to calibrate the initial distribution of soil organic carbon pools against actual measurements of soil respiration. The uncertainties in the original C distribution were measured and reduced via the Bayesian calibration procedure. To derive a revised probability distribution for the parameters, the technique uses two sources of information: (i) prior knowledge regarding the model parameters and (ii) regular data on model O/P variables. The Bayes' theorem which states that the posterior probability of any parameter vector is proportional to the product of its prior probability and its associated data likelihood (eqn (20.1)) is used during calibration.

$$\log(M) = \sum_{s=1}^{u}\left[-0.5\left(\frac{IM - NM}{AB}\right) - \log(AB)\right], \tag{20.1}$$

where *IM* is the mean observed soil winter CO_2 discharges on every sample day, *NM* denotes the model-simulated soil CO_2 discharge, *AB* denotes measurement error or uncertainty, and *n* denotes the number of data points.

Thus, the starting size of the carbon pools incorporates both the initial assumption of the pool sizes and a need for sufficient model performance at the site level. The DailyDayCent model was utilized for a thorough evaluation of ensuing alterations in the carbon balance of Latin America farms from 1985 to 2030, taking into account the consequences of the IPCC Notable Details on Discharges Situations A1B discharge scenario, following fruitful measurement at various sites throughout Latin America (Figure 20.11). The EU27's whole geographical area was divided into approximately 55,000 homogenous units. The model was conducted in every NCU utilizing climate under the A1B discharge scenarios from 1971 to 2000 (baseline) and 2001 to 2028. Under current and predicted climatic conditions, the framework was utilized to generate spatial patterns [21] of NPP, soil heterotrophic respiration, and NEE (Figure 20.12).

Then, by managing DailyDayCent having five various crop rotations using the similar climatic variables, the uncertainties in the current and ensuing carbon balance related to the crop circling option were evaluated. The findings imply that between 1970 and 2000, croplands in Europe were almost carbon neutral; soil carbon losses brought on by ensuing climate alteration neutralized or even partially offset C sequestration brought on by growing enrichments brought on by climate warming and CO_2 fertilization in the

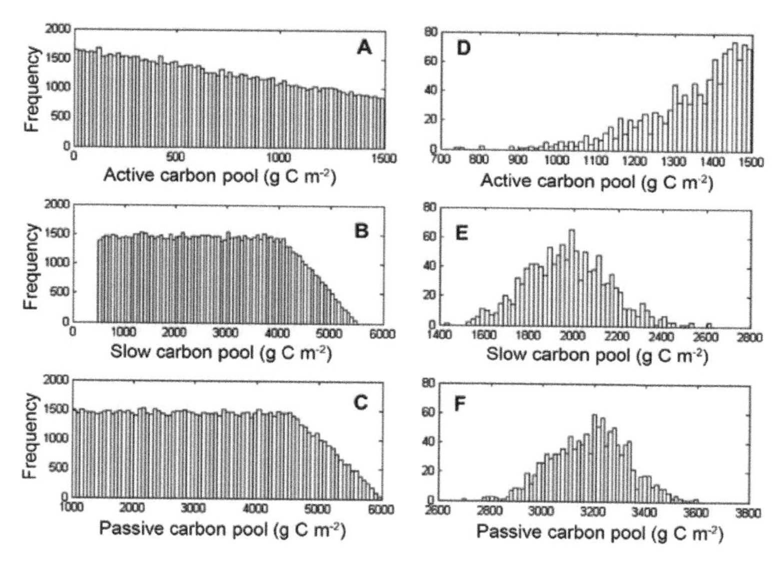

Figure 20.11 Samples from the C1, C2, and C3 pools of the DayCent model's prior (n = 111) and posterior (n = 1214) parameter probability distributions. The graphic shows the posterior parameter distribution of the operative carbon pool, the slow carbon pool, the passive carbon pool, and the prior marginal parameter distribution of each of these pools.

India Ocean area. The A1B scenario predicts that Latin American farming soils would drop carbon and operate as weak sources on the European scale, whereas cultivable systems are a net weak carbon source at the national level; various nations are feeble carbon sinks. The C balance of European agriculture is shown to be sensitive to organizational decisions and climate change, while being near to C neutral today. The study's findings may be utilized to develop more targeted policies at the level of each individual nation, which would assist to lessen these consequences.

20.3.8 Summary

The field of mathematical modeling is very new and developing quickly. As categories like planned, tactical, individual-dependent, bottom-up, top-down, systems, etc., that described the functional elements and qualities of frameworks, evolved, so did our conventional understanding and vocabulary. In line with this classification, statistical representations of physical and biological occurrence basically range greatly in complexity from very simple classical population biology models to complex and dynamic models having many lines of code or more (such as ecosystem models). The ability to analyze and comprehend mathematical models is becoming well understood.

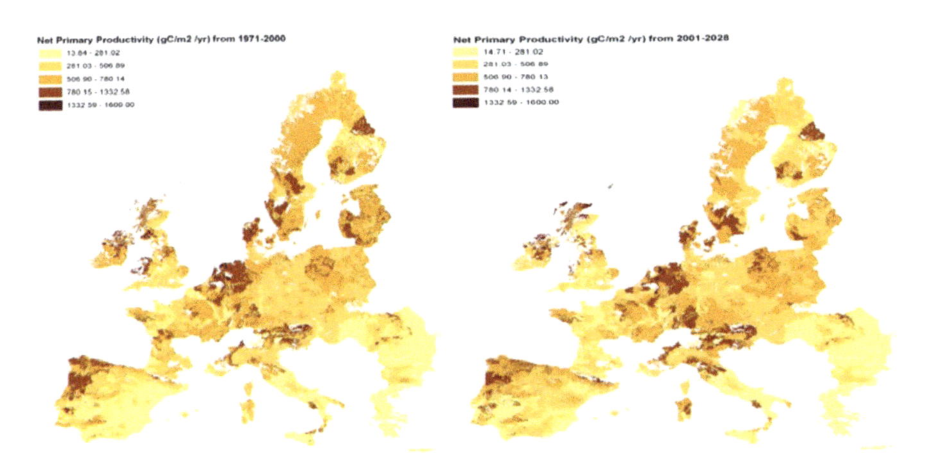

Figure 20.12 The DailyDayCent model's simulation of net primary productivity (NPP). Average yearly NPP, 1971–2000 (left). The NPP's yearly average from 2001 through 2028 (right).

Box. 9. The ELUM project is a condensed example of meta-modeling

The Energy Technologies Institute hired and sponsored the ELUM project to develop a methodology for forecasting the sustainability of bio-energy arrangement throughout the US. The project's goal is to map the US's bio-energy GHG balance from soil respiration up to 2032 by closely integrating field research and computer modeling. The example shows how to incorporate process-based modeling into a user-friendly workflow.

The biomass production data developed for Miscanthus, short-rotation forestry, SRC, and first generation crops were utilized to evaluate GHG alterations following land-usage alteration for different US bio-energy crops. Site-level model simulations were checked against experimental data from the ELUM project.

The result from several ECOSSE simulations utilizing predicted climatic data for the whole United Kingdom is included in the project's final output. These were then included in a look-up table that non-expert users may consult to get location-specific estimates of the GHG fluxes brought on by changing the use of land to raise bio-energy crops. In place of the more time-consuming and inaccurate alternative of using a simple soil organic matter framework, a look-up table was utilized.

The program examines the impacts of various transitions, analyzes geographical and temporal findings, and generates O/Ps in a design which may be accessed in a GIS [22] so that users may post process data in accordance

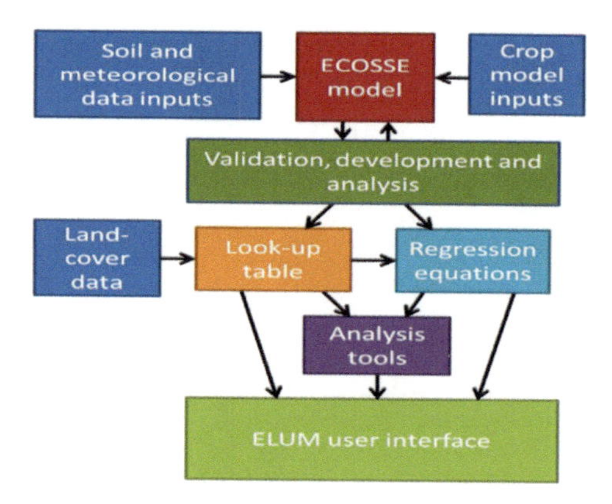

Figure 20.13 Representation of the creation of the ELUM user interface.

with their claims. Users of the tool probably comprise of the below-mentioned people:

- Leaders in the fields of agriculture, the environment, and energy
- Climate and soil scientists; producers and providers of energy; dealers and investors in energy or carbon credits; landowners and supervisors

In actuality, the user input is keyed to search the look-up database for the nearest-neighbor pre-run simulations. Figures 20.13 show the procedure as well as a mock-up of the interface.

In the last section, we looked at several modeling techniques and applications. It may be predicted that more model complexity will result in more accurate modeling, but at the expense of usability. In circumstances where input data can be stated with precision, detailed, process-based models perform well. However, due to their complexity, professional users are required to understand them before they can be translated into a form that can be used by policymakers or laypeople, for example. But none of the aforementioned statements is always true. Even with highly accurate models, performance at the site level is often unpredictable, necessitating a protracted calibration and evaluation procedure. Without additional reliable deterministic ways to verify reliability, this restricts the use of such models to enterprises in C markets. As a result, empirical models that are entirely data driven and transparent in terms of their methods due to their relatively basic structures are still useful today. Simple EF methods may now have more use since they may be included into a protocol like the IPCC's national inventory reporting procedures and the estimate of every EF have a natural relevance as a gauge of the significance of the emission sources.

To get beyond few of the traditional limits of process-dependent models, there have been a number of innovative attempts recently. For instance, the CENTURY model's implementation in the COMET-VR platform, meta-analyses like that of various researches on calibrated DNDC applications in Europe, or the ELUM model interface mentioned here that uses a pre-run look-up table depending on ECOSSE simulations along with point sensitivity analyses.

Various concepts and implementations are addressed in this chapter along with their worth to diverse user groups. Statisticians are familiar with the adage "All models are false—some are useful." The authors must keep in mind that models are merely attempts to explain certain features of actual systems, and that a model's worth should not be only determined by its accuracy.

References

[1] N. Mangra, F. Behmann, A. Thakur, A. Popescu, G. Suciu Jr, G. Giannattasio, R. Uppal, W. Montlouis, "5G Enabled Agriculture Ecosystem: Food Supply Chain, Rural Development, and Climate Resiliency," in 5G Enabled Agriculture Ecosystem: Food Supply Chain, Rural Development, and Climate Resiliency , vol., no., pp. 1–40, 30 Jan. 2023.

[2] I. Hupont, M. Micheli, B. Delipetrev, E. Gómez and J. S. Garrido, "Documenting High-Risk AI: A European Regulatory Perspective," in Computer, vol. 56, no. 5, pp. 18–27, May 2023, doi: 10.1109/MC.2023.3235712.

[3] N. Mishra, I. Jahan, M. R. Nadeem and V. Sharma, "A Comparative Study of ResNet50, EfficientNetB7, InceptionV3, VGG16 models in Crop and Weed classification," 2023 4th International Conference on Intelligent Engineering and Management (ICIEM), London, United Kingdom, 2023, pp. 1–5, doi: 10.1109/ICIEM59379.2023.10166032.

[4] I. B. Matveev, S. I. Serbin and K. Wolf, "Plasma-Assisted Ammonia Combustion—Part 1: Possibilities of Plasma Combustion of Ammonia in Air and Oxygen," in IEEE Transactions on Plasma Science, vol. 51, no. 6, pp. 1446-1450, June 2023, doi: 10.1109/TPS.2023.3273462.

[5] T. T. H. Ngoc, S. Pramanik, P. T. Khanh, "The Relationship between Gender and Climate Change in Vietnam", The Seybold Report, 2023, DOI 10.17605/OSF.IO/KJBPT

[6] P. T. Khanh, T. H. Ngọc, and S. Pramanik, Future of Smart Agriculture Techniques and Applications, in Advanced Technologies and

AI-Equipped IoT Applications in High Tech Agriculture, Eds, A. Khang, IGI Global, 2023.
[7] T. H. Ngọc, P. T. Khanh and S. Pramanik, Smart Agriculture using a Soil Monitoring System, in Advanced Technologies and AI-Equipped IoT Applications in High Tech Agriculture, Eds, A. Khang, IGI Global, 2023.
[8] A. Wang et al., "Identification and Evolution of Soil Organic Carbon Density Caused by Coastal Rapid Siltation Based on Imaging Spectroscopy," in IEEE Journal of Selected Topics in Applied Earth Observations and Remote Sensing, vol. 16, pp. 4287–4300, 2023, doi: 10.1109/JSTARS.2023.3269078.
[9] S. K. D'Mello and B. M. Booth, "Affect Detection From Wearables in the "Real" Wild: Fact, Fantasy, or Somewhere In between?," in IEEE Intelligent Systems, vol. 38, no. 1, pp. 76–84, 1 Jan.–Feb. 2023, doi: 10.1109/MIS.2022.3221854.
[10] I. P. Handayani and N. Folz, "Adaptive Land Management for Climate-Smart Agriculture," 2021 IEEE International Conference on Health, Instrumentation & Measurement, and Natural Sciences (InHeNce), Medan, Indonesia, 2021, pp. 1–7, doi: 10.1109/InHeNce52833.2021.9537265.
[11] S. Pramanik and S. Bandyopadhyay, Identifying Disease and Diagnosis in Females using Machine Learning, in Encyclopedia of Data Science and Machine Learning, Eds. John Wang, IGI Global, 2023, DOI: 10.4018/978-1-7998-9220-5.ch187.
[12] S. Pramanik and S. Bandyopadhyay, Analysis of Big Data, in Encyclopedia of Data Science and Machine Learning, Eds. John Wang, IGI Global, 2023, DOI: 10.4018/978-1-7998-9220-5.ch006
[13] R. Bansal, B. Jenipher, V. Nisha, Jain, Makhan R., Dilip, Kumbhkar, S. Pramanik, S. Roy and A. Gupta, "Big Data Architecture for Network Security", in Cyber Security and Network Security, Eds, Wiley, 2022, https://doi.org/10.1002/9781119812555.ch11
[14] R. Anand, J. Singh, D. Pandey, B. K.Pandey, V. K. Nassa and S. Pramanik, "Modern Technique for Interactive Communication in LEACH-Based Ad Hoc Wireless Sensor Network", in Software Defined Networking for Ad Hoc Networks, M. M. Ghonge, S. Pramanik and A. D. Potgantwar, Eds, Springer, 2022, https://doi.org/10.1007/978-3-030-91149-2_3
[15] S. Pramanik, An Adaptive Image Steganography Approach depending on Integer Wavelet Transform and Genetic Algorithm, Multimedia Tools and Applications, 2023, https://doi.org/10.1007/s11042-023-14505-y.
[16] R. Jayasingh, J. Kumar R.J.S, D. B. Telagathoti, K. M. Sagayam and S. Pramanik, "Speckle noise removal by SORAMA segmentation

in Digital Image Processing to facilitate precise robotic surgery", International Journal of Reliable and Quality E-Healthcare, vol. 11, issue 1, DOI: 10.4018/IJRQEH.295083, 2022

[17] S. Ahamad, V. Veeraiah, J.V.N. Ramesh, R. Rajadevi, Reeja S R, S. Pramanik and A. Gupta, Deep Learning based Cancer Detection Technique, Thrust Technologies' Effect on Image Processing, IGI Global, 2023

[18] B. K. Pandey, D. Pandey, V. K. Nassa, A. S. George, S. Pramanik and P. Dadheech, Applications for the Text Extraction Method of Complex Degraded Images, in The Impact of Thrust Technologies on Image Processing, Nova Publishers, 2023.

[19] V.Veeraiah, D.J. Shiju, J.V.N. Ramesh, Ganesh Kumar R, S. Pramanik, A. Gupta, D. Pandey, Healthcare Cloud Services in Image Processing, Thrust Technologies' Effect on Image Processing, IGI Global, 2023

[20] M. Sinha, E. Chacko, P. Makhija and S. Pramanik, Energy Efficient Smart Cities with Green IoT, in Green Technological Innovation for Sustainable Smart Societies: Post Pandemic Era, C. Chakrabarty, Eds, Springer, 2021, DOI: 10.1007/978-3-030-73295-0_16

[21] B. K. Pandey, D. Pandey, V. K. Nassa, A. S. Hameed, A. S. George, P. Dadheech and S. Pramanik, A Review of Various Text Extraction Algorithms for Images, in The Impact of Thrust Technologies on Image Processing, Nova Publishers, 2023.

[22] P. K. Mall, S. Pramanik, S. Srivastava, M. Faiz, S. Sriramulu and M. N. Kumar, FuzztNet-Based Modelling Smart Traffic System in Smart Cities Using Deep Learning Models, in Data-Driven Mathematical Modeling in Smart Cities, IGI Global, 2023, DOI: 10.4018/978-1-6684-6408-3.ch005

[23] Y. Lu et al., "Remote-Sensing Interpretation for Soil Elements Using Adaptive Feature Fusion Network," in IEEE Transactions on Geoscience and Remote Sensing, vol. 61, pp. 1-15, 2023, Art no. 4505515, doi: 10.1109/TGRS.2023.3307977.

[24] B. Lv et al., "Achievement of Possibly Maximum Photosynthetic Performances for Multi-Primary Laser Lighting for Indoor Farming," in IEEE Photonics Journal, vol. 15, no. 4, pp. 1–6, Aug. 2023, Art no. 6700106, doi: 10.1109/JPHOT.2023.3300050.

Index

About the Editors

Sabyasachi Pramanik is a Professional IEEE member. He obtained a Ph.D. in Computer Science and Engineering from Sri Satya Sai University of Technology and Medical Sciences, Bhopal, India. Presently, he is an Associate Professor, Department of Computer Science and Engineering, Haldia Institute of Technology, India. He has many publications in various reputed international conferences and journals, and has written online book chapter contributions (Indexed by SCIE, Scopus, ESCI, etc.). His research is in the fields of artificial intelligence, data privacy, cybersecurity, network security, and machine learning. He serves as an editorial board member of many international journals. He is a reviewer of journal articles from IEEE, Springer, Elsevier, Interscience, IET, and IGI Global. He has reviewed many conference papers, and has been a keynote speaker, session chair and technical program committee member for many international conferences. He has authored a book on Wireless Sensor Network. He has edited books from IGI Global, CRC Press, Springer and Wiley Publications.

Dr. Sandip Roy is a Professor at the Department of Computer Science & Engineering School of Engineering JIS University Kolkata, India. He obtained his Ph.D. in Computer Science & Engineering from University of Kalyani, India in 2018. Dr. Roy received an M.Tech. degree in Computer Science & Engineering in 2011, and a B.Tech. in Information Technology in 2008 from Maulana Abul Kalam Azad University of Technology, West Bengal (formerly known as West Bengal University of Technology). He was a post-doctoral fellow in the Computer Science and Engineering department of Srinivas University, Mangalore, India between January 2020 and January 2021. He also served as research assistant with different collaborative industry projects of Simplex Infrastructures Ltd., Bharti Airtel Ltd., etc. He has authored over 100 papers in peer-reviewed journals and conferences, and was a recipient of the Best Paper Award from ICACEA in 2015. He has also authored 8 books and also granted 15 patents. His main areas of interest are data science, internet of things, cloud computing, green computing, and smart technologies.

Dr. Rajesh Bose is currently employed as a Professor at the Department of Computer Science & Engineering School of Engineering JIS University Kolkata, India, and is a **Consultant** at R&D Division of Simplex infrastructures Ltd. He graduated with a B.Eng. in Computer Science and Engineering from Biju Patnaik University of Technology (BPUT), Rourkela, Orissa, India in 2004. He went on to complete his degree in M.Tech. in mobile communication and networking from Maulana Abul Kalam Azad University of Technology, West Bengal (formerly known as West Bengal University of Technology – WBUT), India in 2007. He completed his Ph.D. degree from the University of Kalyani in Computer Science and Engineering in 2018 and completed a Post-Doctoral Fellowship in Focusing Artificial Intelligence and Machine Learning in Clinical Practice for an Effective Prognosis at Srinivas University, Karnataka. He also has several global certifications under his belt: CCNA, CCNP-BCRAN, and CCA (Citrix Certified Administrator for Citrix Access Gateway 9 Enterprise Edition), CCA (Citrix Certified Administrator for Citrix Xenapp 5 for Windows Server 2008). His research interests include cloud computing, IoT, wireless communication and networking. He has published more than 200 refereed papers, 4 books and 12 granted patents. Under his guidance 3 Ph.D. scholars have received their Ph.Ds. With over 18 years of experience in industry and education institutions, he is presently recognized by IEEE as an active IEEE Access reviewer. He is experienced in conceptualizing, designing, implementing, and managing projects in Cloud Computing, Datacenter and IoT in the civil construction engineering sector as well as in academic research.